KRAKATAU

KRAKATAU

The Destruction and Reassembly of
an Island Ecosystem

Ian Thornton

HARVARD UNIVERSITY PRESS
Cambridge, Massachusetts, and London, England

First Harvard University Press paperback edition, 1997

Library of Congress Cataloging-in-Publication Data

Thornton, I. W. B. (Ian W. B.)
    Krakatau : the destruction and reassembly of an island ecosystem /
Ian Thornton.
        p.   cm.
      ISBN 0-674-50568-9 (cloth)
      ISBN 0-674-50572-7 (pbk.)
        1. Island ecology—Indonesia—Krakatoa.   2. Krakatoa (Indonesia)—
    Eruption, 1883.   3. Krakatoa (Indonesia)   I. Title.
    QH186.T49   1995
    574.5′267′095982—dc20
    95-4922

For Annie

CONTENTS

ACKNOWLEDGMENTS

In Indonesia I am known as *Profesor yang cinta Krakatau,* "the professor who loves Krakatau." In my home institution, the Professor of Biochemistry calls me "Krakatau Thornton," and in the ring of the phrase I think I can detect undertones of "Crackpot" or "Crackers Thornton." I suppose that there is some truth in both epithets; Krakatau, I confess, has become something of an obsession, in the sense that it has filled my mind since I first visited the archipelago with my wife in 1982.

Together with zoologist colleagues at La Trobe University, in 1984 I began a decade of study of the animals of the Krakatau islands in Sunda Strait, between Java and Sumatra. So far we have made six expeditions to this remarkable archipelago. We were assisted by the Melbourne *Age* newspaper, which made a public appeal for funds. Notable donations were made by CRA Pty Ltd. and Mr. Dick Smith of *Australian Geographic* magazine. I wish to record my thanks to all those whose financial support in the early days made our work on the islands possible—institutions, members of the public, and the faculty and alumni of our own institution—and in particular to one anonymous donor of a most welcome large round sum. Later expeditions were largely financed by grants from the Australian Research Grants Council.

I am particularly grateful to the fifty or more fellow-expeditioners, from a dozen or so institutions, who at various times have worked as teams on the islands. Our expeditions have always been joint affairs between the Indonesian Institute of Science's (LIPI) Centre for Research and Development of Biology, Bogor, and La Trobe University's School of Zoology (but often including scientists from other countries). I am indebted to LIPI for permission to carry out research in Indonesia and to the PHPA (Directorate of Forest Protection and Nature Conservation) for allowing us to work in the Krakatau national park. Dr. Axel Ridder, of Krakatau Carita Beach Hotel on the west coast of Java, provided logistic support and an appropriate base from which to mount the expeditions.

Indonesian biologists who joined several expeditions are Tukirin Partomihardjo of the Bogor Herbarium, whose knowledge of the plants of the islands is unsurpassed; Rosichon Ubaidillah, entomologist, like Tukirin an indefatigable field worker; S. (Bas) van Balen, who can recognize the birds literally with his eyes shut, so well does he know their vocalizations; Darjono, another young ornithologist; Dhamayanti Adidharma, mite specialist; and Sudarman Husen Kartodihardjo, entomologist. Like Abbas Suleiman, Achmad Saim, Agus H. Tjakrawidjaja, Asep Adhikerana, Ichsan U. Din, Lucia Fidhiany, Machfudz Djajasasmita, Mohammed Amir, Eli Mirmanto, and Ramlanto, who each joined expeditions, they became not just companions but also good friends of the visiting scientists.

My expedition colleagues from La Trobe University include Peter Frappell, Tim New, Neville Rosengren, David Walsh, Seamus Ward, and Richard Zann; Peter Rawlinson, also of this group, died tragically on Anak Krakatau's lava fields during our April 1991 expedition. Other participating Australian scientists were Jane Edwards, George and Jenny Ettershank, Stephen Graves, Caroline Gross, Mark Harvey, Darrel Kitchener, Helen Malcolm, Bryan Smith, Chris Tidemann, Mark Walker, and Alan Yen. Although my Indonesian colleagues flatter me with the saying, *naga tua, tenada muda* ("old dragon, young power"), in recent years the need for auxiliary, really young power has increased, and the following La Trobe University students have played an important role in our investigations: David Britton, Geoffrey Brown, Simon Cook, Neil Hives, Paul Horne, Mark Hutchinson, Michael Jaeger, Bruce Male, David McLaren, Rod Mclellan, Natasha Schedvin, Evan Schmidt, and Patrick Vaughan. Expeditioners from other parts of the world have made particularly important contributions: Bryan Turner, insect ecologist, of London University; Stephen Compton, and Sally Ross, fig biologists, of Rhodes University, South Africa; Wolfgang Nentwig, spider ecologist, of the University of Berne, Switzerland; Geoffrey Davison, ornithologist, of the National University of Malaysia; and Mark Bush, University of Hull, U.K.

I welcome the opportunity to register my appreciation to all the above for their work, their cooperation, their great companionship. Without them this book could not have been written.

I am grateful to several reviewers, including Ed Wilson, David Steadman, and Stephen Self, for critically appraising the manuscript. Stephen Self and

Guy Camus provided valuable advice on the archipelago's volcanology, and I owe much to their recent reviews for some of the coverage in Chapter 3. Richard Price, my geological colleague at La Trobe, zoologists Alan Wright, Richard Zann, Seamus Ward, Tim New, and George Stephenson, and Sturla Fridriksson read early drafts of chapters and provided valuable advice. Comments from Ed Wilson and Jared Diamond have helped my thoughts on the biogeography of Krakatau, and Ibkar Kramadibrata, John Flenley, Hideo Tagawa, Junichi Yukawa, and Rob Whittaker, leaders of other Krakatau study teams, have cooperated splendidly. Needless to say, views expressed do not necessarily concur with those of any who have shared with me their experience and expertise, and any errors remain my responsibility. Several illustrations were originally prepared by Jenny Browning, and the photographs, unless otherwise indicated, are by David Walsh (copyright La Trobe University). Frances Pizzey and June Cheah provided effective administrative support, and Alan Marshall, and then Johanna Laybourn-Parry, kindly put up with my continued presence in the School of Zoology long past my official retirement. To all the above, I offer my thanks. I have been fortunate to have worked with such colleagues for so long.

This book is dedicated to my wife, Ann, who instigated and encouraged my involvement in Krakatau research from the start and who accompanied me on several visits, including an overnight stay on Rakata's summit. I could have had no better supporter in this venture, nor any companion who would have put up with "Krakatau Thornton" so readily or with such understanding.

SUNDA STRAIT
AND ADJACENT LANDMASSES

SUMATRA

Teluk Betung

G. Ratai
▲ 1681

Beneawang

5°30'

Semangka
Bay

1106

Lampung
Bay

G.
Ba
12

20

20

1162

Ketimbang

20

Lagundi I

Sebu

50

Vlakke Hoek
(Tg Babi)

100

Sebesi

Krakatau Is.

6°

100

20

50

20

S U N D A   S T R A I T

6°30'

50

I N D I A N
O C E A N

Panaitan I

100

50

20

Second Point
(Tanjung Alangalang)

Welcome
Bay

First Point
(Tanjung Layar)

Ujung
Kulon

0   10   20   30   40

Scale (km)

G. Payung

Java Head

Bathymetry in fathoms   Height in metres

20

7°   104°30'   105°   105°30'

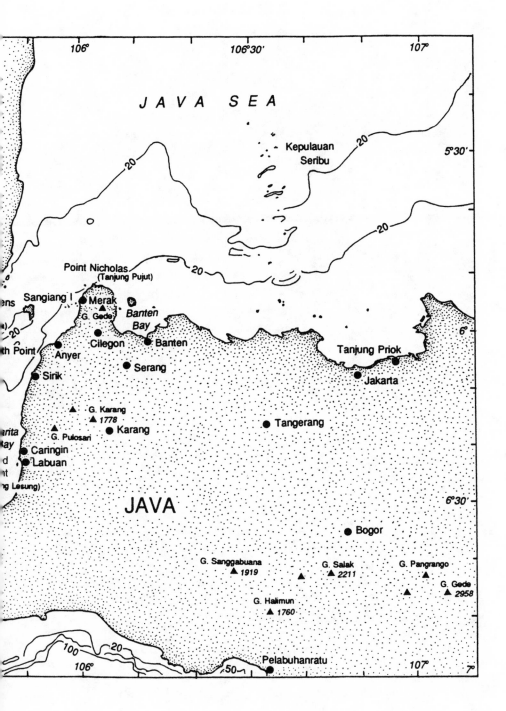

1 ONE TOUCH
OF NATURE

One touch of nature makes the whole world kin.
William Shakespeare, *Troilus and Cressida*

ℰ

Just after 10:00 A.M. on the twenty-seventh of August 1883, an event occurred that literally, as well as figuratively, shook the world. In Sunda Strait, between the islands of Java and Sumatra (Figure 1), the island of Krakatau exploded with the force of more than ten thousand Hiroshima-type atom bombs.

The sound produced was the greatest ever recorded in terms of audible range; explosions were heard in Sri Lanka, Central Australia, and on the island of Rodriguez in the Indian Ocean. The air waves, which had amplitudes greater than those of the largest atomic bomb tests, made four journeys around the globe to Krakatau's antipodes in Colombia, and three return journeys. The tsunamis (marine pressure waves, often incorrectly referred to as tidal waves) that were generated were detected as far away as the English Channel, Alaska, and South Africa. One huge tsunami, with an advancing front in some places as high as a seven-story building and traveling at the speed of a train, surged through the bays and inlets of Sunda Strait about half an hour after the explosion, devastating over 160 towns and villages and killing more than 36,000 people. The eruption column—the rising plume of steam, gases, ash, and fragmented material—penetrated the stratosphere to an estimated height of 40 km (25 miles), and a band of fine particles and aerosols circled the earth for months, gradually broadening to encompass more northerly latitudes. The earth's average annual temperature was lowered by about 0.5° C

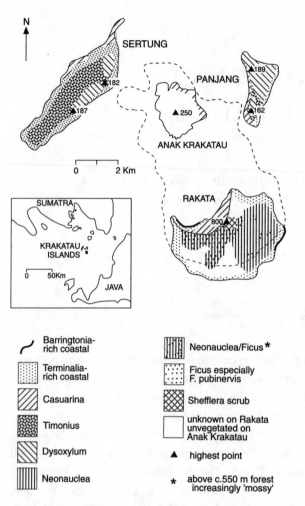

Figure 1. The Krakatau archipelago in July 1993, showing the distribution of main vegetation types. The pre-1883 island of Krakatau is shown as a dashed outline.

and unusual atmospheric effects, such as remarkable sunsets and blue and
green tints to the sun and moon, were seen in the northern hemisphere for
months or years afterwards.

Two-thirds of Krakatau Island, originally 11 km long and clothed in tropical rain forest, disappeared. In its place a submarine depression over 366 m
deep had been created, and the southern half of what had been its southern
volcano, Rakata, was left standing. The Rakata remnant and its two close
neighbours, Sertung and Panjang, were enlarged by a cover of ash at temperatures of hundreds of degrees centigrade to depths of tens of meters. It is

thought that not a plant, not a blade of grass, not a fly, survived after the thick blanket of hot ash had settled on the three islands.

There was then no telephone or radio but the wireless telegraph had been invented some decades previously, and for the first time news of a great natural event was almost immediately available to countries around the world. The event indirectly affected such large areas of the earth and elicited such widespread interest that in January 1884 the Royal Society of London set up a Krakatoa Committee to report on the eruption and its related phenomena, and French and Dutch investigatory teams visited the islands. (Indonesia was at that time the Dutch East Indies.) The unusual meteorological and climatic effects of the great belt of ash circling above the earth, noted in many parts of the world, were soon related to the Krakatau eruption.

It has been suggested (although of course it is impossible to prove) that some of the finest paintings of the great British artist, J. M. W. Turner, may have been inspired by the remarkable sunsets and afterglows seen in London for six months after the great eruption of Tambora, on the Indonesian island of Sumbawa, in 1815. Tennyson could well have been similarly influenced by the meteorological phenomena resulting from Krakatau's eruption when he wrote in the poem *St. Telemachus*, published in 1892:

> Had the fierce ashes of some fiery peak
> Been hurled so high they ranged about the globe?
> For day by day, thro' many a blood-red eve,
> . . .
> The wrathful sunset glared

Because the eruption's effects were detected over parts of the planet many thousands of kilometers away, arousing worldwide interest, the Indonesian volcanologist John Katili, in a Krakatau centenary lecture in Jakarta in 1983, used the quotation from Shakespeare that I have taken as the title of this chapter. This single natural event was seen to have important and obvious effects on the earth's land, its oceans, its atmosphere, its climate, and, at least in one area, on its living organisms. Moreover, in some cases the links between these were clearly evident. As a result of this "one touch of nature," the modern holistic concept of the planet was dramatically demonstrated over a century ago.

In the years since, the Krakatau event and its associated phenomena have been the focus of attention of a wide spectrum of scholars and researchers, including historians, sociologists, geologists, volcanologists, geomorphologists, oceanographers, meteorologists, climatologists, seismologists, botanists, zoologists, ecologists, and biogeographers. Interaction of scientists has been not only between disciplines but also between generations, as investigators have built their studies on the fine pioneer work of their predecessors, sometimes making significant advances as a result.

There were no on-site witnesses of the event, and many of its basic features have been gradually clarified only after decades of volcanological research. The Dutch mining engineer R. D. M. Verbeek had visited Krakatau in 1880, three years before the eruption. He returned in October 1883, six weeks after it, to begin a series of investigations which led to a classic monograph on its volcanology (Verbeek 1884a,b, 1885). His outstanding work has provided the basis and inspiration for many later investigators, and several of the conclusions he made more than a century ago still have much support from today's earth scientists. R. Vincent and G. Camus, two French volcanologists who have recently worked on the Krakataus, have described him as "a pioneer of *modern volcanology*" (their stress) for his quantitative approach to volcanic phenomena (Vincent and Camus 1986, p. 174). Stephen Self, another volcanologist who has studied the Krakatau eruption intensively in the field, has recently written that "Verbeek's remarkable treatise on Krakatau . . . must be considered one of the most significant contributions to volcanology" (Self 1992, p. 120).

Verbeek concentrated on the petrology, the chemical and physical nature of the volcanic deposits. Some forty years later another fine Dutch geologist, C. E. Stehn, made an intensive study of the stratigraphy of the deposits, the arrangement and order of deposition of the layers of volcanic products (Stehn 1929). Stehn's visits were in 1922 and in 1928, when he was also monitoring the first signs of the birth of the new island Anak Krakatau. In spite of these and other detailed on-site investigations, however, several important questions concerning those crucial twenty-four hours in August 1883 are still under discussion by the world's volcanologists. The century-old event still has a special place in volcanological research.

The Krakatau group of islands also has a special interest to biologists. Two Dutch scientists were the principal investigators of the colonization processes occurring on the Krakataus: K. W. Dammerman, Director of the Bogor (then called Buitenzorg) Zoological Museum in Java, and W. M. Docters van Leeuwen, Director of the Government Botanical Gardens, also at Bogor. All subsequent biological workers on the Krakataus owe a great deal to these two great pioneers. As is true of the work of their geological counterpart, Verbeek, many of the conclusions in their classic monographs are regarded as valid today, and many of their predictions have been proved correct (Docters van Leeuwen 1936, Dammerman 1948).

An evolutionary biologist asked to name the most interesting islands in the world would almost certainly choose another volcanic archipelago, the Galápagos, which was colonized by living things a couple of million years ago from continental sources over 600 km (375 miles) away and now provides fascinating examples of the results of evolutionary processes. Krakatau's biological interest, however, like that of the volcanic island Surtsey in the North Atlantic, is of a different nature.

Surtsey emerged from the sea at a position about 50 km (30 miles) from the south coast of Iceland in 1963. It is of course too young to have any evolutionary interest. Its interest is ecological, and Sturla Fridriksson and other Icelandic biologists grasped the opportunity of carefully monitoring the colonization of this biologically sterile, virgin island by living things, and the very early stages of the assembly of a cold-temperate community of plants and animals (Fridriksson 1975, 1995). On Surtsey an ecosystem is beginning to form from scratch, from a standing start, naturally and on a large scale. Colonization has been slow, and only the simplest of communities are now being developed, some thirty years after Surtsey's emergence (Thornton 1984, Fridriksson and Magnusson 1992).

The Krakataus were barren on August 28, 1883. It is generally believed that no plant or animal life on the islands survived the cataclysm. The resulting physical changes created an entirely new ecological situation, a *tabula rasa*, or clean slate. The surviving islands were reshaped by the action of sea, wind, and rain and progressively recolonized by plants and animals from the mainlands of Sumatra and Java. As successful colonists were integrated into island communities over the years, a tropical forest ecosystem was gradually reassembled by natural means.

Approaching the Krakataus by sea from Java, the visitor rounds the northeastern point of Rakata and enters virtually another world. Rakata's massive northern cliff—800 m (2,600 ft) high and almost sheer from the sea to its soaring, often mist-shrouded peak—looms over the water that covers Krakatau's submerged caldera 200 m (650 ft) below the surface. Just as the great rock face dominates the eye, so the great power unleashed in the astounding event of August 1883 dominates the mind. Some 5 cubic km of rock crashed into the sea, leaving the cliff as a clean scar. A small, jagged spur of pre-1883 lava,

Bootsmansrots, protruding at an unusual angle a few meters above the water of the caldera, is another stark reminder of the great explosive eruption.

Except for the great cliff and a more recent northern extension of Sertung, Rakata and its two lower, companion islands are now clothed in tropical forest from the shore to their highest points. Some of the trees on Rakata are now 35 m (100 ft) tall, with girths at breast height of more than 2 m (7 ft) and huge buttresses many times higher than a man. From the boat one may see a white-bellied fish-eagle perched on a tree at the shoreline, or a hawk-eagle patrolling the ridge of Sertung. In the forest, the sounds of pigeons, doves, bulbuls, whistlers, even woodpeckers, may be heard. Rats scuttle through the litter and huge monitor lizards occasionally crash away into the undergrowth. On a tree limb sits a termite nest twice the size of a basketball with a round opening through which a collared kingfisher feeds its young. This blue and white bird protests the intrusion of human visitors with a loud, raucous call, now the most frequent and penetrating sound on the islands' coasts. One may be lucky enough to see the reticulated python or, more likely, the beautiful, gliding, paradise tree snake.

At dusk, around six o'clock, large green cicadas suddenly fill the forest with sound, and about twenty minutes later, just as abruptly, the sound ceases. Around dusk, too, one may see the twisting, flitting insectivorous bats, as well as the rather larger, less maneuvrable fruit bats beginning their search for figs and other forest fruits. For an hour or more after dusk, fireflies may decorate forest glades with hundreds of flashing points of light. At night the forest is silent save for the occasional screech of the barn owl, and if one is sitting quietly and knows where to sit, one may see this white, ghostly bird (*burung hantu* in Indonesian—ghost bird)—see it, but not hear it, as it flies fairly low and absolutely silently, searching for rats or lizards in more open areas.

All this, and more, has been assembled, in a hundred years, on barren islands that are 44 km (27 miles) from the nearest shores of Java and Sumatra, and some 16 km (10 miles) from the nearest island, Sebesi, which itself was very seriously affected by the eruption. Over the years an interlocking, functional community has been reassembled from sources outside the islands. Obviously, other forces have been at work, more gradual but no less powerful than those which began this remarkable chain of events. These positive, constructive forces are the remarkable dispersal powers of animals and plants, and the ability of some species to become integrated into a community that changes and grows in complexity as the colonists arrive and become established.

This process has taken place in an environment more complex physically than that of Surtsey because three islands were involved, one of them 800 m high, almost five times the height of Surtsey. Moreover, whereas Surtsey is cold-temperate, the Krakataus lie near the equator in the humid tropics (mean annual temperature 26.4° C, mean annual rainfall 2,500–3,200 mm), and their benign, warm humidity is in sharp contrast to the harsh climatic conditions of the North Atlantic encountered by colonists to Surtsey. The immediate source areas are the biologically rich islands of Java and Sumatra, about as distant as is Iceland from Surtsey. The reservoir of potential colonists on these islands is much greater than that on Iceland, which was wiped clean by recent Pleistocene glaciation and thus itself carries a much impoverished source biota. Colonization has thus proceeded at a much faster rate on the Krakataus than on Surtsey.

To add to the complexity, and thus the fascination, of the Krakatau case study, in 1930 a second-order event occurred. In the center of the three-island archipelago, from the waters covering Krakatau's submarine caldera, a fourth island, Anak Krakatau (Krakatau's Child), was born after a gestation period of almost fifty years. This emergent, virgin island, of which Surtsey is the cold-temperate counterpart, has grown by periodic eruptions and is now 2 km (1.25 miles) in diameter, and some 270 m (over 900 ft) high. In its turn, it has also received plant and animal colonists, largely from its older companion islands, and has developed its own embryonic biological community. The changes in this community as species are added year by year and as one temporary phase of the succession gives way to the next can be followed within an ecologist's lifetime. Anak Krakatau's emergence has provided a second opportunity for studying one of the central questions of ecology, the way communities of living things are assembled. So the Krakataus have given us not merely a single natural experiment, but a nested pair of them.

The early work by Dutch investigators has been augmented by succeeding generations of biologists, and we now have data for a period of more than a century. Since the Krakatau biota, particularly that of Anak Krakatau, is a simplified one, many general ecological questions can be examined there more conveniently and profitably than in the more complex ecosystems of mainland areas. More importantly, perhaps, the Krakataus have become a classic case study of several aspects of ecological change. These include recovery of a tropical forest ecosystem from extreme disturbance, community assembly and succession in the humid tropics, and the colonization of islands. Island biologists working on the Krakataus seek to answer questions such as: How did living things return to the islands, when, and in what sequence? How did they become organized into communities, and what were the different stages of development as the biota gradually became more complex and mature? Has the process been similar on all the islands and, if not, how can the differences be explained? Have the islands yet reached a "steady state" in terms of their biotic makeup, and have all components of the biota approached this at the same rate? If not, what is the reason for the differences? To what extent does chance play a part in these processes and to what extent are they deterministic and predictable? Can predictions be made about the future development of the island communities and the time needed for the acquisition of an equilibrium state, if one is ever to be achieved? These are some of the questions that will be considered in succeeding chapters, but first we shall consider what exactly transpired during the 1883 eruption that set these processes in train, and why it should have occurred at the particular site of Krakatau.

2 THE DAY THAT SHOOK THE WORLD

The Midgard Serpent will blow so much poison that the
whole sky and sea will be spattered with it . . . In this din the
sky will be rent asunder . . . The sun will go black, earth sink
into the sea, heaven be stripped of its bright stars, smoke
rage, and fire, leaping the flame, lick heaven itself.

The Deluding of Gylfi, Icelandic saga (quoted in Fridriksson 1975)

Awakenings: May and June 1883

In May 1883, earth tremors were felt at Java's First Point, on the Ujung Kulon
peninsula, and a series of earthquakes was experienced at Ketimbang in Sumatra, about 40 km (25 miles) north of Krakatau.

On May 19, after over 200 years of dormancy (see Chapter 3), Krakatau's
northern volcano, Perbuatan, again erupted, exhibiting small to moderate
Vulcanian or sub-Plinian activity. For the next few days, booming explosions
were heard in Anyer and in Palembang and Bengkulu some 250 km (156
miles) away in Sumatra. Air-pressure waves, often accompanying explosions,
rattled windows and stopped clocks in Jakarta (then Batavia) and Bogor (then
Buitenzorg). Ash fell in the Sunda Strait area and floating pumice was extensive.

On May 27 the *Gouverneur-Generaal Loudon* arrived at Krakatau with a
group of 86 volcano watchers. On board were a photographer named Ham-

burg, who captured the dramatic dark cloud of the eruption column over the island, and a mining engineer, J. Schuurman, who had been sent to investigate and report his findings to Verbeek (Verbeek 1885, p. 17–23). As the visitors rather foolishly stood at the summit of Perbuatan's cone on the eastern rim of a crater 980 meters in diameter and 42 meters deep, steam, ash, and pumice fragments up to 10 cm in diameter and large pieces of green-black, glassy rock were being shot into the air. The rock was similar to that of Panjang and Polish Hat (Figure 4a), and Stehn (1929) believed that older material may have been included in the ejecta.

On the climb to the crater the visitors saw trees projecting through the ash as bare stumps several meters high, the branches seeming to have been torn off by force. No leaves or branches could be found in the ash, however, and the wood was dry, without signs of charring. Schuurman ascribed this unusual damage to a whirlwind; volcanologists now would probably suggest a base surge.

Ash fell up to 375 km (228 miles) from the eruption, and unusual atmospheric effects (for example, solar halo, blue moon) were reported almost 3,000 km (1,875 miles) to the northwest. Self (1992) later ascribed these effects to ash or volcano-produced aerosols that had reached the upper troposphere and stratosphere in eruption columns up to 20 km (13 miles) high, and believed that pyroclastic flows may have taken place at this time.

On June 16 an explosion was heard at Anyer and a thick cloud of smoke blanketed Krakatau for several days. When the cloud cleared about a week later, two dense eruption columns were visible, the larger one issuing from the Perbuatan area of Krakatau. In June and August fields of floating pumice covering many square kilometers were reported in the Indian Ocean.

These evidently sporadic eruptions of Perbuatan were the first act of the drama played at Krakatau in 1883. There followed a fairly long intermission, but the final act reached a climax of dramatic intensity that even those in the audience that witnessed the first could not have envisaged.

Crescendo: August 1883

A Dutch topographical engineer, Captain H. J. G. Ferzenaar, was sent to reconnoiter the Krakatau group in early August. Trees grew on the peak of Rakata but in devastated areas there were tree stumps, again without sign of branches or leaves. Volcanic effluvia prevented Ferzenaar from mapping the western, downwind half of the island, but his topographical map identifies three active craters, one on Perbuatan and now two more, on Danan, from which vapor columns rose, with many fumaroles (vents issuing gas and steam). The northernmost part of the main Danan crater appeared to have fragmented or collapsed recently. The extent of volcanic activity convinced the wise captain that "Measuring there is still too dangerous; at least I would not like to accept the responsibility of sending a surveyor . . . I consider a survey on the island itself inadvisable" (Verbeek 1885, pp. 26–27). He was the last person to set foot on the island of Krakatau.

As the Dutch man-of-war *Prins Hendrik* passed close to Krakatau's north coast on August 12, H. McLeod (who was to be the first to set foot on the archipelago after the eruption) observed a 30-m-wide crater from which a column of ash and steam reached a height of 3,400 m (over 2 miles), as measured later from a suitable distance. Further ash eruptions were witnessed by ships during the next ten days. On August 22 the *Sunda,* after passing the erupting volcano, sailed into a layer of pumice in Semangka Bay that was so thick that a bucket lowered overboard was filled with pumice rather than water. Activity was again intensifying. Explosions were heard in the afternoon of the twenty-fifth on the *Prinses Wilhelmina,* near Jakarta, and in the evening ash was falling at Teluk Betung 80 km (50 miles) away in Sumatra and 300 km (190 miles) away, south of Java's First Point.

Climax: August 26–27, 1883

The wisdom of Captain Ferzenaar's decision was soon amply demonstrated. Just over two weeks after he left the island, the great paroxysmal events began. Within a day they reached their climax.

The Plinian phase

At around 10:00 A.M. (all times are local time) on Sunday, August 26, fine ash began to fall in Jakarta, and at 1:00 P.M. a series of violent explosions was heard in Bogor and Jakarta and the telegraph master at Anyer noted oscillations in sea level. An hour later there was a tremendous explosion, and Krakatau's eruption column soared to a height of 26 km (16 miles). Pyroclastic flows and surges swept into Lampung Bay, Sumatra. By 3:00 P.M. the explosions were increasing in intensity; sounds like cannon fire and thunder were heard 600 km (375 miles) away. The first explosion recorded on the pressure gauge at the Jakarta gasworks was at 3:34 P.M., and three minutes later the first small tsunami, with a negative onset (that is, the first displacement being a lowering of sea level), was recorded at Tanjung Priok, Jakarta's port. Latter (1981) later ascribed the wave to subsidence of part of the island. At 4:30 P.M. small tsunamis were noted on each side of Sunda Strait, and at 5:00 P.M. detonations were heard over a wide area of Java and ash and pumice began to fall in the strait. Major explosions were recorded at Jakarta at 4:54 and 5:20 P.M.

The first large tsunami was apparently generated at 5:45 P.M. on August 26 and was propagated in all directions; explosions were heard in Australia with an origin time at Krakatau of about 5:30 P.M. A marked recession of the sea at Tanjung Priok and at Ketimbang may indicate a subsidence, perhaps of the northern part of the island, at 7:00 P.M. At 7:30 P.M. a large tsunami surged into the town of Merak, washing away the Chinese camp and smashing boats, and another arrived at Ketimbang at the foot of Gunung (Mount) Rajabasa in Sumatra. Tsunamis and small explosions continued at intervals of about 10 minutes or so throughout the night. At midnight in Daly Waters, Northern Territory, Australia, 3,252 km (2,032 miles) east-southeast of Krakatau, explosions resembling the blasting of rock lasted for a few minutes, waking the residents, and at 1:55 A.M. on August 27 explosions were heard in Singapore and Penang, 1,400 km (875 miles) to the north-northwest. A tsunami originating at 12:30 A.M. reached Teluk Betung and Vlakke Hoek in Sumatra and partly submerged the village of Sirik, almost the nearest point on Java to Krakatau, yet it was noted as only a slight oscillation at Anyer, some 10 km to the north. Verbeek believed that these highly localized waves had been formed as ejecta fell into the sea at particular places around the island, as Latter would agree decades later.

At Krakatau, ash accumulated on Sertung and Panjang to depths of up to 20 m, and heavy falls of ash and pumice occurred within a 20-km radius. In the eight hours to 2:00 A.M., August 27, a 1-m-thick layer of pyroclastics, including hot pumice, fell on the *Berbice,* almost 100 km west-southwest (downwind) of Krakatau, and static electricity, the phenomenon known as St. Elmo's Fire, played in her rigging. Self (1992) estimated, from Verbeek's data, that the total bulk volume of fallout during the first phase to 5:00 A.M., August 27, was between 8 and 12 km^3, probably the latter, the product of 4–5 km^3 of dense magma.

The Peléan, ignimbrite-producing phase

On the next morning, August 27, there was a dramatic change in the character of the eruption. The paroxysmal, ignimbrite-forming phase began, during which pyroclastic flows were formed. These are fast-moving, turbulent, incandescent masses of solid fragments cushioned like a hovercraft on their own escaping gases. The fragments, mostly fine pumice, are known as ignimbrite.

At 4:30 A.M. wet ash fell at Java's First Point and at 4:56 A.M. an air wave arrived at Jakarta that must have left Krakatau at 4:43 A.M. At 5:30 A.M. there was a thunderous detonation, louder than any previous ones, and the eruption column reached a height of 30 km. Self (1992) regarded this as a co-ignimbrite ash cloud rising above pyroclastic flows. It was later found that most of the total volume of Krakatau's eruption products was emitted during the next eight hours as pyroclastic flows forming 4–7 main deposits composed of unwelded ignimbrite, extending for about 15 km around Krakatau (Self and Rampino 1981, Sigurdsson, Carey, and Mandeville 1991).

A huge tsunami left Krakatau at 5:46 A.M. and crashed into the town of Anyer after 6:00 A.M., destroying the port and killing almost all the inhabitants. Probably the same wave also severely damaged Teluk Betung and Ketimbang in Sumatra, where it beached the government gunboat, *Berouw,* and destroyed coastal villages. At 6:44 A.M. there was a second huge explosion and another large tsunami, and the column rose to a height of 40 km (25 miles).

By 9:00 A.M. it was becoming dark, even at Jakarta, and at 9:06 A.M. another very large tsunami was generated, one that swept into what were now the remnants of Anyer and Merak. At around 10:00 A.M., during heavy ash and pumice fall and in total darkness, the largest explosion of all, and the largest tsunami, marked the eruption's climax. The immense wave swept into the low foreshores of Java and Sumatra bordering Sunda Strait, demolishing what was left of Ciringen, Merak, Anyer, Teluk Betung, and almost 300 other towns and villages. The official death toll was 36,417, although hundreds of bodies were swept out to sea; probably over 40,000 people lost their lives. The *Berouw* was lifted from her stranded position on Teluk Betung's beach and carried nearly 3 km (2 miles) inland behind a small hill, 9 m above sea level; the 28 crew all perished. At Benkulu the onset, at 11:00 A.M., was negative and 50 m of beach was exposed. There was great destruction and loss of life also at the head of Semangka Bay, where the wave arrived at 11:30 A.M. Between 12:10 and 12:30 P.M. the sea reached its highest levels in the Havenkanaal at Jakarta and Tanjung Priok, about 2 m higher than normal, then receded to its lowest level, 3.15 m below normal, by 1:30 P.M. Two more waves, each of slightly less amplitude, were registered at Tanjung Priok at 2:30 and 5:30 P.M., the first stranding the *Prinses Wilhelmina,* which was then in port.

In the open ocean the sea waves traveled at over 700 km/hr (438 mph), crossing the Indian ocean to reach Bombay, Sri Lanka, Mauritius, Rodriguez, and Port Elizabeth in South Africa, 7,500 km (4,688 miles) distant. They reached Auckland over 7,700 km away, stranding vessels in port, and at Fremantle and Derby, Western Australia, the wool crop, baled on the docks ready for loading, was spoiled when the wave swept over it. At greater distances the tsunami's effects were smaller but detectable, as, for example, in Panama, San Francisco, Alaska, and Le Havre in the English Channel, 18,000 km (11,250 miles) away.

The detonation of the 10:00 A.M. explosion was heard in Singapore, Sri Lanka, Perth, and the island of Rodriguez at a distance of 4,700 km. An enormous mass of pumiceous ignimbrite was ejected; the deposit on southern Sertung and Rakata was first reported to have been 100 m thick. The finer particles of ejecta rose into the stratosphere to a height of between 60 and 80 km (37–50 miles). The ash cloud obscured the sun and darkness enveloped Sunda Strait and the surrounding regions, where the temperature fell by 5° C.

Shortly after the great explosion a mud rain began in the Sunda Strait area and lasted for three or four hours. On the *Gouverneur-Generaal Loudon* the mud was half a foot deep on the deck within 10 minutes. A fourth large explosion at 10:52 A.M., registered on the gasworks' pressure gauge at Tanjung Priok, was hardly heard or was inaudible in western Java yet was clearly audible in Sri Lanka. Throughout the eruption there was frequent lightning, often in the eruption column. At 11:10 A.M. on August 27, lightning struck the lighthouse at Java's First Point, burning convict laborers through their irons.

At some time after noon (the time is unfortunately imprecise) an important but fateful incident, unique in the eruption sequence, took place. What Verbeek called "the burning ashes of Ketimbang" were experienced about 40 km north of Krakatau. Following the 10:00 A.M. explosion, the family of Ketimbang's Dutch Controller, the Beyerinks, were in their hillside cabin, which was surrounded by about three thousand local people, when clouds of hot tephra swept several kilometers inland. As Mrs. Beyerink, the Controller's wife, later recalled: "Someone burst in shouting 'shut the doors, shut the doors' . . . Suddenly it was pitch dark. The last thing I saw was the ash being pushed up through the cracks in the floor boards, like a fountain . . . I felt a heavy pressure, throwing me to the ground. Then it seemed as if all the air was being sucked away and I could not breathe" (Furneaux 1964, p. 106). Mrs Beyerink and her three children, one of whom died, were badly burned by hot ash, and about a thousand people outside the hut perished. Simkin and Fiske (1983a, p. 81) quote the Rev. J. E. Tenison-Woods as reporting in 1884 that one of the people rescued at Ketimbang said: "I am sure I was burnt mainly by the fire that spurted out of the ground as we went along. At first, thinking only of the glowing ash showers, we endeavoured to shelter ourselves under beds . . . but the hot ashes came up through the crevices of the floor, and burned us still more." Quite clearly the tragic event was not the result of a simple fall of hot ash. Horizontal flows of considerable force had evidently traversed some 40 km of sea from Krakatau. The event has all the hallmarks of a fast-moving, ignimbrite-producing, pyroclastic flow (earlier known as a *nuée ardente*, or "glowing cloud").

At about 4:30 P.M. a tremendous explosion was heard in the Sumatran port of Krui, which was still in darkness, and both Verbeek (1885) and later Escher (1919a) believed an explosion occurring at 4:35 P.M. caused the last sea wave in August. Activity then diminished. Explosions were again heard in Bogor from 7:00 P.M. and in Jakarta from 8:00 P.M.; they increased in volume until 10:00 or 11:00 P.M., then declined. By the morning of August 28, Krakatau was again quiescent.

The Air and Sea Waves

Sounds and air waves

Because the eruption's accompanying air waves were fortuitously recorded on the Jakarta gasometer's pressure gauge (there was no barograph there), the times of the explosions can be determined with some precision. The large 10:00 A.M. explosion of August 27 was also registered at barographs in various parts of the world, which means that its energy can be calculated. The four great explosions of August 27, 1883, were at 05:30, 6:44, 10:02 (or 9:58), and 10:45 A.M.; these registered the greatest movement on the Jakarta gasometer pressure gauge.

Explosions were heard in Java and Sumatra up to 10:00 A.M., but afterwards sounds were heard only at fairly distant locations. The huge volume of pyroclastics put out at the time of the 10:00 A.M. explosion probably effectively blanketed the area that it covered, much as a snowfall muffles sound. Moreover, the velocity of sound in the atmosphere decreases, then increases, with altitude. Since sound waves from large surface explosions are reflected to earth from atmospheric air layers, the sound may focus at a zone some distance from its source, while only weak sound waves, diffracted along the ground, reach closer areas. A similarly anomalous silence was experienced during the Mount St. Helens lateral blast: witnesses enveloped by the blast generally did not even hear the trees falling around them (Keiffer 1981). Keiffer pointed out

that although in clear air sound can travel for about 10 km before it is attenu-
ated to one-third of its strength, the small particles within the blast cloud
would have absorbed and scattered the acoustic energy, increasing its attenuation more than a thousand-fold. Within the cloud, sound would have reached
equivalent attenuation within less than 10 m.

The Krakatau explosions were heard over an area comprising one-thir-
teenth of the earth's surface, from Rodriguez to Manila and as far south as
Perth and Alice Springs in Australia (see Figure 2). Of the major explosions,

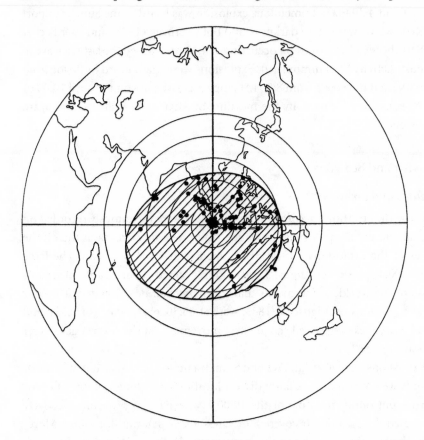

Figure 2. The earth's "Krakatau hemisphere." Black dots mark the places from which
sounds of the 1883 explosions were reported, the most distant of which, the island of
Rodriguez, is almost 4,700 km (3,000 miles) from Krakatau. The approximate range
of audibility of the eruption is indicated by the cross-hatched area; its eccentric shape
is probably due to the westward movement of the lower atmosphere in the region of
the trade winds. The concentric circles centered on Krakatau are spaced at intervals
of 10° "latitude" from the Krakatau "pole."

the first was heard in Alice Springs, Australia, the second and fourth in Sri Lanka, and the third in Sri Lanka, Perth, and Rodriguez, 4,700 km (3,000 miles) away, the only recorded instance of sounds being heard at such a distance from the source. Sounds were heard at greater distances westward than eastward of Krakatau, probably owing to the general direction of the prevailing trade winds. In many cases, such as at Diego Garcia, Burma, Singapore, Manila, Timor, Port Blair, and places nearer to Krakatau, the sounds were originally thought to be ships in distress.

The phenomenon of large air waves passing round the earth's surface had not previously been observed as a result of volcanic eruptions. The Royal Society subcommittee concerned with air waves and sounds collected the records of 46 continuously recording barographs. Using the records of the ones nearest to Krakatau, at Shanghai, Calcutta, Bombay, Mauritius, and Melbourne, Strachey (1884, 1888) calculated the time of origin of the third, great air wave at Krakatau (9:58 A.M.) and its velocity (674–726 miles per hour, or 1,078–1,161 km per hour). Strachey's estimate was corroborated by the time of the trace of the explosion on the gasometer pressure gauge at Jakarta, which indicated a time of origin at Krakatau of 10:02–10:07 A.M.

Strachey provided projections of the earth's Krakatau hemisphere and its antipodean hemisphere that marked the progress of the air wave of the 10:00 A.M. explosion on August 27 round the earth (Figure 3). At 5:00 A.M. on August 28 it returned to its point of origin, having traveled round the earth to Krakatau's antipodes, near Bogotá, Colombia, and rebounded, the first of three such round trips, and one more to Colombia, to be recorded. This was of considerable interest to meteorologists in showing how air waves are propagated in the atmosphere, and decades later the great Krakatau air wave was frequently cited in studies of man-made explosions.

Sea waves (tsunamis)

Tsunamis are long-period pressure waves in the sea, their wave lengths (crest to crest) often exceeding 100 km. The speed of horizontal motion is proportional to the square root of the water depth (and also depends on the wave's amplitude); as a wave approaches shore it slows, its amplitude increases, and the distance between waves decreases. As a result of this telescoping effect, waves that were quite small on the open sea can become high, devastating waves at the shoreline. Moreover, local topography is important; bays and estuaries have a funneling effect, increasing the height of the wave, so wave

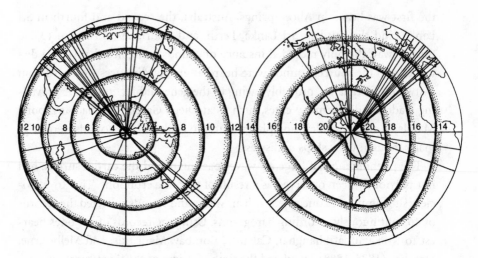

Figure 3. The passage of the atmospheric pressure wave around the earth to Krakatau's antipodes in Colombia, the first half of one of three round-trip journeys and one more to Colombia to be plotted. Isochrons are in hours from inception of the wave.

height may vary substantially between places on the shore only a few kilometers apart.

Reports of witnesses were often (understandably) imprecise as to the timing and extent of waves, but all agreed that the wave after 10:00 A.M. on August 27 was the largest. This was also the wave that traveled over great distances. The reception of the waves was registered and measured at two tide gauges in Indonesia, one of them at Tanjung Priok, and at gauges as far away as Alaska and the English Channel. The Royal Society committee (Wharton and Evans 1888) examined the records of the world's automatic tide gauges. Explosions at 1:36, 2:21, 9:16, and 9:58 A.M. all seem to have had accompanying tsunamis, the last one being the great tsunami. The final explosion, at 10:45 A.M., may or may not have generated a tsunami—there is insufficient evidence to decide.

The Aftermath

Morphological changes

As daylight was returning at 10.00 A.M. on August 28, 24 hours after the climactic event, the *Gouverneur-Generaal Loudon,* on her way to Jakarta from Teluk Betung in Sumatra, reported that two-thirds of Krakatau Island had disappeared (see Figures 4 and 10).

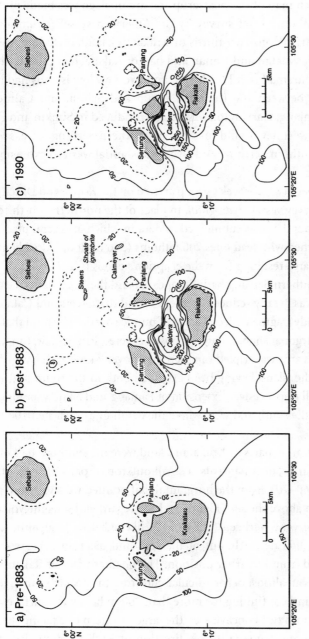

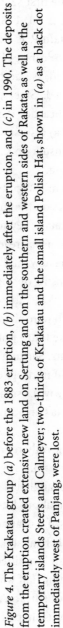

Figure 4. The Krakatau group (*a*) before the 1883 eruption, (*b*) immediately after the eruption, and (*c*) in 1990. The deposits from the eruption created extensive new land on Sertung and on the southern and western sides of Rakata, as well as the temporary islands Steers and Calmeyer; two-thirds of Krakatau and the small island Polish Hat, shown in (*a*) as a black dot immediately west of Panjang, were lost.

The first sketch map of the post-eruption archipelago was made on September 10, 1883, on a brief survey by the American vessel *Juniata* under Captain P. F. Harrington. Two-thirds of Krakatau Island, including the entire volcanoes Perbuatan and Danan and half of Rakata, had indeed disappeared, as had the small islet Polish Hat, and two new low islands had been formed: Steers, between the Krakatau group and Sebesi, and Calmeyer, northeast of Panjang. Harrington's sketch (reproduced in Simkin and Fiske 1983a) also shows a new island just east of northern Sertung, an eastward peninsula on southern Sertung, and a very substantial western extension of Rakata.

In mid-October 1883, Verbeek's group arrived on the *Kediri,* and Dutch surveyors on the survey vessel *Hydrograaf.* In place of the northern two-thirds of Krakatau they found a new, submerged caldera in which at several places no bottom was reached with lead lines 200 fathoms (366 m) long. Already much of the new land area resulting from the eruption's deposits had been lost. The peninsula on southern Sertung had disappeared and the large western extension of Rakata was greatly reduced. Verbeek found that Steers and Calmeyer, which were already fragmented, consisted of pumice overlain by mud that had cracked into polygonal shapes. Huge deposits covered Panjang, Sertung, and Rakata. Sertung's area was tripled (an increase of over 7 km^2) by a layer of ash 20–40 m deep, the island having been extended both in the south and west. There was a small northwestern extension of Panjang, and 5 km^2 was added to Rakata's area in the south and west where the depth of ash reached 100 m in places.

The only remaining parts of Krakatau island were the southern half of the Rakata volcano and Bootsmansrots, a small outcrop of parallel slabs of andesite angled at 60–70° from the horizontal that pointed westward and projected about 5 m above the sea (Figure 5). Judd regarded this as the volcanic plug of a Danan vent. Verbeek (1884 p. 12) likened it to "a gigantic club, which Krakatau lifts defiantly out of the sea," and believed it represented Danan lava tilted from its earlier position. The northern face of Rakata's remaining half was an almost perpendicular cliff from the peak, 813 m high, to the sea surface, and continuing for more than 200 m below the surface. The cliff was shaped into two concave arcs, the larger, western one facing the gap between Sertung and Panjang in the direction of Krakatau's missing part, and the smaller, eastern arc facing toward Panjang. This great scar, almost a section of the Rakata volcano through a plane near its center, must have been

Figure 5. Bootsmansrots and the southern half of the Rakata volcano (background) are the only remnants of Krakatau Island.

a magnificent sight in the early years after 1883; it is still an impressive feature today (Figures 5, 6).

The distribution of ash and pumice

Sea lanes in the bays and inlets of the Sumatran shore of Sunda Strait were not cleared of pumice until December. Most pumice floated out to the Indian Ocean, and well over 130 sightings were reported by ships in 1883, 1884, and

Figure 6. Landslides sometimes occur in the late afternoon on Rakata's great northern cliff scar (seen from Anak Krakatau, 1984).

1885. Extensive pumice rafts sometimes stretched to the horizon on all sides. The *Bothwell Castle,* for example, in mid-ocean, steamed through pumice for 2,150 km (1250 miles). A substantial amount remained afloat long enough to reach Natal in South Africa, over 8,000 km (5,000 miles) away, just over a year after the eruption. Some pumice rafts near Krakatau were thick enough for seamen to walk on, and some supported tree trunks. In December 1884, trees with trunks up to five feet in diameter and huge buttress roots jammed with pumice made landfalls in the eastern Caroline Islands (Kusae), 6,500 km (4,100 miles) northeast of Krakatau (Simkin and Fiske 1983a). Simkin and Fiske quote a letter to the Royal Society suggesting that pumice rafts may also have carried more macabre cargo over such distances. A quantity of clean human skulls and bones was found in Zanzibar, 6,170 km (3,900 miles) away, stranded all along the beach at high-water mark in July 1884, almost a year after the eruption.

Verbeek gathered information on the fall of ash from land observations and ships' reports, His resulting distribution map of ash falls showed a boomerang-shaped area, mainly to the west, the result of northeasterly and southeasterly trade winds prevailing during and after the eruption. Ash deposits fell away with distance. Very fine ash particles, carried high in the atmosphere, fell outside this area, and eleven ships to the west of Krakatau reported receiving such falls. The most distant report was from off the horn of Africa, 6,000 km (3,750 miles) west-northwest of Krakatau, 10–12 days after the eruption. The *Medea*, off the western coast of Australia and some 1,600 km (1,000 miles) south-southeast of Krakatau, received an ash fall on the night of August 30–31, indicating that not all the ash cloud moved westward.

Atmospheric effects

After the eruption there were widespread reports of blue or green suns and moons just after rising and before setting, and intense red skies at sunset and sunrise. These sightings occurred in a fairly regular sequence from east to west, and the vivid red skies reappeared in India and Sri Lanka after circling the earth for 12 days, indicating that the dust cloud causing the effect traveled at a speed of about 3,200 km (2,000 miles) per day. Verbeek had little doubt that Krakatau's eruption was responsible for these phenomena, although at the time other explanations were offered, such as ice crystals and a cloud of cosmic dust. He realized that extremely fine ash and dust particles would remain airborne for a long time and would be carried for vast distances. Verbeek believed that solid ash particles must have been responsible for the blue or green suns and moons, since they were observed largely during the first month after the eruption and only at places close to the equator. The remarkable sunsets were reported over such a wide area and for such a long period that he did not think that sufficient ash would have remained in the atmosphere to account for them. He ascribed these effects to water vapor condensing and freezing in the stratosphere, possibly with ash particles as condensation centers.

These effects received very wide attention, of course. Simkin and Fiske suggested that three-quarters of the world's population (then 1.4 billion) must have been aware of the remarkable sunsets and sunrises. The atmospheric effects were the subject of the major part of the Royal Society's report (Symons 1888), and their possible causes were the topic of lively discussion in scientific journals and newspapers.

From the time that the spectacular red skies first appeared, the Reverend Sereno E. Bishop, in Honolulu, made detailed observations that were valuable contributions to their understanding. He was the first to document the westward movement around the equator of the high-altitude eruptive products (what he called the Equatorial Smoke Stream) as evidenced by their atmospheric effects "announcing the great convulsion around the earth, and belting the globe with fiery skies . . . the vast cloud mass . . . , having shed its heaviest dust, spreads far abroad, and begins its slow march of months across the continents to cover the whole globe with the same wonderful glares" (Bishop 1884a, p. 110). He noted that "the after-glows are seemingly more brilliant than the primary ones . . . a vast blood-red sheet covers the west . . . As it sinks and rests low on the horizon . . . it precisely simulates the appearance of a remote and immense conflagration" (Bishop 1887, p. 63). Indeed, on November 27, 1883, fire engines were called out in New York state and Connecticut when afterglows in the western skies falsely signaled extensive fires in the distance.

Russell and Archibald (1888) plotted the progress of the main sky phenomena westward from August 26 until they reappeared in the Indonesian area on September 9 and recorded the gradual broadening of their distribution to include higher latitudes by the end of November. They noted that the height of the dust veil, as calculated from the duration of the twilight glows, gradually declined, from about 32–40 km (20–25 miles) to about 18–20 km (11–13 miles) by January 1884, as material in the cloud gradually subsided.

A broad, opalescent circle surrounding the sun was first described by the Rev. Bishop in December 1883 (Bishop 1884b, p. 260): "the very peculiar corona or halo extending from 20° to 30° from the sun . . . may become visible around the globe, and give ample opportunity for investigation." Noted 10 days after the eruption, Bishop's ring, as it is now known, was indeed widely evident; it reached maximum brightness about a year later and was last seen in 1886. It occurred again after several notable eruptions in the present century. Like the afterglow, Bishop's ring can provide information (from its radius) on the size of particles in the dust veil. A purplish patch in the shape of a trapezium was also reported fairly often in mid-latitudes from 20 to 50 minutes after sunset, occasionally repeated 40 minutes later, at an elevation of about 20° from the horizon. Following the 1982 El Chichón eruption, Lamb (1983) described both Bishop's rings and purple patches (at 22–27 and 52–57 minutes after sunset, respectively) in England in January 1983. His observations led him to conclude that there was a main layer of particles at a height of 18–20 km and a less dense layer at about 35 km. He suggested that particles of diameter 0.8 to 10 mm would have produced the ring of 20° radius that was seen after August 1883, and that appearance of the ring may be limited to times when only one of the stratosphere's overlapping dust layers is involved, probably at one particular stage of an explosive eruption.

Eruption clouds produced by recent eruptions, such as that of El Chichón, have been monitored by laser radar and satellites and sampled by meteorological balloons and aircraft—all of which were unavailable, of course, in 1883. El

Chichón injected large quantities of sulfate into the stratosphere because of the geological layers through which it erupted, and in January 1983 the highest layer of its dust veil consisted of droplets of sulfuric acid so small that they were not detectable by laser radar. The striking parallel with the atmospheric effects of Krakatau suggested that sulfuric acid droplets played a significant role at Krakatau, also, as suggested earlier by Archibald (Russell and Archibald 1888).

The effect on world climate

Simkin and Fiske noted that the Krakatau eruption has been termed a turning point in history for the science of meteorology. The first evidence of circulation patterns in the stratosphere was gathered by following the path and rate of progress of Krakatau's high ash clouds through their unusual atmospheric effects. The effect on the earth's temperature and climate was used, along with other such examples, in the framing of "nuclear winter" scenarios a decade or so ago, when the possible climatic effects of nuclear war were brought to our attention. General interest has been revived, of course, with each subsequent major eruption.

Benjamin Franklin may have been the first person to link climatic change with a volcanic eruption. In a paper read in 1784 to the Manchester Literary and Philosophical Society in England, when he was the first ambassador of the newly formed United States to France, he ascribed the unusually severe winter of 1783–1784 to a reduction in the sun's normal heating effect by a "dry fog" that prevailed in much of Europe over most of the 1783 summer. "The rays of the sun were indeed rendered so faint in passing through it, that when collected in the focus of a burning glass, they would scarcely kindle brown paper" (Franklin 1784, p. 359). As a possible cause he speculated (p. 360) "whether it was the vast quantity of smoke, long continuing to issue during the summer from Hecla [Laki in Iceland] . . . which smoke might be spread . . . over the northern part of the world, is yet uncertain." Laki's massive fissure eruption, the greatest basalt eruption in history, generated about twice the aerosols of Krakatau.

In the second decade of this century the general question of the effect of volcanic explosions on climate received considerable attention. Great reductions in the intensity of solar radiation were shown to have followed notable volcanic eruptions. Abbot and Fowle (1913) and Kimball (1913), as examples, found that periods of low intensity of direct solar radiation and low temperature in the United States were correlated with periods of great amounts of dust in the atmosphere as a result of the eruptions of Krakatau and nine other major eruptions. Recovery to normal radiation values took from 17 to 34 months. Decades later, after examining climatic records over the previous 100 years, Wexler (1951) suggested that frequent, large volcanic eruptions may affect world climate for very much longer periods.

Rampino, Self, and Fairbridge (1979) warned that dating errors and poor resolution (the precision with which an effect can be distinguished from natural, background variation) make it difficult to determine whether eruptions lead to cooling or vice versa, and that a case could be made equally for climatic cooling triggering explosive volcanism. They found that numerous eruptions coincided with climatic cooling that had begun before the eruptions, and many periods of cooling could not be associated with any large eruption. For example, the famous 1815 Tambora eruption, said to have resulted in the "year without summer" (1816) in North America and Europe when average temperatures were lowered by from 0.5 to 1° C, occurred at a time of pronounced low sun-spot activity (indicating low solar ultraviolet output), and the associated minor glacial advances began before the eruption. Krakatau also erupted after decrease in temperature had already begun.

The Tambora eruption, like Krakatau's, was large-volume, paroxysmal, Plinian, and ignimbrite-forming (Volcanic Explosivity Indices 7 and 6, respectively, on a scale of 1 to 7). Tambora injected over twice the volume of ash into the atmosphere (50 km^3 from Tambora, 20 km^3 from Krakatau) and caused twice the reduction in the earth's mean annual temperature (0.8°C versus about 0.4°C) (Rampino and Self 1982). Agung (Bali, 1963) and El Chichón (Mexico, 1982), were relatively small-volume Vulcanian or sub-Plinian eruptions with Volcanic Explosivity Indices of 4. Each ejected less than half a cubic kilometer of rock, yet they produced climatic effects as great as those of Krakatau and Tambora. This was because their rocks were exceptionally sulfur-rich andesite, and their eruptions injected larger amounts of sulfur volatiles (SO_2, H_2S) into the aerosol layer of the stratosphere than did the relatively sulfur-poor eruptions of Krakatau and Tambora.

Krakatau left evidence of its eruption in two unexpected depositories—ice and trees. Large ice sheets, such as those of Greenland, contain a record of the snows that built them, including microparticles of volcanic fallout. Once calibrated for date, cores of ice sheets can therefore provide information on past volcanic eruptions (Hammer, Clausen, and Dansgaard 1980). After examining an ice core from Greenland at the 70-m level (corresponding to the period 1884–1891), Herron (1982) found that in the three years following the

Krakatau eruption sulfate levels were at least twice the nonvolcanic average; levels peaked at least a year after the eruption, indicating a stratospheric travel time of about a year. Rampino and Self (1982) confirmed not only that the amounts of sulfate from the Tambora, Krakatau, and Agung events were much more similar than the widely differing volumes of matter ejected would suggest, but also that the eruptions were fairly well correlated with reductions in world temperature. They concluded (Rampino and Self 1984) that sulfate aerosol contributes greatly to the blocking of solar radiation, and it now appears to be generally accepted that the long-lived clouds in the stratosphere after volcanic eruptions are sulfuric acid aerosols, and that it is the amount of sulfur-rich volatiles injected into the stratosphere by volcanoes, not the quantity of ash, that determines their climatic impact.

One might find trees an unusual source of information on Krakatau's climatic effect, but those in temperate zones may hold records of climatic variation in their wood. Recent investigations indicate that a drop in temperature, such as that after a volcanic eruption, may have a detectable effect on the growth rings. A distinctive "frost ring" is sometimes formed when below-freezing temperatures during the growing season cause ice formation and dehydration, compressing the outermost zone of the tree's wood before secondary thickening and lignification of the immature xylem cells in the annual ring is complete. La Marche and Hirschboeck (1984) found a remarkable coincidence of frost-ring dates in tree rings from subalpine pine trees in the western U.S.A. with five notable eruptions, including Krakatau. Moreover, the 22 eruptive events from 1500 to 1968 that had a dust veil of at least Krakatau's significance showed a highly significant correlation with the formation of frost rings within the next two years (a frost ring was never found to precede an eruption). In sum, major eruptions are likely to be followed by notable effects of frost. The frost event in 1884, the year after Krakatau's eruption, was particularly notable, as frost rings occurred in more than 50 percent of the trees at all sites examined.

3

THE GREAT ENIGMAS
OF 1883

Ah, but a man's reach should exceed his grasp,
Or what's heaven for?
Robert Browning

In spite of the fact that Camus and Vincent (1983a) published a paper entitled *Un siècle pour comprendre l' éruption du Krakatau* (A century for the understanding of Krakatau's eruption), many aspects of the eruption are still not fully understood. Krakatau guarded its secrets well. After considerable advances in volcanological techniques, subsequent research both at the Krakatau site and at other volcanoes, and the thorough airing of numerous theories, the explanations for various phenomena remain controversial: the way the caldera was formed, the causes of the loudest natural explosion known, and the generation of the largest tsunami known to have been produced by volcanic action. There is no substitute for direct evidence at the time, or, to paraphrase De Nève, post-mortem examinations are no substitute for reports of physicians attending living patients! Because of the nature of the event, however, no physician could get near the patient at the critical time, and in any case the lights had been switched off. The climactic act, the great explosion of around 10:00 A.M., was played in an empty, dark theater. The Krakataus were uninhabited—luckily so, for no one would have survived on the islands; there were no survivors from the 3,000 or so people on the island of Sebesi, 13 km away. Even the great tsunami was seen by very few. At Anyer, for example, the few survivors of the earlier wave had fled, and the great wave arrived in total daytime darkness. The "blackout" lasted for 57 hours within a radius of 80 km (50 miles) and for 22 hours within 200 km (125 miles).

Thus, rather as they must do to study a prehistoric eruption, volcanologists have had to work with indirect evidence: reports of witnesses and entries in ships' logs, the gross morphological changes to the archipelago itself, the airborne cloud of fine particles and its informative atmospheric and climatic effects, and the records of the air and sea waves at barographs and tide gauges around the world. Finally, the eruption products that were deposited on the remnants of the archipelago and on the sea bed have long been available for study. This large-scale detective work, which is still under way, has led to stimulating controversy.

The two most discussed problems of the event are the manner of formation of the submarine caldera and the way the tsunamis, particularly the great tsunami, were generated, and what relationship, if any, there is between the two. Debate on the formation of the tsunamis, perhaps the most puzzling aspect of the eruption, began almost immediately. Wharton and Evans (1888) maintained that collapse of the island would have produced sea waves much greater than those that occurred, and they concluded that the sea waves were caused by a *raising* of the sea bed and by ejecta entering the sea. Verbeek (1885) thought that subsidence of the northern part of the island could not have formed sea waves as large as those that occurred, since after many eruptions it would have been but a hollow shell by the time it collapsed. Most explanations of caldera formation and tsunami generation conform to one of three generic eruption scenarios that have been proposed—explosion, collapse, and a third theory incorporating elements of both.

Explosion Hypotheses

The idea that the eruption was explosive, caused by the reaction between seawater and molten magma, was the generally accepted textbook view for some decades. Escher (1928), drawing a comparison with the 1906 Vesuvius eruption, believed that an immense cylinder 4 km deep and about 800 m in diameter was cored out of the conduit by explosion. The walls of the old cones then slid in along a funnel-shaped gliding plane, their rocky debris forming a thick layer beneath the great pumice deposit of 10:00 A.M., August 27. Yokoyama (1981, 1982, 1987) is the modern proponent of the explosion hypothesis. By a comparison with the 1956 Bezymianny eruption on the Kamchatka peninsula, he calculated that the energy of the 10:00 A.M. Krakatau explosion was suffi-

cient to account for the volume of magma removed or expelled. He put the
origin of the explosion at the deepest point of the caldera, west of Danan. In a
gravity survey Yokoyama and Hadikusumo (1969) had found that the deposits
of low density had a funnel shape, which they considered inconsistent with a
collapse hypothesis but indicative of an explosive event. The seismic reflection
survey of Sukardjono, Lubis, and Sukorahardjo (1985) also showed that the
eruption deposits had a concentric, nondirectional distribution, and this was
confirmed by a recent underwater survey (Sigurdsson et al. 1991). These reports are all consistent with an explosion explanation. Yokoyama thought that
Steers and Calmeyer could have consisted of explosive debris.

By constructing a refraction diagram of the tsunami in the Sunda Strait region, Yokoyama calculated that at the area of its origin near Krakatau the sea
level was raised by 30–40 m. He ascribed the small sea waves to ejecta entering
the sea, as Verbeek had done, but believed that the great tsunami was caused by
a violent explosion, for, he maintained, tsunamis can only be caused by sudden
water movements. Phreatomagmatic eruptions (in which water and magma
interact), however, are characterized by ejecta of very small grain size, and the
Krakatau ejecta did not have this characteristic. Yokoyama suggested that if
magma were separated quickly from water by a seal of chilled lava, the typical
small grain size might not be acquired. He maintained that any explanation of
the great tsunami must account for the following: the tsunami and air wave
were generated simultaneously with the largest explosion; the first change in
sea level was positive at Tanjung Priok and at the 14 more distant ports at
which it was properly registered (which meant that water was displaced upward at the point of origin); and the tsunami spread concentrically. A landslide or avalanche tsunami and a collapse of the island to the north would both
tend to produce a directional tsunami. Only a submarine explosion, he believed, satisfies all the requirements (Yokoyama 1987). All investigators, however, have found that the ignimbrites are relatively poor in lithics (fragments of
dense rock), and explosion theories will continue to be questioned until the vital evidence required, the massive deposits of rock fragments that must have
been produced, is found.

Collapse Hypotheses

Verbeek (1885) believed that the eruption probably became submarine at
10:00 A.M. on August 27 because extensive mud rain was reported only after
that time. On the basis of the time of origin of the great tsunami, he thought

that subsidence of the wall of one or both of Danan and Perbuatan occurred shortly before 10:00 A.M.; such a collapse would allow access to large quantities of seawater and cause submarine explosions. The northern half of Rakata, with a volume of about a cubic kilometer, having been previously undermined and fractured and, unlike the southern half, not resting on an "Ancient Krakatau" foundation, then also crashed into the sea (Figures 4 and 10). Verbeek considered that seawater would have rushed in to fill the newly formed submarine cavity and that the movement of the water would have caused the initial retreat of the sea observed at some places along the mainland coasts. He calculated that the volume of ejected material was 18 km^3, of which at least 12 km^3 fell within 15 km of Krakatau. In his view the earlier sea waves—the one that destroyed Anyer and other, smaller waves—must have been caused by these enormous masses of ejecta falling into the sea. Verbeek considered, however, that it was the slumping of the northern half of Rakata volcano into the sea that created the eastern part of the caldera and generated the great tsunami.

Following his study of the stratigraphy of the deposits, Stehn (1929) concluded that Perbuatan subsided at 4:40 A.M., much earlier than Verbeek had thought, and that the collapse and explosion at 10:00 A.M. occurred at the deepest part of the caldera, west of Danan, and caused the great sea wave. The great mass of pumice was ejected as a consequence of the explosion, and this was when a southwest fracture system (graben) occurred in the sea bed. After this came the subsidence of the northern part of Rakata. Like Verbeek, Stehn regarded the lack of basalts in the rock debris found in the ignimbrite as evidence that the northern half of Rakata, Krakatau's basalt (and easily its largest) volcano, was lost by subsidence rather than by explosion. Stehn concluded that the second great explosion, at 10:52 A.M., which destroyed Danan, ejected blocks of rock and pumice and formed the last layer of the 1883 deposits. Both Verbeek and Escher believed that there was another subsidence, perhaps at 4:35 P.M., and this was when the fracture of the sea bed between Panjang and Rakata occurred. Further subsidence then caused the last tsunami of August 1883. The theories of Verbeek and Stehn thus require the volcano to be collapsing *before* the main explosions and ignimbrite emissions occurred.

Dana (1890) had also suggested that the caldera was formed by collapse, but that the island collapsed into a cavity left by rapid emission of pumice. This view has since received wide support. Reck (1936) noted that Rakata's great arcuate northern cliff is analogous to the scalloped bays along the caldera of Santorini (Thera) in the Mediterranean, which resulted from a foundering event, not from explosion. In a review of calderas and their origins, published after visiting Krakatau, Williams (1941) selected Krakatau as the type example of calderas formed in what is now regarded as the classic manner, by collapse after exhaustion of the magma chamber following violent emission of large volumes of pumice. Again, the low proportion of old rock fragments from Danan, Perbuatan, and Rakata in the ejecta was taken as crucial evidence against the explosive decapitation of these volcanoes. Williams proposed that the form of caldera recognized by Escher (1919b)—two basins separated by a northwest-trending ridge and with 100-m-deep grabens leading out to the southwest and southeast—supports collapse and engulfment as the mechanism of caldera formation. He found no evidence of the series of submarine explosion funnels or thick layer of rocky debris that Escher's theory required. Williams did not think that the fall even of great masses of pumice into the water could have generated a wave exceeding 40 m in height. He considered that the tsunamis were generated by the formation of submarine grabens and by material slumping from land into the sea. He agreed with Stehn that the last two great explosions (at 10:00 A.M. and 10:52 A.M. on August 27) originated to the southwest of earlier vents, at the deepest point of the caldera floor, and that these account for more than 90 percent of all the 1883 ejecta, nearly all of which had been deposited, in the sea and on the islands, as an almost unstratified pumice layer up to 100 m thick. He believed that the 10:00 A.M. explosion was followed almost at once by collapse of much of Rakata's cone into the sea and the initiation of the great wave, as well as the formation of the western submarine graben. Because the upper layer of all deposits included lithic blocks, mostly of andesites like those of Danan, with pieces of coral and shells, Williams supported Stehn's view that the final great explosion, at 10:52 A.M., involved demolition of the remaining part of Danan. There appears to have been no collapse then, for there was no accompanying tsunami. The collapse occurred six hours later, generating the last tsunami and producing the eastern submarine graben. Sigurdsson et al. (1991) later confirmed that on Panjang the massive ignimbrite deposits are overlain by a dark grey pyroclastic flow unit that contains many lithic blocks up to 2 m in diameter—the final deposits of the eruption sequence.

The Dana-Williams model of caldera formation was supported by Self and Rampino (1981) and Self (1992), who considered that the major pyroclastic flows were generated by collapses of the eruption column following the major explosions (Figure 7). The paucity of lithic debris in the deposits, their lack of the typically phreatomagmatic, very fine grain size, and the fact that pyroclas-

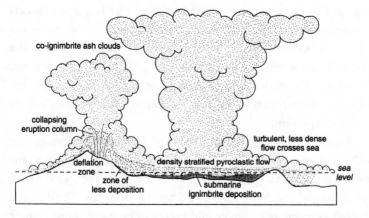

co-ignimbrite ash clouds

collapsing
eruption column

deflation
zone

zone of
less deposition

density stratified pyroclastic flow

submarine
ignimbrite deposition

turbulent, less dense
flow crosses sea

sea
level

Figure 7. Self's representation of the movement of low-concentration pyroclastic flows from Krakatau's collapsing eruption column on the left. After the flows had undergone density segregation and density filtering, the low-concentration part of one flow moved over the sea to Sumatra while the denser stratum moved under the sea surface. The turbulent aerial flow may have been responsible for the "burning ashes of Ketimbang."

tic flow deposits cover southeastern Rakata on the other side of the peak, all indicated to them that the island was not collapsing during the eruption. They identified an area as a possible submarine slump of part of Rakata's northern half into the newly formed caldera, which would mean that the Rakata material could not have been removed earlier by explosion. Moreover, the caldera evidently was not extensively filled with 1883 deposits, supporting the Dana-Williams view that it was formed by subsidence.

Self and Rampino attributed the major tsunamis to the entrance of several cubic kilometers of pyroclastic flows into the sea, citing the great 1815 Tambora eruption from a crater over 15 km inland and 2850 m high. The Tambora tsunami could not have been generated by either collapse or explosion but was probably caused by the immersion of pyroclastic flow deposits at Tambora's base. They acknowledged, however, that slumping, for example of the northern half of the Rakata cone, could have been implicated in the later waves.

Francis (1985) contended that the hypothesis that entry of pyroclastic flows into the sea generated the later tsunamis depended entirely on accurate timings of the air and sea waves. As a result of an innovative and detailed analysis based not on times of origin but on the relative arrival times of air and sea waves at Jakarta, he concluded that it was most unlikely that the great sea wave was produced in this way. He followed Verbeek, however, in suggesting only that this may have been the cause of some of the smaller tsunamis. Francis noted that the second-largest tsunami was generated at 9:20 A.M. along with a fairly large air wave, and this precursor explosion may have destabilized the volcano prior to a major collapse that caused the great 10:00 A.M. tsunami. Francis believed that the evidence supported a scenario similar to that proposed shortly after the eruption by Verbeek—collapse of the volcanic edifice triggering a major explosion.

A curious feature of Krakatau's sea waves had puzzled investigators for many years. Waves reaching points far from Krakatau, often beyond intervening land, but not those arriving on nearby shores and following a direct path from the island, arrived much too early to be normal tsunamis. The discrepancies were often as much as 10 hours, and highly improbable diffraction or refraction around land barriers, and impossible speeds, would be demanded if the observed displacements at the arrival points resulted from simple tsunamis. Both Wharton and Verbeek had ascribed these "distant waves" to coincidental seismic tremors. It was later shown that in these cases the sea wave was excited by the air wave—received energy from it. The air wave passed over land barriers, of course, and was capable of re-exciting the sea wave if there was a sufficiently long fetch. This accounted for both the apparent avoidance of land obstacles and the "early" arrival (air waves travel at about twice the speed of sea waves). The disturbance in the sea, however, was greater than could be accounted for purely by the effect of the atmospheric pulse on the water. It has since been suggested that the wave was amplified by a process of resonant coupling to free waves in the atmosphere with velocities of about 790 km/hr (220 m/sec), the same as the velocity of waves in the deep ocean. This coupling permitted a highly efficient transfer of energy from these atmospheric waves to the ocean, resulting in sea waves with amplitudes several times the hydrostatic values (Ewing and Press 1955, Harkrider and Press 1967).

Latter (1981) showed that the arrival times of many, but not all, sea waves at Tanjung Priok coincided with the arrival there of air waves caused by some of the minor eruptive events. The largest air wave, caused by the 10:00 A.M. explosion, reached the pressure gauge at Jakarta's gas works at 10:08 A.M., when a large sea wave was recorded on the tide gauge at Tanjung Priok. This sea wave should have arisen at Krakatau at 7:40 A.M., but no event took place there to cause it at that time. Latter concluded that it was in fact an air-sea wave, caused by coupling of the large air-pressure wave to the sea. The great tsunami, presumably generated at the same time as the largest air wave, had a much larger amplitude and did not arrive at Tanjung Priok until more than two hours later. Latter's calculations showed, however, that this tsunami should have been generated at Krakatau 13 minutes before the large explosion. He therefore suggested that it arose not at Krakatau but at a site closer to Jakarta by a distance equivalent to the 13-minute discrepancy, the outer edge of Calmeyer, and that it was caused by pyroclastic flows entering the sea. Partly supporting the general hypothesis of Self and Rampino, he concluded that three other tsunamis had been formed in this way.

The detailed 1990 field studies of Sigurdsson et al. (1991) confirmed that the submarine deposits were similar to those on land. They calculated the volume of the present caldera as 8.9 km^3 and were able to correct for subsequent erosion in estimating the dense rock equivalent of the total erupted products as 9 km^3. Like other studies, this one supports van Bemmelen's calculation that when the pumice erupted is re-assessed as magma, its volume corresponds closely to the loss in volume of the island or volume of the caldera (van Bemmelen 1909). This correspondence and the paucity of scattered lithic debris are strong arguments in favor of a collapse explanation.

Debris Avalanche and Lateral Blast

Following field work on the islands in 1981 and 1983, a team of French volcanologists proposed a mechanism of caldera formation combining elements of both explosion and collapse hypotheses (Camus and Vincent 1983a,b; Camus, Diament, and Gloaguen 1992; Vincent and Camus 1983, 1986; Vincent, Camus, and Larue 1984). They compared the Krakatau eruption with that of Mount St. Helens in Washington state, U.S.A, in 1980, the study of which has allowed volcanologists to review several historic and prehistoric eruptions *"avec une nouvelle optique,"* with a new perspective. Both eruptions resulted in a horseshoe-shaped "avalanche crater," hummocky-surfaced deposits of a debris avalanche, and deposits of a directed lateral blast or a zone of devastation caused by it. Following the Mount St. Helens model, they hypothesized that at 10 o'clock on August 27 the 800-m-high Rakata volcano was situated at a structural unconformity, the edge of the prehistoric caldera, and that its disequilibrium in this position led to massive gravitational failure. Only gravitational flank failure, not the collapse of the whole island, could account for the fact that most of the submarine caldera lies to the west of the former Krakatau island. An avalanche, with a flow of debris, occurred, followed by a lateral blast (both directed toward the north by the shape of the newly formed northern cliff of Rakata), thus allowing seawater to reach the magma. Rakata's northern cliff was the southern boundary of the "avalanche caldera" so formed (Figure 8). A submarine eruption with extensive ignimbrite emissions began soon afterward and ended with the collapse of the caldera, according to the Williams model, after over 10 km³ of magma had been ejected. Their hypothesis thus combined entry of large volumes of water with caldera collapse after the emission of ignimbrites. The great 10:00 A.M. tsunami could have been caused by debris flow or, more likely, by lateral blast, in the same way that water was displaced in Spirit Lake by the Mount St. Helens blast in 1980. The tsunami may well have originated 15 minutes before the explosion, as Latter had calculated. Had it been caused by the debris flow, then the succeeding lateral blast of the explosion could have overtaken it and increased its amplitude, thus producing the largest volcanic tsunami known.

Camus's group cited the results of their 1983 bathymetric survey over the sites of Steers and Calmeyer, which showed a hummocky submarine surface of the deposits, marked off from the adjacent smooth relief of what were probably submarine ignimbrite flows. At the base of the west Panjang cliff, they had found hydrothermally altered andesite blocks resting on pre-1883 lava and covered by the thick layer of ignimbrite. This find supported their contention that the tsunami was related to the debris flow. They pointed out that debris avalanches are extremely mobile, the median horizontal distance traveled being about ten times the vertical drop; some have involved more than 20 km³ of material and reached distances of up to 100 km (63 miles) from the eruption point.

Stehn had earlier observed that the curvature of Rakata's great cliff does not correspond to that of the 1883 caldera, and he had also noted the paucity of basaltic rocky debris in spite of the fact that about half Rakata had disappeared. Camus and Vincent now argued that if the great cliff was the result of a debris avalanche flowing to the north, then both these apparent paradoxes were explained.

Verbeek (1885) had described the "burning ashes of Ketimbang" as a lateral blast and had remarked that the devastation ended abruptly northwest of Ketimbang, where the intervening island of Sebesi offered protection from any direct, horizontal effects of the eruption. Camus and his co-workers regarded the event as a Mount St. Helens–type directed lateral blast limited to a fairly narrow sector to the north of the archipelago, which could also have included Lageundi, Sebesi, and Sebuku. Simkin and Fiske (1983b) suggested that the massive area of floating pumice could have provided a "land bridge" for pyroclastic flows not ordinarily able to travel over water. Self and Rampino speculated that the "burning ashes" may have been ash-cloud surges rolling forward

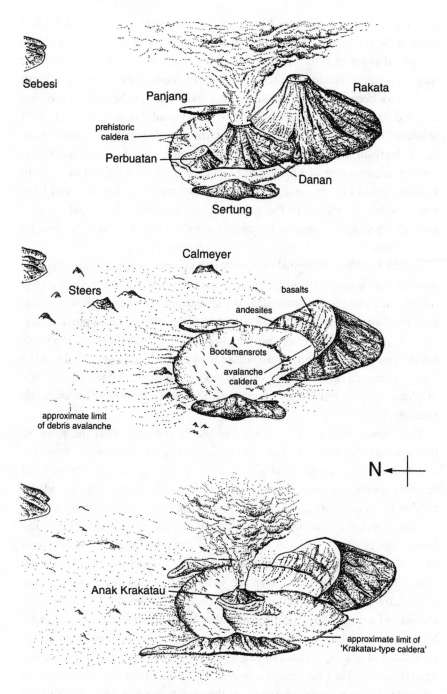

Figure 8. The eruption scenario proposed by Camus and Vincent, represented here by views before the 1883 eruption *(top)*, immediately after it *(middle)*, and after the creation of Anak Krakatau in 1930 *(bottom)*.

over the sea surface for some 40 km, after their parent pyroclastic flows had entered the sea. Similar ash clouds moved over water in the St. Pierre (Martinique) disaster of 1902. Sigurdsson's group also considered that pyroclastic flows spreading (presumably radially) from the base of the eruption column could have been segregated gradually by gravity near to the islands, evolving into a denser, submarine component, which evidently mixed very little with seawater (for it is poorly sorted and contains abundant pumice), and a more dilute, buoyant, aerial component. The upper component could have had an internal temperature of up to 600° C and, perhaps with the aid of a steam cushion, could have traveled at a speed of 540 km/hr (338 mph), and for a much longer distance than the submarine component. This upper component, they believed, accounted for the hot flows reaching Lageundi, Sebesi, and Sebuku, and (presumably in about 4 minutes) Ketimbang.

Camus and Vincent argued that, although submarine, the Plinian eruption may not have been strictly phreatomagmatic in the sense of water having a significant influence on its dynamics. At the very high discharge speed through only several tens of meters of water, the fine grain size typical of phreatomagmatic eruptions may not have been aquired. Stoiber and Williams (1985), after analyzing ash samples from Krakatau held in museums, concluded that the water-to-magma ratio (about 0.13) was too low for the very fine grains typical of phreatoplinian eruptions to have been formed (a ratio of 0.2 is evidently necessary).

The French volcanologists made the important point that for some 17 hours before the 10:00 A.M. tsunami, the eruption was Plinian (or Vesuvian) and its deposits were associated with subaerial pyroclastic surges that settled at temperatures high enough for their upper levels to be incipiently welded. After this tsunami, however, the eruption was ignimbritic (or Peléan) and the massive, unsorted flow deposits of this phase were emplaced within a few hours. The 10:00 A.M. tsunami thus marked a major change in the course of the eruption, and its cause must also explain the simultaneous and dramatic change in the eruption's character and style. They argued that sudden release of pressure owing to flank failure of the volcano was this cause, and that immersion of a debris flow from the resulting avalanche triggered the tsunami. Vincent, Camus, and Larue (1984) had confirmed Stehn's observation that the great mass of ignimbrite on Panjang rested on a pyroclastic fall layer that had been eroded and in some places removed. Like both Stehn and Williams, they attributed this erosion to the passage of the great tsunami. Camus and Vincent (1983a,b) had found that the bottom of the Sebesi channel was "hummocky" (later confirmed by Sigurdsson et al. 1991), a conformation not expected of typical ignimbrite deposits. Like Verbeek, Stehn, and Latter, they argued that the ignimbrite emissions *followed* the tsunami and therefore could not be advanced as its cause.

Finally, Camus has recently noted that if the explosion or debris avalanche occurred before the massive ignimbrite eruption, any resulting lithic material would have been buried beneath the thick ignimbrite deposits, and the short submarine cores taken by Sigurdsson's group (Sigurdsson, Carey, and Mandeville 1991) could not have reached it. Absence of evidence, implied Camus, is not evidence of absence.

ｅ

Crucial questions bear on these interpretations: Could coupled air-sea waves reach Jakarta over the short intervening fetch of sea (see Yokoyama 1987)? Was the island collapsing before the great ignimbrite emissions, or were they the cause of collapse? Did the massive pyroclastic eruption occur just before, or just after, the great tsunami?

Perhaps the above hypotheses of caldera formation and tsunami generation are not all mutually exclusive. It is likely that during the course of so complex an eruption many events cited as evidence for the competing theories occurred—submarine faulting, flank failure, massive collapses of pieces of Krakatau island into the sea, submarine explosions, and the entry into the sea of huge quantities of pyroclastic flows—and some of these may have been occurring simultaneously. The question remains, as in Verbeek's time: Which of them was the predominant cause of caldera formation, and which was the cause of tsunami generation? And are they the same? The answers may still rest beneath the thick blanket of submarine deposits at the bottom of Sunda Strait.

4 WHY KRAKATAU?

> [The All-father] flung the serpent into the deep sea which
> surrounds the whole world, and it grew so large that it now
> lies in the middle of the ocean round the earth, biting its
> own tail . . . the Midgard Serpent is writhing in giant fury.
>
> *The Deluding of Gylfi,* Icelandic saga (quoted in Fridriksson 1975)

Why should the remarkable events outlined in the previous chapters have occurred at Krakatau? Why not, for example, at New York, or even Jakarta? What
is it about this location in Sunda Strait that such cataclysmic eruptions have
occurred there? To answer this question we need to enlarge our perspective, to
"the whole world" in fact.

It is now known that the earth's surface is composed of about seven great
tectonic plates and a number of subsidiary ones. These plates make up the
earth's thin "crust," which covers the much deeper interior layer of the mantle.
A submarine ridge runs round the earth along the floor of the great oceans,
rather like the seam of a tennis ball (or the Midgard Serpent). New ocean crust
is formed as molten rock, or magma, rises in upwelling convection currents of
the mantle and adds new material to the plates on each side of the ridge. The
mid-oceanic ridge thus forms an accretion margin of the plates on each side of
it. Put another way, new ocean floor is being created and is gradually moving
away from the ridge on each side—at a rate of several centimeters per year.

Some plates comprise both continental and oceanic crust, and as the plates
move away from the ridge the continents are carried along, like corks fixed to
a conveyer belt. The deep-sea trenches of the Pacific and Indian oceans, lying
along the edges of continents or along island arcs, are convergent margins of
plates, where moving plates converge and plate material is being consumed.

The crust under the deep oceans is 6–8 km thick and is composed largely of dense, black, fine-grained, basaltic, igneous rock, but the much thicker continental crust is less dense. As a result, where continental and oceanic crusts meet at convergent plate margins, the continental crust overrides the plate with an advancing margin of oceanic crust, which is subducted, passing down beneath the opposing plate along a deep ocean trench, the subduction zone. Thus plate material is being created at the mid-oceanic ridge and consumed at the trenches.

The release of strains and stresses created by the subduction of this crustal material triggers earthquakes, and the friction caused by the movement of such large masses of material results in volcanic eruptions. Volcanoes (and the epicenters of earthquakes) do not occur at random over the earth's surface. They occur predominantly along the margins of the tectonic plates. Submarine volcanic activity occurs along the mid-oceanic ridges (accretion margins) themselves, and occasionally eruptions build up part of the ridge above sea level, as in Iceland, Surtsey, and Tristan da Cunha on the mid-Atlantic ridge. The products of these volcanoes are always basalts. Along the subducting plate margins, as the descending plate slides below the overriding plate, some of the rock melts. The molten magma rises through the crust above and so includes components from the mantle above the plate's descending front as well as material from the oceanic crust itself. When the overriding plate carries a continental landmass, material from the overlying continental crust is also added to the mix. Some of this magma may reach the surface and erupt. The volcanic products in this case are andesites, named after the Andes, a chain of volcanic mountains formed in just this way. The result is often a line of andesite volcanoes, perhaps 250 km (156 miles) beyond the trench and parallel to it, emitting lava products that differ in a number of important ways (notably greater viscosity and higher silica content) from the basalt lavas produced by volcanoes on constructive plate margins.

The accreting margin of the Indo-Australian plate is the ocean ridge running between Australia and Antarctica and into the Indian Ocean. This plate, carrying with it the Australian landmass, has been moving northward for at least 40 million years, and part of India, on its leading edge, has been driven under the Asian plate to form the two-continent-thick Tibetan plateau. Further east, the plate's leading edge is oceanic crust alone, and this is plunging under the Asian plate along a deep trench some 3,000 km (1,875 miles) long, the Sumatra-Java trench. Beyond the trench a line of offshore islands has been formed, as well as a line of volcanic mountains along the spines of Sumatra and Java and on the Lesser Sunda Islands to the east (Figure 9). The Indone-

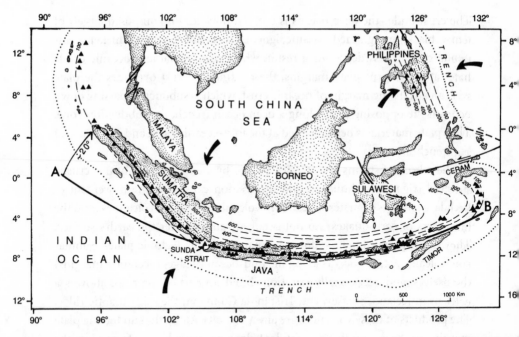

Figure 9. The Indonesian volcanic arc was formed as the Indo-Australian
tectonic plate underthrust the Asian plate along the Sumatra–Java trench
(dotted line). Differences in the angle and rate of subduction under Java and
Sumatra have resulted in the rotation of Sumatra through about 20°
clockwise from the fulcrum of rotation at Sunda Strait.

sian archipelago, situated where two of the earth's great plates are impinging
on one another, is thus the site of a substantial proportion of the earth's great
earthquakes and severe earth tremors. It is also one of the most volcanic regions of the world, with 177 volcanoes, 75 of which have been active since the
year 1600. DeNève (1983, 1985 a,b) estimated that since the Upper Pliocene (4
million years ago), more than a third of the earth's volcanic eruptions have occurred there, and that from 15 to 20 eruptions of various degrees of intensity
happen there every year. Within this great volcanic alignment, as Verbeek correctly observed, Sunda Strait, the setting of the Krakatau eruption, is an area of
weakness.

It had long been appreciated that the orientation of the line of volcanoes
running along the long axes of Java and Sumatra changes sharply at the site of
Krakatau (Berghaus 1837). Verbeek (1881) had also noted a shorter, almost
north–south, line of volcanoes and islands of volcanic rock within Sunda
Strait, comprising Gunung (Mount) Rajabasa on Sumatra, the islands of Tiga,

Sebuku, Sebesi, and Krakatau, and Gunung Payung on the Ujung Kulon peninsula of Java. (A map of the area immediately precedes Chapter 1.) He believed these lineaments to represent fractures in the earth's crust, and observed that the sea in Sunda Strait is deeper to the west of the "Sunda fracture" than it is to the east. He further remarked that Krakatau is situated at the exact intersection of the three volcanic "fracture zones" of Sumatra, Sunda Strait, and Java. It is only since the 1960s that these observations, made before the 1883 eruption, have been satisfactorily explained.

Geologists have shown that the rate of subduction of oceanic crust is faster, and the angle of subduction shallower, along the Sumatran trench than along the Javan trench. The change is abrupt and occurs at Sunda Strait, which is thus a boundary between two segments of subducting lithosphere. A major tectonic break occurs at Sunda Strait (Fitch 1972); the distributions of epicenters of shallow earthquakes resulting from subduction are different east and west of the strait, as are the patterns of faults (which indicate the general directions of major sliding movements between blocks of rock). Katili (1970) has described a great Sumatra fault zone, which ends in Sunda Strait, possibly extending to the Krakatau area. The magma of Krakatau was generated from remelted crustal material (Oba, Tomita, and Yamamoto 1992) and has a wider range of chemical composition than that of typical island-arc volcanoes (Nishimura, Hardjono, and Suparka 1992). The northwest–southeast alignment of volcanoes in Sunda Strait represents this peculiar type of volcanism and, as Verbeek had noted, it crosses the major east–west alignment of the volcanoes of Sumatra and Java at the site of Krakatau.

It has now been confirmed that very powerful tectonic forces are operating at the strait. Extensional, tensional forces are evident in the south, and convergence and compression are detectable in the north, making what geologists describe as a wrench-fault system. The American geologist Ninkovich (1976, 1979) proposed that Sumatra and Java were originally in alignment but, because of the different speeds and angles of subduction along the Sumatran and Javan trenches, Sumatra has rotated clockwise at a rate of 5–10 degrees per million years over the past 2 million years (Figure 9). The Malay peninsula may also have been displaced, in an anticlockwise movement. The fulcrum, or hinge, of Sumatra's massive rotational movement is at Sunda Strait, which is thus being pulled open at the southern end; the volcanic axis noted above is the result of the dilational fractures which are produced in this zone of crustal distension. Ninkovitch's rotation theory has been supported by a comparison of the paleomagnetism of volcanic rocks on each side of Sunda Strait (Yokoyama et al. 1985). When erupted magma cools, it takes on the magnetic polarity of the earth's magnetic field prevailing at the time of its cooling. Yokoyama's group found that samples of rocks some 3 million years old from Sumatra had magnetic declinations (deviance from geographic north) averaging about 10° E, but samples of the same age from Java had declinations averaging about 10° W. They suggested that the original declinations were the same, the present difference being the result of the rotation of Sumatra during the last 2 million years.

Zen (1985) has summarized the tectonic importance of Sunda Strait and the difference between the volcanic geology of Sumatra and Java. In Sumatra there are numerous large depressions indicating volcanic activity that in Quaternary time (the last 2 million years) produced vast quantities of acid pyroclastics. Such volcanotectonic depressions occur where two or more faults exist en echelon and their volcanoes are fed by granitic magma occurring near the surface. Depressions of this type do not occur in Java, whose andesite volcanoes are fed by magmas from deeper levels. Zen stressed that Sunda Strait and the Banten area of West Java (roughly, that part of Java west of Jakarta), are more similar in volcanic characteristics to Sumatra than they are to the rest of Java. The pattern of north–south volcanic lineaments and fault zones in the general area of the strait led him to suggest that Sunda Strait itself may be a submerged, Sumatran-type volcanotectonic depression, with the huge deposits of acid pyroclastics in Lampung (southern Sumatra) and Banten as the products of its vulcanism. Such deposits are not known from more easterly regions of Java. A belt of seismicity with an unusual density of shallow- and medium-depth earthquake foci also coincides with the Krakatau volcanic lineament. The belt is wider in the south of the strait than in the north, and Zen believed this pattern, as well as the fault patterns, to be related to the opening of Sunda Strait in the south by the clockwise rotation of Sumatra.

Sunda Strait, the link between the Java Sea and the Indian Ocean and one of the busiest marine straits in the world, is therefore also the focus of very powerful tectonic forces. And within the strait, one of the most seismically and volcanically active regions on earth, Krakatau occupies a crucial position.

Previous Volcanic Activity

Historical evidence of volcanic activity in the region of the strait is not as clear as we might wish, but Judd (1889) suggested that the Javanese Book of Kings *(Pustaka Raja)* provides evidence of a major eruption there in the year A.D. 416. This suggestion was supported by De Nève (1985b). The narrative relates that:

in the year 338 Saka [= 416 A.D.] a thundering noise was heard from the mountain Batuwara [now called Pulosari, the nearest Javanese volcano to Sunda Strait], which was answered by a similar noise coming from the mountain Kapi [lying west of the Banten region of Java]. A great glaring fire, which reached to the sky, came out of [Kapi]—the whole world was greatly shaken, and violent thundering, accompanied by heavy rains and storms, took place; but not only did not this heavy rain extinguish the eruption of fire of the mountain of Kapi, but it augmented the fire; the noise was fearful, at last the mountain Kapi with a tremendous roar burst into pieces and sunk into the deepest of the earth. The water of the sea rose and inundated the land. The country to the east of the mountain Batuwara to the mountain Kamula [Gunung Gede in Java] and westward to the mountain Raja Basa, was inundated by sea; the inhabitants of the northern part of the Sunda country to the mountain Raja Basa were drowned and swept away with all their property.

After the water subsided the mountain Kapi and the surrounding land became sea and the island of Java [the Sanskrit Yawa-dwipa, referring to central and Southern Sumatra and Java] divided into two parts . . . This is the origin of the separation of Sumatra and Java.   (Judd 1889, pp. 365–366)

Judd was not prepared to say that we should accept the date 416 for this "very grand eruption of Krakatau" and the accompanying subsidence leading to the formation of Sunda Strait. He noted that frequently the tradition of an earlier event is attached to a later, similar one. The narrative does show, however, that in the fifth century there was a belief in the former unity of Java and Sumatra, and that large-scale volcanic activity was occurring in the Krakatau area. Moreover, the separation of Java and Sumatra is associated with an explosive volcanic event of considerable magnitude at a site near to that of Krakatau. There are also suggestions in the narrative of a caldera collapse, resulting in one or more devastating tsunamis, at least as great as those of 1883. As early as 1884 Judd wrote, "It is not improbable that the subsidence accompanying this great outburst gave rise to the depression which forms the strait now separating Java and Sumatra" (Judd 1884, p. 85). If the much larger scale geological processes outlined above are accepted, however, it seems unlikely that the strait originated as recently as the first millennium A.D. De Nève, though, noted that there are no records of a navigable seaway between the Java Sea and the Indian Ocean until the year 1175, when the Arabian navigator Yakut first reported its existence (De Nève 1985b).

Verbeek (1885) believed that the islands Krakatau, Sertung, and Panjang originated as what Judd (1884) called the "basal-wreck" of an even earlier

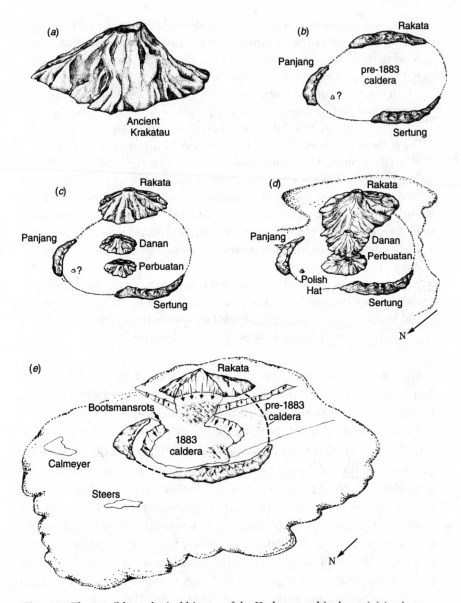

Figure 10. The possible geological history of the Krakatau archipelago: *(a)* Ancient Krakatau; *(b)* remnant islands around the prehistoric caldera following the explosive eruption of Ancient Krakatau; *(c)* renewed volcanic activity within the pre-1883 caldera and the emergence of the islands of Danan and Perbuatan; *(d)* fusion of the three volcanoes as one island, Krakatau; *(e)* the Krakataus immediately after the August 27, 1883, eruption—note the grabens created toward the southwest and southeast. (Modified from P. Francis and S. Self, "The eruption of Krakatau," *Scientific American*, 249: 172–198; copyright © 1983 by Scientific American, Inc. All rights reserved.)

event: the eruption of a large, andesitic, island stratovolcano, "Ancient Krakatau," about 11 km in diameter and 2 km high (Figure 10a). Verbeek believed that this ancient island may have exploded and collapsed—in prehistoric times, but possibly more recently, as recorded in the Javanese chronicles above—leaving three major fragments (Rakata, Sertung, and Panjang) and a smaller one (Polish Hat) around the rim of its submerged 7 km diameter (4.5 miles) caldera (Figure 10b). From studies of sedimentary deposits in the Sunda Strait area, Ninkovich (1979) estimated that the great eruption of Ancient Krakatau occurred about 60,000 years ago. Zen and Sudradjat (1983), however, were skeptical of the hypothesis. They found no trace of this ancient caldera, either from bathymetric features or from the distribution of gravity anomalies, and they questioned the likelihood of a caldera-forming paroxysmal eruption (as in 1883) occurring twice at the same place.

There are records of seven eruptive events in the strait between the ninth and sixteenth centuries, when Krakatau was well known as the "fire-mountain" during Java's Cailendra dynasty. During this period, or earlier, Rakata's basalt cone is thought to have been built up to a height of over 800 meters, and two smaller andesitic island volcanoes developed within the ancient caldera, in line with it (Figure 10c). Escher (1919b) supported Verbeek's suggestion that as these island volcanoes grew they coalesced to form a single compound island, Krakatau, about 9 km long and 5 km wide, comprising the three volcanoes, Rakata (822 m high) in the south, Danan (450 m) to the north, and Perbuatan (120 m), the northernmost (Figure 10d).

It is known that Krakatau island was well forested in the seventeenth century, when timber and sulfur were regularly obtained. In 1620 the Dutch East India Company set up a naval station on the island and later established a shipyard. A period of what was probably Strombolian-type volcanic activity in 1680 lasted for several months, during which andesitic lava was extruded from Perbuatan. No further activity was reported by observers on land or sea until the first signs of re-awakening of the volcano, in May 1883 (Chapter 2).

5   LIFE RETURNS

It would be very interesting to follow step by step the
progress of the development of new life on this land now
dead, but which in a few years, thanks to the intense heat of
the sun and the abundance of equatorial rains, will surely
have been re-covered in its green mantle.

Edmond Cotteau, *En Océanie* (1886)

The 1883 cataclysm substantially changed the archipelago physically. Two-thirds of the largest island, Rakata, disappeared, together with the smallest one, Polish Hat. The remaining islands were considerably enlarged (Sertung to three times its former area) by a thick blanket of hot pyroclastic deposits (Figure 4). These unconsolidated deposits were of course extremely prone to erosion by both sea and rain. The "wetter" season followed soon after the great eruption, and rain and the northeastern sea swells began their work immediately.

Reshaping of the Archipelago

In early 1884 the French government sent a team of official investigators and René Bréon, W. C. Korthals, and Edmond Cotteau visited the Krakatau group on the twenty-sixth of May (Bréon and Korthals 1885). The new islands Steers and Calmeyer and the islet east of Sertung had already disappeared as a result of wave action; Verbeek, however, later stated that Steers was still above water as a sandbank at low tide in August and September 1884. A British Admiralty chart dated 1887 shows Steers and Calmeyer as "mud and pumice almost awash" and the island east of northern Sertung and the promontory on its southeast merely as "breaks." The Sebesi Channel (between the Krakataus and Sebesi), apart from the remnants of Steers and Calmeyer, has the notations "breaks" and "channel foul" and many depths therein are shown as 2 fathoms or less.

The west monsoon (November to March), the period when both rain and wave erosion is greatest, followed within a few months of the eruption. The first post-eruption photographs, taken in 1884, show that rain erosion had rapidly formed large alluvial fans and steep-sided, V-shaped gullies separated by narrow, sometimes knife-edged ridges on all three islands. On Rakata the gullies radiated from the summit, and on Sertung and Panjang they ran from the islands' longitudinal ridges to their west and east coasts, respectively. Their courses virtually fixed the future morphology of the islands for decades to come; as plant colonization proceeded the valleys were stabilized, the morphological pattern thus being "frozen." In some cases, however, rain erosion broke through the narrow inter-gully ridge, joining two gullies and leaving the intervening ridge isolated. Climbing along such a ridge is a frustrating exercise; one is left exhausted, faced with either a difficult descent and ascent to the next, which one hopes *will* continue, or a return to shore and a new start. A gully may be no safer bet; it too may end—at a tuff cliff.

The newly formed coasts were unprotected by coral reefs or sand beaches, and the sea quickly ate into the thick blankets of deposits to form pumice cliffs, particularly on the south and west of Sertung and Rakata, which are in the direct path of prevailing swells. As the cliffs receded, the openings of gullies were cut back until the valley bottoms, often filled with rain-borne rubble, were exposed well above the base of the new cliff, forming "hanging" valleys (Figure 11). Material captured by marine erosion was moved along the coast by longshore drift and often deposited elsewhere on the island. The British Admiralty chart dated 1887 shows that already the large new promontory on western Rakata had been reduced by about two-thirds, but a small peninsula and several islets and "breaks" indicate its former extent. The present isolated cliff stacks of tephra off the south end of Turtle Beach are remnants of this once extensive western blanket of deposits; presumably they have survived because they rest on a lava basement. Escher's map shows that although between 1886 and 1919 some 2–3 km^2 of this new western area had disappeared, a rounded, lobate foreland, Krakatau South, had been formed at Rakata's southernmost point (Escher 1928). Krakatau South is shown in five maps and charts dated from 1903 to 1940. A fine vertical aerial photograph of Rakata (Hoogerwerf 1953), however,

Figure 11. "Hanging" valleys backing Turtle Beach on Rakata. The peak of Rakata is in the background.

reveals that by 1946 it had been replaced by a shallow embayment (now called Handl's Bay). Van Borssum Waalkes (1960) noted that in 1951 Krakatau South had been much reduced since the 1930s and that a new spit was being formed between what are now known as Owl Bay and South Bay. The area of the "new spit" has now become reduced to a few hectares (Figure 12).

Thus there has been a generally west-to-east migration of material along Rakata's south coast in stages: the large western extension was reduced first and fed the formation of Krakatau South, which in turn was eroded to form a smaller "new spit" at the eastern point. The erosion in the west slowed, presumably as lava basement was encountered, but has still been detectable over the last few decades, and there has evidently been a reduction in the turnover of material at each stage (Bird and Rosengren 1984, Thornton and Rosengren 1988).

Movement of material also became obvious on Sertung. The first stages of the formation of a northern spit were noted by Penzig (1902) in 1897. He remarked that sand and pumice from the western side of the island were being deposited on the northeast, and that the northern, level locality was enlarging rapidly. A northern spit was first noted and mapped in 1908. By 1919 it had extended, and an area of sea entrapped at its base had formed a shallow lagoon.

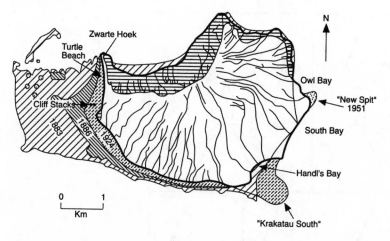

Figure 12. The changing size and shape of Rakata over a century. The 1984 coast is shown by the thick black line, inside of which horizontal hatching represents pre-1883 lavas. The extent of the island in 1883, 1886, and 1924 is indicated by the different cross-hatching patterns. Backer (1929) provides a map of the Krakataus (plate III) labeled "in 1908," in which the lobate foreland "Krakatau South" is shown.

Escher recognized and illustrated a pattern of a prograding eastern shore and an eroding, retreating western one, and, moreover, he appreciated the implications for the vegetation. He predicted "crumbling of the low west coast, shrinkage of the lake, and acquisition of new land toward the north" (Escher 1919b, p. 208), and finally the elimination of the lake. In 1924 the lagoon was open to the sea but by 1928 it was closed and, as Escher had predicted, smaller than in 1919. The spit had indeed elongated northward between October 1921 and July 1924, and a second small shallow lagoon had been formed near its tip. In 1933 the spit had extended even further to the north, and Dammerman (1948) records the northern lagoon as having increased in size. By 1940 Escher's predictions were fulfilled; the basal lake had disappeared and the northern lagoon was much reduced in area. Vertical aerial photographs in 1946 revealed a further extension of land. The spit now extended for over 3 km north of the island proper, had become very narrow (less than 100 m wide) in its middle section, had no lagoon, and had migrated several hundred meters eastward. The British Admiralty chart of 1956 shows that in the decade since the aerial photographs were taken the spit had been breached and its isolated distal portion dispersed (Figure 13).

The spit's eastward migration resulted from the imbalance of two processes acting simultaneously, rapid marine erosion of the west coast and accretion of

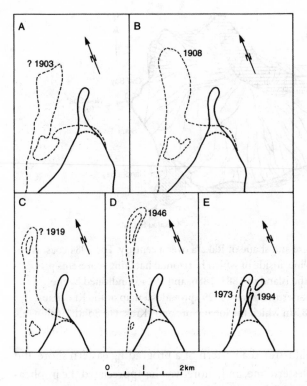

Figure 13. Changes in the shape, size, and position of the Sertung spit since the 1883 eruption. The bold outline common to each figure represents the spit's coastline in September 1984. In May 1994 the spit was fragmented and its distal portion isolated (second bold outline in *E*). The resulting islet persisted for only 4–5 months; in October it was awash and in November its position was indicated by surface breaks only.

the east coast by the emplacement there, in the lee of the swells, of submarine deposits and material eroded from the western tuff cliffs (Figure 14). Rosengren and Suwardi (1985) showed that the pattern of longshore drift was determined by the wave refraction patterns generated by westerly swells and that the spit's shape was determined by interference patterns of three sets of waves. By 1982 there had been considerable further eastward migration and the spit was then only about 2 km long. Rates of recession and accretion were found to be up to 2.5 m and 2 m, respectively, per month. In 1986 Rosengren found that the spit had narrowed further and moved eastward, the rate of movement in places being 8–9 m in 3 months—over 30 m per year. The feature was 600 m (about 650 yards) east of its 1946 position and 1 km east of its 1906 position.

Figure 14. Casuarina trees on the west coast of Sertung's spit (seen from northwest) are killed by ring-barking as loose pumice scours their bases on this high-energy beach. Note the change in vegetation near the island's former northern bluff at the spit's base *(upper right),* where casuarina woodland gives way abruptly to mixed forest. Panjang, Anak Krakatau, and Rakata are visible in the distance.

The shortening and eastward migration may result from changes in near-shore submarine topography—as shoals and banks formed in 1883 were gradually dispersed, thereby changing erosion patterns—and possibly also from a gradual reduction in the supply of sediment available for, and provided by, longshore drift (Rosengren 1985). In 1993 the spit was about 1 km (just over half a mile) long and averaged less than 100 m (110 yards) wide, being less than half that width at its narrowest point, which was devoid of vegetation and overwashed in high seas. By May 1994 the distal section was isolated and by November the resulting islet had been dissipated.

Erosional change on Panjang, which was not enlarged by the 1883 eruption to anything like the same extent as Sertung, has been limited to the reduction of the newly created northern extension and some recession of the southeast coast, with deposition in the northeast (Bird and Rosengren 1984).

As exemplified by the above summary, the study of Krakatau since the great eruption is a study of change, and the biological changes have been no less re-markable than the physical ones. Dutch biologists of the time, particularly the botanists, fully appreciated the opportunity recognized by Cotteau and attempted to monitor, "step by step," the reassembly of a community on the archipelago. Although we may wish that surveys had been made earlier, particularly the zoological ones, as well as more frequently and more systematically, we must be thankful that they were made at all, and, moreover, as thoroughly and extensively as they were made. Without the basis they provided it would be difficult, if not impossible, to understand the unique biological process taking place on these islands.

Before examining the way a community was reassembled on these changed and still changing islands, we need to consider the standard against which the community may be measured; what do we know about the Krakatau community before the great 1883 eruption? Unfortunately, the short answer is very little.

The Pre-1883 Biota

Reports in general terms from sailors and observers on passing ships and from rare visitors to the islands, two small collections of plants and molluscs, and a scrap of evidence from wood carbonized in the 1883 eruption, constitute the only information available on the pre-1883 biota.

There is some indication that damage to the biota from the 1680 eruptions was not extensive. Vogel had written of his surprise in February 1680 at finding the island, which was green and covered with trees when he had last seen it in September 1679, now burnt and arid and projecting large, incandescent blocks into the air at four places (Vogel 1690). Yet Hesse wrote that, only three months later, it was covered in tall trees and wilderness like the other islands in the strait (van den Berg 1884). In comparing these contradictory reports, van den Berg concluded that Vogel had exaggerated the size of the 1680 eruption, which could not have caused much damage to the biota, and De Nève (1985b) discovered from early records that by March 1681 a pepper crop was being harvested for sale. Whatever the damage may have been, the subsequent lull in volcanic activity over the next two centuries would have permitted considerable recovery.

On the first voyage of the famous Yorkshire explorer, James Cook, the *Endeavour* spent a day and two nights at Krakatau in January 1771. Joseph Banks recorded in his journal, "At night Anchord under a high Island call in the draughts *Cracatoa* and by the Indians *Pulo Racatta* . . . This morn when we rose we saw that there were many houses and much Cultivation upon Cracatoa, so that probably a ship might meet with refreshments who chose to touch here" (Beaglehole 1963, p. 233). In January 1777, on his third and last voyage, Cook again found a village and cultivated fields on the island. A year after his death on the island of Hawaii in February 1779, the *Resolution* and *Discovery* called at Krakatau on their return to England. A shore party rediscovered the village near a small stream and a thermal spring in the south of the island. Krakatau was described as covered in forest apart from some small clearings for rice cultivation, and the human population was said to be not large (Cook and King 1784). The *Resolution* anchored in the narrow roads between Krakatau and Panjang for four or five days, and the expedition artist, J. W. Webber, had time to make drawings of village houses and the 1780 lowland vegetation (Joppien and Smith 1987). Tukirin Partomihardjo of the Bogor Herbarium, Java, suggested to me that the distinctive palm in Figure 15 may be *Licuala spinosa* (which recolonized by 1982), and the equally distinctive plant on the right *Cordyline terminalis,* which has not been recorded on the Krakataus. On the extreme right is a large grass looking much like *Saccharum spontaneum,* and Dr. Partomihardjo suggested that the fern in the left foreground within the tree buttresses might be *Nephrolepis hirsutula.* These two plants recolonized the islands by 1920 and 1987, respectively.

Van Breugel, administrator of Banten, recorded that Krakatau was no longer a source of pepper in 1787 but was still inhabited (Simkin and Fiske 1983a), and Horsburgh stated that villagers were raising goats and poultry and growing fruit, and ships visited for water, firewood, and food (van den Berg 1884). Some time later a small penal settlement was established there, and when this was disbanded the island remained uninhabited. Junghuhn (1853) wrote that Rakata's mountain was covered in forest from its base to the peak. When Verbeek arrived in 1880, the islands were covered in dense vegetation (average annual rainfall is over 2,600 mm), apart from several small lava flows near Perbuatan's 1680 craters, which were bare or only very sparsely vegetated (Verbeek 1885).

We have a little information about the flora before 1883 from a small collection made by in 1856, which lacked precise information on localities (Teijsmann 1857). The collection comprised six species (Docters van Leeuwen 1936): *Dendropthoe pentandra* and *Viscum articulatum* (both Loranthaceae), the legumes *Intsia amboinensis* and *Mucuna gigantea,* the tree *Dysoxylum arborescens* (Meliaceae), and the orchid *Dendrobium uncatum.* The first two are common parasitic mistletoes in the Java lowlands, and the *Dysoxylum* is a forest tree of the pantropical mahogany family; the others are beach and coastal plants. One further piece of information comes from a study of carbonized

Figure 15. John Webber's sketch of a clearing on Krakatau in 1780,
made on the return journey of Cook's last expedition.

wood found under the 1883 pyroclastics in 1979. From its cellular structure
this was identified (Baas 1982) as *Macaranga tanarius,* a small, fast-growing,
early-successional, heliophilous tree known to the Indonesians as *tutupan,*
meaning "cover," because of the speed with which it covers open areas. The
Macaranga, Dysoxylum, and the two legumes have been found on the islands
since 1883, and several other species of *Dendrobium,* but not *D. uncinatum,*
have been recorded. Loranthaceae have not been found since the eruption.

Thus from this fragmentary information we know that some plants have recolonized and others have not. Our knowledge of the animals is equally poor, and confined to land molluscs. The mollusc specialist von Martens (1867) identified five species from Krakatau, all fairly large land snails, two quite conspicuous species of *Cyclophorus,* a species each of *Chloritis* and *Hemiplecta,* and *Amphidromus inversus.* None has recolonized the islands.

In spite of the paucity of information on the biota, it seems safe to assume that before 1883 the islands were forested from the shore to the peak of Rakata, with only the small areas of 1680 lava in the north of Krakatau being more sparsely vegetated. The British tropical forest botanist P. W. Richards believed "there is every reason for supposing [the pre-1883 vegetation] was mostly tropical rain forest similar to that now existing in the neighbouring parts of Sumatra" (Richards 1952, p. 270). If so, then only a fraction of the biota has recolonized.

The Development of Grassland

The first scientist ashore after the eruption was Verbeek, who stated that in October 1883 he found no sign of life. Nine months after the eruption Cotteau visited the islands and wrote, "In spite of all my searching I could find no sign of plant or animal life on the land, except a solitary very small spider; this strange pioneer of the revival was in the process of spinning its web!" (Cotteau 1886, p. 126, my translation).

Although in the two years after the eruption some observers on ships reported that the land was bare, devoid of vegetation, others claimed to be able to make out scattered plants with their telescopes. Verbeek, for example, thought he could see grasses on Rakata from his ship after only a year. On hearing these reports, Treub visited the islands with Verbeek, Brun, and others, from June 17 to 24, 1886, probably staying 4 days on the archipelago. A botanist, Treub was the first biologist to visit the islands after the eruption, and his report drew wide attention to the questions of dispersal and recolonization (Treub 1888). From the ship Treub could distinguish plants near the summit of Rakata with field glasses, but ashore he was able to investigate only the coastal area of the west side of Zwarte Hoek (Black Corner) on the northwest of Rakata, the place at which Cotteau had landed in 1884. Here Treub found 11 species of ferns, mostly away from the actual coast, 15 species of flowering plants, mostly beach plants, and 2 species of mosses. He found no lichens but 6 species of "blue-green algae," or cyanobacteria, formed a gelatinous film over the ash substrate. Treub believed that this hygroscopic layer provided a medium for the germination of fern spores and facilitated the growth of their minute prothalli, the first stage of fern growth. He concluded that ferns, which were the most numerous plants away from the coast both in species and individuals, would always precede colonization by flowering plants in such situations.

On the coast Treub found *Ipomoea pes-caprae,* a convolvulaceous ground creeper of sandy beaches throughout southeast Asia and the western Pacific; *Scaevola taccada,* a halophytic beach shrub; *Calophyllum inophyllum* and *Hernandia peltata,* littoral trees; *Wedelia biflora,* a common halophytic composite found on or behind beaches; *Cerbera manghas,* a small tree of mangrove swamps and sandy areas; and *Erythrina orientalis,* a tree often found on rocky or sandy beaches. Also on the shore were a grass, *Pennisetum macrostachyum,* which grows in sunny, barren areas in Java, and two sedges (Cyperaceae). The *Scaevola, Pennisetum,* and *Wedelia biflora* also occurred further inland, together with an island beach shrub, *Tournefortia argentea,* a grass of fairly open places, *Neyraudia madagascariensis,* three species of composites that have seeds with a parachute-like pappus, and the ferns and mosses.

Treub assumed that all plant life had been extirpated by the eruption and was convinced that the flora he encountered was a new one that had colonized the islands from outside the archipelago and was not derived from survivors. Although this assumption was challenged over the next few decades (see Chapter 6), this first visit by a biologist appeared to establish the fact that 26 species of vascular plants had colonized the island within three years of the eruption. Further, and surprisingly at that time, it showed that 18 wind-dispersed plants (three composites, two grasses, and the ferns and mosses) had arrived and become established as early as the sea-dispersed species (the beach plants).

The mollusc specialist A. Strubell also visited the islands in 1889 but could find only marine molluscs (Dammerman 1948). Not one of the five species of land molluscs reported in 1867 by von Martens was seen.

In September 1896 J. G. Boerlage and W. Burck made the first post-eruption visit to Panjang (Docters van Leeuwen 1936). During their one-day stay they met Lt. A. K. Nolthenius, an officer of the Triangulation Brigade of the Topographical Service, which was setting up a survey marker. Nolthenius made a small plant collection in the latter half of 1896. The combined plant collection comprised about 53 species, including the first fig species to be found, *Ficus septica, Ficus fulva, Ficus hispida* and *Ficus padana,* and *Macaranga tanarius,* but, in contrast to Treub's finding on Rakata ten years earlier, there were only four species of ferns.

Emil Selenka, a vertebrate zoologist, had called in at the Krakataus whilst on a round-the-world journey with his wife in 1889. In the resulting travel book they reported the presence of spiders, flies, bugs, beetles, butterflies and "gigantic lizards enlivening the peaceful scene" in the shade of a 20-cm-thick *Casuarina* tree, among scrub and coconut trees almost 4 m high (Selenka and Selenka 1905, p. 139). The "gigantic lizards" must have been the large monitor, *Varanus salvator,* one of the world's longest lizards, related to the famous Komodo dragon. The monitor was again present in 1896, and birds and insects were mentioned without being identified (Muller 1897).

Plant associations and the "grass-steppe"

In March of the following year Penzig, Treub, Pak Idan, Boerlage, M. Raciborski, and G. Clautriau spent half a day on Rakata at Zwarte Hoek's west side, the same area previously visited by Treub. For the first time Sertung (as well as Panjang) was also explored; the party collected for about two hours on Panjang's northern beach area and made a similarly brief visit to southern Sertung.

A remarkable change had occurred since Treub's first visit eleven years earlier. Rakata's interior hills and valleys were now covered in a dense growth of grasses the height of a man or more, what Penzig (1902) called a grass-steppe. This consisted of wild sugar cane or *glagah (Saccharum spontaneum),* which can grow to a height of 3 m, along with the grasses *Pennisetum macrostachyum* and *Neyraudia madagascariensis,* with scattered trees. Numerous trailers that are usually beach plants *(I. pes-caprae, Canavalia rosea, Vigna marina, Cassytha filiformis)* occurred in the dense grass, severely hampering the party's progress. On the higher slopes the grassland comprised *alang alang* grass (*Imperata cylindrica,* a grass which in this region is the first cover after clearing or fire) and *Pogonatherum paniceum* ("bamboo-grass"), together with ferns. Pteridophytes (ferns and their allies; 14 species) now formed a minority of the flora of 69 species of vascular plants. Eight of the littoral plants and *S. spontaneum,* which was also found on the beach, had not been found by Treub.

Definite plant associations were now recognizable. On the beaches was the association of low-growing, sand-binding plants, many of them creepers, characteristic of sandy shores in Indonesia and the west Pacific and named after its most usual member, *Ipomoea pes-caprae.* This was backed in places by coastal trees and shrubs, including *Casuarina equisetifolia, Terminalia cat-*

appa, Barringtonia asiatica, Calophyllum inophyllum (on Panjang), *Hibiscus tiliaceus, Morinda citrifolia, Cerbera manghas,* and *Pandanus tectorius.* This association is known as the *Barringtonia* formation, or, perhaps more appropriately on the Krakataus since *Barringtonia* is not its main component there, *Terminalia* forest. *B. asiatica* is a large littoral tree with beautiful, large, white, night-opening flowers, which are pollinated by moths and nectarivorous bats, and large, heavy, four-sided, buoy-like fruits, which will stay afloat for many months. *Terminalia catappa,* the "Indian almond," is a distinctive coastal and near-coastal tree with a pagoda-like branching habit and leaves that turn reddish in the dry season, features which make it one of the most easily recognized trees on the islands, even from a distance. *Casuarina equisetifolia* is a pioneer tree of open, disturbed ground and has narrow, pine-like leaves. It is intolerant of shade and thus cannot self-regenerate in the same area. Treub had found fruits of *Barringtonia, Terminalia,* and *Pandanus* in the drift zone in 1886, and *Pandanus* was now numerous on the beach. Three species of orchids were found, all having self-fertilizing flowers. *Ficus fulva* and *Ficus padana* occurred on both Rakata and Panjang, and *Ficus hispida* and *Ficus septica* on Panjang.

In 1905 Rakata's southeast coast was visited for the first time (as well as Zwarte Hoek), by T. Valeton. Several species of trees that were to become important in the later development of the islands' vegetation were first found on this visit: two further fig species, *Ficus fistulosa* and *Ficus hirta* at Zwarte Hoek, *Guettarda speciosa* on the southeast coast, and, at an unspecified location, a wind-dispersed tree that later came to dominate Rakata's forests, *Neonauclea calycina* (Rubiaceae). This is a fast-growing, heliophilous tree that can reach about 30 m in height with a girth of about 2 m, and its translucent foliage forms a rather dense crown. It has abundant, small, winged seeds. In Java the species is usually associated with submontane rain forest rather than coastal areas and is found up to an altitude of 1,200 m.

Because of illness Treub was unable to lead the April 1906 expedition, and A. Ernst headed a group which included C. A. Backer of the Bogor Herbarium and two visitors, A. A. Pulle and D. H. Campbell. They spent a day on the islands. A coastal woodland had now developed on southeast Rakata, mainly consisting of casuarinas, now from 2 to 15 m tall and covered in climbers, along with scattered coastal trees, including the figs *F. fulva* and *F. fistulosa,* which for the first time were seen to be in fruit. Inland, ferns had further decreased in number, and trees and shrubs were scattered in the savanna grassland of the gradual slopes. The party hacked through the *Saccharum* grassland, where they were attacked by black and red ants, but could reach a height of only 40–50 m, failing to attain the lower ridges and ravines in which patches of trees could be seen. The vegetation on Zwarte Hoek's eastern beach was less developed than that on its west, which had been investigated previously. Sertung's beach vegetation was similar to that on Rakata; there were several casuarina woods, and the interior was a grass steppe with occasional trees and shrubs.

It was not until 25 years after the eruption that the islands were surveyed for animals. A four-day visit was made to Krakatau in May 1908 by the zoologist E. Jacobson with members of the Topographical Service, and, at Treub's request, Backer was again included. Zwarte Hoek and Rakata's southeast coast were visited, and a few hours were spent on the north and northeast sides of both Sertung and Panjang (Backer 1909; Jacobson 1909). The coastal casuarina trees on southeast Rakata were now up to 35 m tall, and a zone of dense *Saccharum* many hundreds of meters wide, with open spaces and herbs, ferns, and a few trees, notably *Pipturus,* backed the beach vegetation or fringed the coastal forest where it occurred. Jacobson and Backer reached the wooded ravines seen two years earlier, at a height of 400 m. About ten species of plants were found there that had not been seen on the lower slopes, notably the shrub *Cyrtandra sulcata* and the seventh fig species to be recorded, *Ficus montana* (=*quercifolia*). The day after the two biologists had left the expedition, its leader, A. F. Herderschee, and photographer, J. Demmeni, climbed to 475 m, where they bivouacked. Next morning Herderschee was suffering badly from rheumatism, and Demmeni climbed to the summit alone. On the higher, steeper slopes *Saccharum* became sparser, ferns making up the main vegetation, and Demmeni's photograph of the summit showed *Saccharum,* ferns, and two shrubs—which Docters van Leeuwens (1936) later thought to be dwarf *Ficus fistulosa*—but no trees. It is unfortunate that the biologists had left; more precise information on the summit region at this time would have been of great interest; three decades later Backer himself was to criticize expedition leaders for not investigating Rakata more extensively.

Jacobson documented the fauna of a vegetation still dominated by grassland. He carried out no light-trapping, nor did he make thorough examinations of the litter, apart from searching for earthworms. Butterflies, which were not very numerous, evidently were not collected as assiduously as on later expeditions, and collecting may have been biased toward Diptera (two-winged flies). In just three days on the islands over 200 species of animals, 192 of them insects, were recorded. No mammals were found, and Jacobson particularly noted that no bats were seen on any night or evening, but there were 13 species of non-migrant land birds.

In the beach area the savannah nightjar, *Caprimulgus affinis,* was present, along with the collared kingfisher, *Todirhamphus chloris* (Figure 16), and a migrant, the common sandpiper, *Tringa hypoleucos.* Neither the nightjar nor the kingfisher requires trees for nesting, the former nesting on bare ground, the latter in holes in banks and tuff cliffs or in termite nests (Figure 17) so lack of trees would not have prevented their establishment. Both are insectivorous, the kingfisher also taking crabs. Jacobson also observed a second kingfisher on Rakata, the small blue, *Alcedo coerulescens,* which has not been recorded since. In the *Ipomoea pes-caprae* association the large carpenter bee, *Xylocopa latipes,* was visiting the flowers of *Canavalia rosea,* and there were 11 species of chrysomelid leaf beetles, a leaf-eating coccinellid (ladybird) beetle, and the sweet-potato weevil, the larvae of which bore into *Ipomoea* tubers. A small black pentatomid bug and, on Sertung and Panjang, a fulgorid (lantern fly), both plant-sucking insects, were also present. Potter wasps (Eumenidae), which provision their cells with caterpillars, and digging wasps (Sphecidae and Pompiliidae), which largely select spiders and Orthoptera as provisions for their larvae, were also collected.

Figure 16. The "supertramp" collared kingfisher, an early and successful colonizer.

Figure 17. Nest of the termite *Nasutitermes matangensis,* used as a nest site by the collared kingfisher. The entrance hole is made by the kingfisher, and the termite nest may be an active or an inactive one.

In the *Terminalia* forest the pink-necked green pigeon, *Treron vernans,* an obligatory fruit (largely fig) feeder, and the emerald dove, *Chalcophaps indica,* which is a ground feeder on seeds and fallen fruit (and possibly also insects), were seen. The white-breasted wood-swallow, *Artamus leucorhynchus,* was feeding on flying insects caught on the wing above the tops of tall trees near the coast, and the edible-nest swiftlet, *Collocalia fuciphaga,* was also taking insects on the wing. Two fruit- and insect-eating bulbuls were present, the yellow-vented, *Pycnonotus goiavier,* and the sooty-headed, *Pycnonotus aurigaster.* The latter, like the blue kingfisher, then occured in Java but not Sumatra, and, also like the kingfisher, was never again recorded. The beautiful, yellow and black, black-naped oriole, *Oriolus chinensis* (Figure 18), which feeds on fruit and insects in high treetops, was on all three islands.

Nests of the leaf-sewing ant, *Oecophylla smaragdina,* notorious for its ready and painful bite, were already evident, and shiny green dolichopodid flies of the genus *Agonosoma* were conspicuous, as was the black and yellow swallow-tail butterfly, *Troides helena,* its caterpillar feeding on the vine *Aristolochia.* The fruits of the yellow-flowered small tree, *Hibiscus tiliaceus,* were already being

Figure 18. The black-naped oriole, an insectivore-frugivore of high treetops, probably colonized soon after trees were established.

attacked by the pyrrhocorid bug, *Dysdercus crucifer.* The *Casuarina* woodland, then as now, appears to have been less rich in animal species than the *Terminalia* forest; Jacobson found the larva of a longicorn beetle (Cerambycidae) in a casuarina branch.

The *Saccharum* and *Pennisetum* grassland had already reached its climax, and mixed forest was beginning to replace it. The lesser coucal, *Centropus bengalensis,* was quite common in this habitat, and the two bulbuls and the long-tailed shrike, *Lanius schach,* were also present. The large-billed crow, *Corvus macrorhynchos,* was also found; it is an omnivorous bird of fairly open country but requires trees for nesting. Ants were already plentiful both in numbers and species. On Rakata there were 20 species; *Crematogaster dohrni* was abundant in the *Saccharum* fields, their large, round carton nests built on the grass stalks. A *Tetramorium* species, which also makes papery nests round twigs and stalks, was also very common, along with *Plagiolepis longipes,* and a painfully stinging myrmicine ant, probably *Tetraponera rufonigra,* made its presence felt in the grass fields. The ant genus *Camponotus* was represented by 3 species and *Polyrhachis* by 7. *Polyrhachis* species tend aphids, scale insects, membracids, and fulgorids for their secretions, but Jacobson collected only 1 scale insect (on Rakata), 1 aphid (*Aphis malvae* on Panjang), 1 fulgorid (on Panjang), and a membracid on Rakata and Panjang. Lycid beetles and fireflies (lampyrid beetles, *Pyrophanes appendiculata*) were present. The lampyrids feed on snails, and a very small (2 mm long) snail, possibly *Gastrocopta pediculus*, was often abundant in litter, as was the tree-dwelling snail, *Amphidromus porcellanus*, in the ravines up to the height Jacobson reached (370 m). In the Rakata grassland a jassid and a coreid bug, and a click beetle (Elateridae), the larvae of which feed on grass roots, were found, and on the grasses of Sertung a species of "cuckoo-spit" bug (Cercopidae). Of the sphecid wasps in this habitat, Jacobson found species of *Crabro* and *Bembix*, which hunt for flies, and *Larra* and *Tachytes*, which provision their larvae with Orthoptera, particularly crickets; none of these has been recorded since. Jacobson found acridids (short-horned grasshoppers) and gryllids (crickets) to be abundant, represented by 5 and 4 species, respectively, but only 1 species of long-horned grasshopper (Tettigoniidae) was found. Only 1 skipper *(Polytremis lubricans)*, of a butterfly family (Hesperiidae) often associated with grasses, was collected. There were 3 species of termites, all tree nesters. Jacobson's spider material was later lost but contained about 28 species, yet only 1 spider-hunting pompilid wasp was found.

Myriapods were abundant. Rakata was "teeming" with the 15-cm-long (6 in) centipede *Scolopendra subspinipes,* and a large millipede, a species of *Spirostreptus,* was also abundant and found up to 370 m. Scorpions *(Chaerilus variegatus),* whip scorpions *(Thelyphonus caudatus),* and pseudoscorpions *(Chelifer birmanicus)* were present, the pseudoscorpions under the bark of decaying casuarinas. There were 3 species of wood lice, two of them cosmopolitan species. Jacobson's search for earthworms in the soil was unsuccessful, although he found one immature specimen of a species of *Pheretima* in a decaying tree trunk.

The common house gecko, or *chechak, Hemidactylus frenatus,* found near baggage piled on Rakata's beach and also near the surveyors' base camp on Panjang, was thought to have been introduced accidentally. A young specimen of the monitor was caught as it swam near the boat, at anchor off Panjang. The python was already present, although not seen by this party. In his report of work on volcanic gases in September 1905, Brun (1911) included a photograph of a Rakata ravine with a note that at the end of the ravine a large "boa-constrictor" was found. This of course was the reticulated python, *Python reticulatus,* found on later visits and still present eight decades later (Figure 19).

At the end of 1917—or earlier according to Backer (1929)—Mr. Johann Handl, a German pumice collector, took up residence on the south coast of

Figure 19. Peter Rawlinson with a reticulated python
on northern Sertung, 1990.

Rakata, where he built a house and planted a garden. He and his entourage
lived there until 1921. There was only one important ecological consequence
of his stay, the introduction of the black rat, *Rattus rattus,* to the island.

The Second Twenty-Five Years: Forest Formation

The Cyrtandra *highlands and the decline of grassland*

The First Netherlands Indies Natural Science Congress was to be held in
Jakarta in October 1919, with a symposium on Krakatau and a congress excursion to the archipelago. A preparatory visit was made to Rakata (4 days)
and Sertung (2 days) in April by Docters van Leeuwen, M. Bartels, A. L. J.
Sunier, and Handl's son. Bartels's brief was the birds and mammals, Sunier's
other land animals and the fauna of Sertung's brackish lake, and Docters van
Leeuwen's the plants and plant galls.

Sunier discovered the first bat to be seen on the islands, *Cynopterus brachyotis angulatus,* a small dog-faced fruit bat, on both Sertung and Rakata. On

Rakata he also found the first true earthworm, an undescribed species of *Pheretima*. Most of his effort, however, was put into an investigation of the lagoons. The mangrove-fringed (*Lumnitzera racemosa* and *Excoecaria agallocha*) lagoon at the southeastern end of the spit was completely enclosed by the silting up of its outlet. It contained marine forms, including sharks, rays, sawfish, 15 species of bony fish (many of them euryhaline forms), and over 30 species of marine invertebrates. A number of aquatic insects were also present, including 5 species of water beetles, chironomid larvae, 5 mosquito species, water striders (gerrids), *Hydrometra*, backswimmers (corixids), naucorid bugs, and 10 species of dragonflies, chiefly brackish-water forms. As a result of migration of the spit, the lagoon became smaller and heavy rainfall caused the salinity to decline significantly and the marine species disappeared. The aquatic insect fauna also declined, apart from mosquitoes and dragonflies, in which species numbers increased.

Bartels recorded an additional 12 species of non-migratory land birds but two of Jacobson's species (the small blue kingfisher and the sooty-headed bulbul) were not found, making a total of 23 species. Among these were two grassland species not seen subsequently, *Geopelia striata*, the zebra dove, which feeds on grass seeds, and *Centropus sinensis*, the greater coucal (heard once on Sertung). In addition, the following were found for the first time: the pied imperial pigeon, *Ducula bicolor*, a specialist frugivore; white-breasted waterhen, *Amaurornis phoenicurus*, which feeds on invertebrates and vegetable matter; white-bellied fish-eagle, *Haliaëtus leucogaster;* and brahminy kite, *Haliastur indus.* Also noted for the first time and occurring on both islands visited were: the Asian koel, *Eudynamys scolopacea*, and orange-bellied flowerpecker, *Dicaeum trigonostigma*, which feed on figs, berries, and insects, the former laying its eggs in nests of the large-billed crow and the oriole, which act as foster parents of koel young; and the Asian glossy starling, *Aplonis panayensis*, a fruit and insect feeder. There were also two sunbirds that feed on nectar and insects, the olive-backed, *Nectarinia jugularis*, and plain-throated, *Anthreptes malacensis*, and four insectivores: the Pacific swallow, *Hirundo tahitica*, an aerial feeder; the pied triller, *Lalage nigra* (foliage); the magpie robin, *Copsychus saularis* (trunks and bark); and the mangrove whistler, *Pachycephala grisola* (branches). Several species that Jacobson had found only on Rakata were found on Sertung by Bartels (1919).

Although it is possible that some of the birds recorded by the specialist Bartels (and later by Siebers) may have been present but missed on Jacobson's brief 1908 visit, others were almost certainly new immigrants. The white-bellied fish-eagle, brahminy kite, and Pacific swallow are easily recognized in flight, and the koel, white-breasted waterhen, and magpie robin are recognizable by their calls or song. Gregarious species like the pied imperial pigeon and the glossy starling would also almost certainly have been recorded by Jacobson had they been present in 1908, and it is unlikely that the two sunbirds or the woodpecker (see below) would have been missed.

Docters van Leeuwen (1923) noted that in 1919 Handl's neglected garden on Rakata was being crowded out and at least ten species of weeds were present. Most were confined to the immediate area and had not spread. On Krakatau South the *Ipomoea pes-caprae* association was thriving, with several beach shrubs, and a mature casuarina forest extended to its north and west, within which were scattered *Macaranga tanarius, Pipturus argenteus, Ficus septica, Ficus fulva, Premna corymbosa,* and *Guettarda speciosa.* Climbers such as *Aristolochia tagala, Mucuna gigantea, Cayratia trifolia,* and *Ipomoea gracilis* covered the trees, the last two even reaching the tops of tall casuarinas. Further to the east, *Terminalia* forest formed a narrow strip behind the upper shore, and further inland was a casuarina wood with intrusions of *Macaranga, Premna, Pipturus,* and *Ficus.* Areas of grassland had scattered small patches of the mixed woodland species.

The expedition attempted to reach the summit from the east point of the island by following a ravine, but at a height of 100 m an unscaleable wall of tuff was reached. Here there were about 16 species of ferns on the ravine walls and a young woodland of *Macaranga, Ficus fistulosa, Premna,* and *Neonauclea calycina,* with the low shrub *Cyrtandra sulcata* in very moist places.

On the beach the large bee, *Apis dorsata,* visited *Ipomoea pes-caprae* flowers, while *X. latipes* and scoliid wasps visited the pea-like flowers of *Canavalia.* Smaller bees attended flowers of the grass *Ischaemum muticum.*

The mountain was climbed on the following day. Beyond the point previously reached, *Cyrtandra* bushes gradually became more common as understorey in the young woodland, which consisted of *Neonauclea, Radermachera* and *Ficus fistulosa,* with an occasional large *Ficus variegata. Macaranga* and *Pipturus* were absent. A ravine wall was covered with a film of cyanobacteria in which the prothalli and young plants of the fern *Ceropteris* were growing. Trees were more sparse above 400 m and the *Saccharum* and ferns that Demmeni had found widespread in the uplands in 1908 had been replaced by *Cyrtandra,* which filled the slopes and ravines, only the ridges carrying grasses and ferns. Above 500 m a dense stand of *Cyrtandra* covered the whole upper part of the mountain, including the ridges, epiphytic mosses festooning the branches. There was one huge *Ficus retusa*, visible from the boat, and occasional *Ficus fistulosa*. The summit was clothed in a dense stand of *Saccharum* mixed with stunted *Cyrtandra*, the fern *Nephrolepis biserrata*, and a few stunted *F. fistulosa*. Docters van Leeuwen remarked upon the large number of annoying flies on the summit. After a 5-hour climb, the descent took 2 hours, and Docters van Leeuwen (1936, p. 146), who is said to have climbed Mt. Pangranggo (3,000 m) in West Java more than a hundred times, described it as "the most strenuous excursion I have ever made in the tropics."

On Sertung the hills were covered in grassland with scattered trees, and there was a young casuarina woodland with a few young *Macaranga*, *Pipturus*, and *Premna* already within it.

At the congress it was decided that investigations of the Krakatau biota should be made at regular intervals, and Docters van Leeuwen volunteered to take on the botanical and Dammerman the zoological studies. In this way the congress provided a great stimulus to renewed studies of the colonization of the Krakataus.

Between 1919 and 1922 Dammerman made six visits, in December 1919 (8 days, with the bacteriologist Groenewege), April 1920 (7 days, with Docters van Leeuwen), September 1920 (7 days, with J. Siebers, ornithologist at the Bogor Museum), April 1921 (1 day, with H. Boschma, a visiting marine biologist), October 1921 (3 days) and January 1922 (7 days, with Docters van Leeuwen). Dammerman concentrated his efforts on Rakata and Sertung, both of which had been declared nature reserves by 1919. In this period 770 species of animals were recorded, 621 on Rakata and 335 on Sertung (Dammerman 1948). This is more than three times the number found by Jacobson, but of course the amount of collecting time was much greater. Moreover, Dammerman used light-trapping and soil and litter sifting, methods not employed by Jacobson.

A second species of fruit bat, *Cynopterus horsfieldi*, was found in 1920 and although both fruit bats were numerous, insectivorous bats were never seen. *Rattus rattus* was present on Rakata, both in the area of Handl's house and at Zwarte Hoek. There were now 29 species of non-migratory land birds, and a sparrow hawk of the genus *Accipiter* and the Sunda woodpecker, *Picoides moluccensis*, were first recorded. A second gecko species, *Lepidodactylus lugubris*, was found on both Rakata and Sertung, and for the first time the beach skink, *Emoia atrocostata*, was found, fairly abundantly, on northern

Sertung only. The python was recorded on both Rakata and Sertung, many years before rats were known on the latter island. Handl complained that it took his chickens.

Two species of fig-pollinating agaonine chalcid wasps were found on Rakata in 1920, together with two of their parasites, as well as chalcid parasites of mantises, ants, dipterous larvae, and butterfly larvae. Two species of *Halictus* bees were visiting *I. pes-caprae* flowers on the beach and now a second carpenter bee, *Xylocopa confusa*, visited *Canavalia*. The large bee, *Apis dorsata*, which nests high in tall trees, was visiting both *Ipomoea* and *Ischaemum* flowers, and 5 soil-nesting species of the genus *Nomia* collected grass pollen for their broods. A number of insect pollinators, therefore, were well established. Two species of leaf-cutting bees, *Megachile umbripennis* and *Lithurgus atratus*, were found for the first time. Sphecid digging wasps (4 species not found previously), which collect caterpillars for their developing young, were found among the beach plants, and 5 species of pompilid digging wasps, which provision their nests with spiders and Orthoptera, were found. Two more species of potter wasps (Eumenidae) were found in the *I. pes-caprae* association, and 3 species of mutillid wasps ("velvet ants"), which parasitize the eumenids. Two further species of scoliid wasp parasites of beetle larvae were also present in this habitat. Eight species of carabid ground beetles were collected, whereas only one had been discovered in 1908. Thus by 1920 parasitic and predaceous insects formed an important component of the fauna.

Of the butterflies, representatives of the family Lycaenidae were found for the first time on this visit (6 species), as were Satyridae (5 species) and Pieridae (3). A second papilionid (swallowtail) species was feeding on *Aristolochia* vines. There were now 6 skippers (Hesperiidae), compared with one in 1908.

Apis dorsata was also present in the mixed woodland, and social wasps of the genera *Vespa* (1 species) and *Polistes* (3 species) were collected for the first time. Wood-boring beetles of the groups Cerambycidae (5 species), Scolytinae (3 species), Anthribidae (1), and Platypodinae (1) were also first recorded; several of the cerambycids were probably boring into fig trees, and one was probably attacking *Casuarina*. Pierid butterflies, the larvae of which feed on tree-like legumes such as *Albizzia, Cassia,* and *Erythrina,* were present in the woodland (3 species), as were satyrids (5 species) and lycaenids (7). Several families of moths, notably Geometridae (5 species), Arctiidae (4), Noctuidae (25), Pyralidae (21), and a hawk moth (Sphingidae), were first encountered, but since light-trapping had not been carried out in 1908 the sudden appearance of large numbers of moth species may have been an artefact of methodology.

Fungus gnats (Mycetophilidae) had increased from 1 to 5 species, and gall midges (Cecidomyidae) were recorded for the first time (10 species). The only mosquito in 1908 had been the day-biting *Aedes albopictus*, which breeds in small natural and artificial bodies of water, such as those in leaf axils and tree holes and in upturned containers. By 1921 4 more species were present, 3 more of *Aedes*, breeding in crab holes and brackish water, and *Culex sitiens*, a common tropicopolitan, brackish-water species, collected only on Sertung. Tipulids (crane flies) had also increased from 1 to 7 species, but of the 8 species of Lauxaniidae found in 1908, only 1 was present in 1919.

The green cicada, *Dundubia rufivena*, was first encountered in 1920, and fulgorid and jassid homopterans (sap-sucking bugs) had increased from 1 to 19 species and 1 to 8 species, respectively. Coccids (scale insects) were first found in the 1919–1921 period, when there were 10 species found on 8 species of plants. Thrips were not mentioned by Jacobson, but perhaps were overlooked; in the 1919–1924 period 13 species were found, mostly in the flowers of beach and coastal plants. Acridid grasshoppers and crickets, which were poorly represented in 1908, and tettigoniids, of which only 1 species was present, by 1921 were represented by 9, 9, and 8 species, respectively. A mantis, *Hierodula patellifera*, was first recorded in 1919. Dragonflies also increased from 2 species in 1908 to 12 in the 1919–1921 period, largely because of the brackish lagoon on Sertung and Handl's wells on Rakata. The ant lion, *Myrmeleon frontalis*, was first found in 1920, its pits being found first around Handl's house on Rakata. Whereas Jacobson had collected about 28 species of spiders, between 1919 and 1922 Dammerman collected some 52 species.

The tree snail, *Amphidromus porcellanus*, that had been found on Rakata in 1908 was now less abundant, but 5 smaller species were found by sifting humus, litter, and dead leaves. Dammerman made a special study of the soil and litter fauna, and this probably partly accounts for the very greatly increased representation of ants (from 21 to 34 species), beetles, including scarabaeids (from 2 to 8 species), carabids (1 to 9 species), staphylinids (0 to 19 species), pselaphids (0 to 6 species), and tenebrionids (0 to 19 species).

The development of Rakata's Neonauclea forest, 1922–1929

Docters van Leeuwen (with his son and Dammerman) stayed on the archipelago for six days in January 1922 (Docters van Leeuwen 1936). Groups of trees within the grassland were now extensive, coalescing in many places, and the vegetation now consisted of patches of *Saccharum* in a young forest, rather than patches of trees in a grassland. Handl's house was deserted and dilapidated and the garden overgrown with *Macaranga*. Most of the cultivated plants, as well as the weeds, had disappeared. Near the area of the house Docters van Leeuwen's eleven-year-old son made the first capture of a python, a specimen 2 m long; the gut contained rat hairs. The old casuarina wood near the landing stage was choked by *Ipomoea gracilis*, and the climbers *Aristolochia tagala, Mucuna*, and *Tinospora glabra* were much more common than previously.

A second climb to the summit was made over two days, from Handl's house, through a few strips of grassland (in which ants were much less common than on Ernst's visit) and a monotonous mixed wood of 15-m-high *Ficus (ampelas, fistulosa, fulva, septica), Pipturus*, and, higher up, *Neonauclea* and *Ficus montana*. At around 400 m *Saccharum* and the *Nephrolepis* ferns were rarer than two years previously, *Cyrtandra* and *Ficus ribes* more common. At 600 m the *Cyrtandra* was still dominant, but *F. ribes* and *F. fistulosa* were frequent, and at 700 m *F. pubinervis, Villebrunea*, and, in a few ravines, the shrubby *Ficus subulata*, occurred. The summit was reached "going on all fours" after a climb of seven and a half hours. The party descended along the route, now overgrown, that had been cleared two years previously, and no new plants were seen. Docters van Leeuwen had the impression that the *Cyrtandra* belt had shifted upward somewhat since his last visit, and that in the lowlands the grass steppe had regressed.

As an adjunct to a visit to Ujung Kulon, Dammerman and Docters van Leeuwen made a four-day excursion to the Krakataus in July 1924, spending a day on Sertung and three on Rakata. The casuarinas at Krakatau South were now up to 2 m in circumference and much reduced in number, the foreland's southern shore having been eroded. Handl's house was a ruin and just a few cultivated plants survived. Up to a height of 400 m *Macaranga* was still common, with scattered groups of *Ficus montana*, and young *Ficus pubinervis*. Higher up, *Neonauclea* had increased in numbers and the trees were up to 20 m tall. From about 600 m to 750 m *Cyrtandra* was still predominant, but now with more intrusions of *Ficus ribes, F. fistulosa, Villebrunea*, and *Neonauclea*, and *Ficus subulata* covered many ridges. From 750 m to the summit *Cyrtandra sulcata* clumps alternated with patches of ferns and the grasses *Saccharum* and *Imperata*.

Docters van Leeuwen next visited the Krakataus in May and June 1928 for ten days, with his wife. At the usual campsite on the southeastern side of Rakata, the *Saccharum* had now almost completely disappeared and the remains of Handl's house had been claimed by the forest. Krakatau South had eroded further, and there were only a few old casuarinas. On the morning of the summit climb Docters van Leeuwen was bitten by a giant *Scolopendra* centipede but although he "felt dozy," he began the climb. The mixed forest was much denser in the ravines, and large trees of *Macaranga, Neonauclea, Ficus fistulosa, F. fulva,* and *F. variegata* grew on ridges as well as in the ravines. After two hours, at a height of 250 m, his foot was so swollen and painful that the climb was abandoned. Two days later the attempt was repeated. Up to about 400 m, *Macaranga* and *F. fulva* were the dominant trees, above this *Neonauclea* was dominant. Ridges covered with *Saccharum* were now rare, the young, dense *Neonauclea* forest extending up to about 700 m. The *Cyrtandra* stand had retreated to the slopes above this height, and higher up it was now combined with more trees, particularly *Neonauclea, Villebrunea, F. fistulosa,* and *F. ribes.* At the summit itself was one *Neonauclea* plant and individuals of *F. fistulosa, Melastoma affine,* and *Villebrunea rubescens.*

The highlight of the post-congress excursion of the Fourth Pan Pacific Science Congress (held in East Java) on May 12, 1929, was a visit to the Krakataus. Krakatau South was found to be almost completely eroded away and its casuarina wood and most of the *Terminalia* forest were gone.

A visit was made to Panjang on July 9, 1929, by Docters van Leeuwen and Stehn. On this island, too, grassland had largely disappeared, having been crowded out by the fern *Nephrolepis.* Shrubs of *Melastoma* and *Lantana* grew in open places, and groves of trees, *Macaranga, Ficus (fulva, septica,* and *fistulosa), Premna, Trema,* and *Pipturus,* were frequent. The tree *Timonius compressicaulis,* which was to become an important component of the islands' forests, was also common, in contrast to Rakata, where it was found only in the Zwarte Hoek area. *Morinda citrifolia* was spreading up to the ridge, and there were occasional specimens of *Terminalia catappa.* Ravine walls were covered with the grass *Poganotherum paniceum,* and on the ground were *Psilotum nudum, Melastoma, Leucosyke, Neonauclea,* and a few terrestrial orchids. A visit to the southern end of the island showed much the same vegetation.

The expedition climbed to the summit of Rakata (Docters van Leeuwen's fifth ascent) from the southern beach. At around 400 m the *Macaranga–F. fulva* forest merged into a *Neonauclea* forest, and at 700 m the *Neonauclea* became mixed with *F. fistulosa* and occasional *Cyrtandra sulcata.* The thick *Cyrtandra* cover was now confined to the region above 750 m, and now many trees were mixed within it. On the summit itself were *Cyrtandra* bushes with *Neonauclea* among them, 3–4 m high, and *Ficus (ribes, fistulosa, montana), Homalanthus, Leucosyke, Vernonia cinerea, Saccharum, Imperata, Melastoma,* and, for the first time, *Saurauia nudiflora.* The beach on the western side of

Zwarte Hoek had been much eroded, and it was in a large ravine in this area, with prominent *Macaranga* and *F. fulva*, that Docters van Leeuwen found groups of *Timonius compressicaulis*.

The southern part of Sertung was also explored, and here too *Saccharum* had almost completely disappeared. Carbonized stems of plants were found in the walls of ravines, including one 60 cm in diameter; all were carbonized to the core.

The vegetation had become much denser on all three islands, and the ravines and ridges were in deep shade when Docters van Leeuwen, Dr. and Mrs. Ernst, and F. W. Went, visited the islands for four days in February and March 1931. On Rakata's northern cliff small *Macaranga* trees could be seen growing a few hundreds of meters below the summit. There was dense forest at Zwarte Hoek and on the southeast side of the island. The vegetation of northern Sertung was composed of the same species as before but was now so thick and entwined with lianes, dense growths of *Stenochlaena*, and patches of *Caesalpinia bonduc*, that the party could make little headway and constantly lost its bearings. In areas seriously damaged by Anak Krakatau's 1930 eruption there were large patches of high *Saccharum*.

Dammerman arranged for the British spider specialist W. S. Bristowe to visit the Krakataus for four days in February 1931, along with two volcanologists, one of them W. A. Petroeschevsky, who had just returned from a damaging eruption of Mt. Merapi in Java. Before being called away to Merapi, Petroeschevsky had built a rough hut with a concrete floor and corrugated iron roof on Panjang's ridge, as a base from which to monitor the activity of Anak Krakatau. The party spent two days on Panjang, and half a day on each of Rakata and Anak Krakatau. Bristowe collected almost a hundred species of spiders on Panjang, including two domestic species. In a later popular account of this visit he made a number of interesting comments on aspects of Panjang's fauna in 1931.

Bristowe remarked (1969, pp. 147, 148) on the "brilliant little sunbirds hovering round their blossoms." These almost certainly would have been the olive-backed sunbird, *Nectarinia jugularis*, recorded on the island in the following year by Dammerman. His reference to "night-time bats" establishes the fact that bats were present on the island before the first species, *Rousettus amplexicaudatus*, was recorded in 1933. Rats and house geckos, also mentioned by Bristowe, were already known on Panjang. *Rattus tiomanicus* was recorded only on this island, as common, in 1928, and the house gecko, *Hemidactylus frenatus*, was present as early as 1908, although it had not been seen since. Another, rather less common house gecko, *Lepidodactylus lugubris,* was found on the island almost two years after Bristowe's visit and *H. frenatus* was not seen then; it is possible that *L. lugubris* was introduced with Bristowe's party. Bristowe saw a 2-m-long (6 ft) python, and the species had often been seen on the island by staff of the Volcanological Survey. He also noted a thysanuran, then known only from Rakata but probably present on the other islands, a springtail, *Mesira calolepis,* common on shore in drift debris, and a bronze-black tiger beetle. The last species was undoubtedly *Cicindela holoserica,* and Bristowe's comment is the first record of this family of predaceous beetles on the Krakataus. He remarked also on the land crabs and "a few kinds of land snails."

In May 1932, with his sister-in-law assistant, Miss C. C. Reijnvaan, Docters van Leeuwen made his last expedition to the islands. They visited Anak Krakatau for a few hours and spent two days on Panjang, where ten species of weeds were found in the area recently vacated by the volcanologists. On the Panjang hills *Saccharum* culms were mostly dead; there were occasional fields of *Imperata* and *Nephrolepis,* but the grasses had largely been replaced by tree cover, *Macaranga* and *Ficus fulva* predominating. Near the northern beach occasional clumps of casuarinas were densely festooned with *Ipomoea gracilis* and *Mucuna gigantea,* many of them fallen, having been killed. A dense, mixed forest had largely replaced the *Terminalia* forest, and many of the new immigrants were species with edible fruits.

Faunal change with forest formation, 1932–1934

Dammerman's second intensive series of faunistic investigations began in 1932. In August he visited Sertung and Rakata with Went, in order to assess the effect of Anak Krakatau's July eruptions on these islands. Further expeditions were made in November 1932, January, April–May, October, and December 1933, and April and August 1934 (Dammerman 1948).

By 1933 "Krakatau was almost entirely covered by trees even up to a great altitude and the grass jungle, at any rate as covering large areas, had largely disappeared" (Dammerman 1948, p. 149). Several groups of animals had declined both in numbers and diversity since the series of visits about a decade earlier. For example, fulgorid bugs had decreased from 19 to 8 species, jassids from 8 to 4 species, crickets from 7 to 2, coccinellid beetles from 6 to 2, tipulid flies (whose larvae feed on grass roots) from 6 to 3, and lygaeid bugs from 16 to 9 species. In contrast, other groups, notably wood-boring beetles, had increased in representation, scolytines (shot-hole borers) from 3 to 10 and cerambycids from 5 to 8 species. Two other families of wood-boring beetles, Bostrychidae and Buprestidae, first made their appearance during this series of surveys, as did passalid, clerid, and mordellid beetles and aradid bugs. Of the Lepidoptera, the skippers, many of which feed on grasses, had declined from 6 to 4 species, and there had been a large turnover of species (some species becoming extinct, others newly colonizing) in the families Lycaenidae, Geometridae, Noctuidae, and Pyralidae. Mosquito species had increased from 5 to 10, and prosobranch molluscs from 2 to 5, both probably the result of the development of Sertung's lagoons at this time. The land molluscs had increased also, however, from 6 to 12 species.

A third species of gecko, *Cosymbotus platyurus*, had arrived, and in 1933 a second snake ("by its small size and dark appearance we could be quite sure it was not a young python") was seen on Panjang (Dammerman 1948, p. 349). This could have been the paradise tree snake, *Chrysopelea paradisi*, which was found on all the islands in the 1980s expeditions. A crocodile, *Crocodylus porosus*, 2.8 m long, was shot near the brackish lagoon on Sertung in 1924; claws of the monitor were found in its gut.

In 1928 the first insectivorous bat was found on the islands, a single female of the large, horseshoe bat, *Hipposideros diadema*, at 600 m on Rakata, and in 1933 bats of a third frugivorous species, a rousette, *Rousettus amplexicaudatus*, probably seen by Bristowe two years earlier, were found roosting in a Panjang cave.

Four more resident land birds were present, two flycatchers—the mangrove blue flycatcher, *Cyornis rufigastra*, and the golden-bellied gerygone, *Gerygone sulphurea*—the red cuckoo-dove, *Macropygea phasianella*, and a second swiftlet, the glossy swiftlet, *Collocalia esculenta*. The long-tailed shrike, *Lanius schach*, first seen in 1908 and quite common in 1919 and 1920 and breeding on two islands, was absent. It is a bird of open country habitats and has not been recorded since.

ℰℕ

Up to the 1930s the assembly of communities of animals and plants was substantially similar in the lowlands of all three of the older islands (Docters van Leeuwen 1936, p. 265), the differences between Rakata and the other two islands being largely attributable to Rakata's much greater height providing greater heterogeneity of its physical environment and thus its habitat diversity.

The vegetational succession in the interior lowlands was through cyanobacteria and ferns (seen only on Rakata), to grassland, to *Macaranga–Ficus fulva*, species-poor, heliophilous mixed forest. This was succeeded on Rakata alone by *Neonauclea* forest, first in the uplands, and on Panjang the tree *Timonius compressicaulis* became common. The change from grassland to forest was perceptibly faster in southern and southeastern Rakata than on the rest of the archipelago. By the 1920s changes in the fauna were recognized, accompanying the change in habitat. Some species usually associated with grasses and open country, such as several skippers, lycaenids, and the shrike, were lost, and many forest inhabitants, such as wood-boring beetles, fruit bats, and frugivorous birds, colonized. Colonization by the last two groups, which are plant dispersers, led to an increase in animal-dispersed plant species, which in turn permitted further colonization of their dispersers.

The emergence of Anak Krakatau in 1930 and its ash falls, both then and periodically since that time, have undoubtedly affected the development of the biota on its older companion islands Sertung and Panjang, but not on Rakata, which was shielded from any such effects by its great cliff scarp. Moreover, the effects on the vegetation of Sertung and Panjang were different. Before exploring this topic, however, we must first consider a great debate which began as a result of the surveys treated early in this chapter, a discussion of great importance with respect to the origin and assembly of Krakatau's ecosystem. The controversy became known as "The Krakatau Problem."

6 THE "KRAKATAU PROBLEM"

And no one has a right to say that no water-babies exist, till
they have seen no water-babies existing; which is quite a
different thing, mind, from not seeing water-babies, and a
thing which nobody ever did, or perhaps ever will do.

Charles Kingsley, *The Water-babies* (1863)

Treub never doubted that the 1883 event had extirpated all life on the islands
and that the flora he encountered in 1886, what he called "Krakatau's new
flora," was indeed entirely new, all the species having arrived from outside the
archipelago.

In the first place it is necessary to prove that the present flora must be considered as new and that it is not derived from remains of the luxuriant vegetation present on the island before the eruption. Nothing is easier than to
furnish this proof. At the time of the eruption the trees felled or smashed by
violent outbursts must have been half carbonized, in view of the extremely
high temperatures that certainly prevailed over the whole island. After that
Krakatau had been covered, from the summit right down to sea level, in a
layer of burning ash and pumice. This layer had a thickness varying between
one and sixty meters. In those conditions it is clear that no vestige of the flora
would have been able to exist after the cataclysm. The most persistent seed
and the most protected rhizome must have perished.   (Treub 1888, p. 214,
my translation)

By far the majority of subsequent writers, particularly those who worked on the islands, agreed that a natural experiment in recolonization and biotic reassembly from a "clean slate" was in progress. Indeed, this was what had directed international attention to Krakatau following Treub's paper.

Over the years, however, a handful of biologists who had not visited the archipelago voiced misgivings. They based their concerns on pieces of evidence which, they believed, threw into question the theory that life was completely extirpated in 1883. They received little attention, general scientific opinion being overwhelmingly on the side of the big battalions, the principal Krakatau scientists such as Verbeek, Treub, Penzig, Ernst, Backer, Docters van Leeuwen, and Dammerman. Finally, in 1929, Backer changed his views completely and published a 300-page monograph in which he mounted an attack on the theory of total extirpation that could not be ignored (Backer 1929). The "Krakatau Problem" became a topic of keen international debate for the next few decades.

Early Criticisms of the Extirpation Theory

J. P. Lotsy was the first to question formally Treub's assumption of total destruction (Lotsy 1908). He pointed out that in 1905, only 22 years after the eruption, Valeton had found a large specimen of *Cycas rumphii* growing on Rakata near the coast. Cycads were thought to be very slow-growing, and Lotsy believed that this plant was too large to have been less than 22 years old and must have been a survivor of the eruption.

The idea of survival was taken up more generally by Lloyd Praeger (1915). Although accepting that many plants of the new flora were colonizers from overseas, he suggested that many seeds must have been buried by floods and landslides before the eruption, some to a considerable depth, and within two months many of these could have been exposed by rain erosion cutting deep valleys and gorges into the ash, in places exposing the old soil surface. Seeds able to survive high temperatures may then have germinated. The zoologist Michaelsen (1924) suggested that earthworms could have survived. Some species occur meters below the surface and can live below ground in humus-rich soil without surfacing. Michaelsen described a hitherto unknown species of *Pheretima*, first found, in large numbers, in localized areas of Rakata in 1919. He regarded this as a surviving Krakatau endemic (a species occurring only on Krakatau), rather than a post-1883 colonizer. In the following year the zoogeographer Scharff (1925) suggested that insect larvae deep underground could also have survived the eruption, and adult insects, spiders, snails, and even lizards could have been sheltered from the hot ash in rock crevices and overhangs. Scharff argued that the tsunamis would have covered considerable stretches of the island several times and thus cooled the deposits in these areas. If Krakatau's vegetation were derived from surviving seeds and spores, one would expect the new vegetation to appear simultaneously throughout the island. This, he said, was precisely the case. He also noted that Ernst (1908) had remarked that the first plants of the littoral vegetation to appear were found furthest from the beach, which is not what one would expect if they had been sea-dispersed. While not denying the possibility of long-distance dispersal, Scharff hesitated to believe, without very convincing evidence, that an island entirely deprived of life could be provided with a host of animals and plants solely by "accidental" means.

The Critique of C. A. Backer

It was Backer, the iconoclast and acknowledged plant systematist, who mounted the most extensive and detailed attack on the theory of total extirpation. After his second visit to the islands with Jacobson in May 1908, he had been so convinced of the total destruction of the biota that he had written, "All researches have fully confirmed the presumption that the old flora has been totally destroyed, so that the new flora must exclusively consist of introduced elements" (Backer 1909, p. 191). Twenty years later, having completely changed his views, he regretted his earlier compliance: "all of us—like everyone then—on the mere authority of Treub took the total destruction of the vegetation by the eruption of 1883 for a well-proven fact, not liable to any doubt. Often since I have severely blamed myself for this levity" (Backer 1929, p. 143). His extensive and detailed critique (Backer 1929) included criticisms, both implicit and explicit, of the planning, experience, specialist knowledge, and record-keeping (among other aspects) of the botanical investigators on expeditions up to and including 1921. In his view, these and other failings rendered the whole question of the elimination of the biota not only insoluble but not even worth consideration.

Apart from the substantive issues, the discussion of the "Krakatau Problem" is interesting in two other ways. First, it shows that iconoclasts can stimulate debate, thought, and further research, even if they do not eventually win the day. It is likely that the program of intensive and regular field studies on the Krakataus by Docters van Leeuwen and Dammerman was instigated at least partly because they knew that Backer's broadside was being loaded and would surely be fired, and Docters van Leeuwen's 1936 monograph on the islands' vegetation may have benefited from the controversial context in which it was written.

Second, the controversy showed that, as in other occupations, quirks of personality can intrude upon and add piquancy to scientific debate. Backer wrote without sparing previous investigators, some of whom were his colleagues, and without mincing words. As examples, of the Treub expedition Backer wrote: "the new flora of the island has never yet been investigated by persons who had made a sufficient study of the flora of the Dutch East Indies and were able to recognise *on the spot* all or nearly all plants" (p. 8); "the eminent savant Treub, who was no florist and whose knowledge of tropical plants was very limited" (p. 9); "the harsh words must be spoken: taken as a whole the botanical investigation of Krakatao in 1886 was a *failure;* because of the superficiality it could not be otherwise" (p. 83). He was equally severe on the Penzig expedition (p. 89): "in short, this trip may not be called a scientific oecological investigation made by experts, but was a mere excursion of persons interested into the problem but not seriously trying to solve it."

Over 70 pages were devoted to detailed comments on Docters van Leeuwen's expeditions. Concerning Docters van Leeuwen's first ascent to the summit (an enterprise not attempted by Backer himself) he commented (p. 213): "a period of six hours [ascent only] was much too short for a thorough investigation of the dense vegetation along the way; very many plants may have been over looked; the data acquired cannot but be very incomplete. Yet a great many valuable observations might have been made by an able botanist." And on Docters van Leeuwen's description of the vegetation at Zwarte Hoek (p. 218): "It appears his aim was not to study seriously the development of the vegetation but only to hastily collect as many species as possible . . . Collecting some plants and vaguely speculating on them should, in the absence of all serious investigations, not be considered identical with trying to solve difficult oecological problems. The sooner there comes an end to this childish delusion, the better it is. The money misspent on costly botanical excursions and publications of this kind could be used much better." Backer's own critique, be it noted, was privately published. His final verdict on Docters van Leeuwen's 1919 expedition, coming by now as no surprise to the reader, was that "the botanical expeditions to Krakatao in 1919, when considered from a scientific view-point, were a complete failure" (p. 272). Docters van Leeuwen's survey of Sebesi's flora in 1921, made in order to compare it with that of Krakatau, was dismissed quite simply (p. 278): "The results of this trip are not of the slightest importance for the solution of the problem of the revegetation of Krakatao, and will therefore be passed over here."

Backer's polemic style is well illustrated by his final summation (p. 286):

> Let me conclude this paper by comparing the litterature [sic] on the revege-tation of Krakatao to "the colossal skeleton brontosaur" some American au-thor speaks of, "that stands fifty-seven feet long and sixteen feet high, the awe and admiration of all the world, the stateliest skeleton that exists on the planet. There was one small bone and the rest was built out of plaster of Paris. If the builders had not run short of this stuff the monster would have become much bigger." I have tried to find the very few bones in these wagon-loads of rubbish and will finish by stating once more, that . . . the Krakatao-problem can neither now nor in the future either be posed or solved and is of no importance at all for Botanical Science.

Such forthright, often personal, criticisms were rebutted by those concerned who were still living, Ernst and Docters van Leeuwen. Other observers, follow-ing the debate from overseas, were drawn into the personal issues simmering under the surface of this controversy. Thus, Sir Arthur Hill wrote that it was "unfortunate that he [Backer] has written his account with so much acrimony, not unrelated, it would appear, to the fact that he was 'Formerly' as he states on the title page, 'Government Botanist for the Flora of Java' . . . It is . . . much to be regretted that it is marred by the aspersions that he has cast on distin-guished botanists, both dead and living, whose contributions to the Krakatao eruption history have come to be regarded as classics." Sir Arthur continued, using a particularly apt metaphor, "it is to be regretted that he has chosen the medium of a scientific work as a vent for the eruption in print of his own pent-up feelings" (Hill 1930, p. 627). One of Britain's foremost plant ecologists, Tur-rill (1935), who accepted that the flora was essentially a new one, also suggested (p 442) that "Backer's arguments are weakened by a dogmatic and unnecessarily polemic style which, rightly or wrongly, leads the reader to sus-pect a personal bias against one or more of the previous investigators."

Backer's monograph was widely reviewed, and his personal criticisms may have deflected the attention of some reviewers from the substance of his case. He was generally acknowledged to have had a thorough knowledge of the flora, and he made several valid points. He asserted that the first surveys were of short duration (they were) and were very localized and confined to lowland areas (they were), that the intervals between surveys were too long (true), that the investigators' knowledge of the flora was inadequate (certainly less than Backer's), and that no attention was paid to recording ecological information either when specimens were collected or in the reports (certainly true of Treub's survey). Since nothing was discovered about the upland areas of Rakata in these surveys, Backer argued that to dismiss entirely the possibility of survival on the higher slopes was unwarranted. He believed it possible that in Rakata's steeper interior the "new" vegetation may have arisen from upland survivors, although he acknowledged that this could apply neither to lowland coastal areas of Rakata, many of which were "new" terrain (land formed after 1883), nor to the low islands of Sertung and Panjang. Backer cited Verbeek's report that the ash layer was thinner on the steeper slopes than on the more gradually sloping lowlands, and argued that the ash would have been eroded within one or two months by the heavy rains, in September and October, at the end of the east monsoon. If so, some roots and rhizomes, for example, that were below the soil surface at the time of the eruption may not have been killed. Like Scharff (and independently), Backer suggested that projecting rocks on the sides of deep ravines could have protected seeds, spores, rhizomes, roots, and in some cases small plants from the deposits of ash, and the rains at the end of the monsoon would have favored revegetation by these survivors. Backer also drew attention to the fact that, when digging a well near the coast, Handl found incompletely carbonized portions of trees, according to Handl at the bottom of the ash layer.

Although Treub had seen plants near the summit from his boat in 1886 and Haeckel, passing the south coast of Rakata in 1901, wrote of narrow brown ridges radiating from the top of the cone, piercing its green mantle, Backer pointed out that it was not until April 1906 that the upland ravines were even approached by shore parties. On that visit Backer (1929, p. 28) could see from the boat that "several ravines were clothed with woods running up high against the cone; the upmost ridges and the top bore scattered trees." He cited Treub's observation in 1886 that Sertung and Panjang, much lower islands, were devoid of vegetation. If the plants found on Rakata in 1886 were colonists, why had they not also colonized Sertung and Panjang? And why did the vegetation of these lower islands appear so much later? The answer, Backer argued, was that some of Rakata's "new flora" represented survivors on its upland slopes. His point was that the possibility that some of Rakata's (and only Rakata's) new vegetation was derived from upland survivors should not have been dismissed entirely. Since it was dismissed, investigations on the possibility of survival were not made, and by the 1920s it was too late to make them.

Because the early surveys were brief, localized, superficial, and insufficiently frequent, Backer argued that the presence of a species in one survey and its absence in the next should not necessarily be interpreted as an extinction. Similarly, absence in one and presence in the next should not necessarily be interpreted as a colonizing event. This is valid criticism, certainly when data are based on short surveys of limited scope, and it applies in some degree to surveys made much more recently. If investigators overlook species, they will obviously be led to make spurious estimates of species turnover and to distort the records of extinctions and immigrations by including erroneous records of absences, what is termed "pseudoturnover."

The 1929 discussion of "The Case of Krakatau"

The appearance of Backer's book was timed to coincide with the Fourth Pan-Pacific Science Congress, held in Jakarta in 1929. A joint meeting of the divisions of earth sciences and biology was held to discuss "The Case of Krakatau" (Pacific Science Congress 1929). Backer did not attend the congress.

Docters van Leeuwen opened discussion by seeking opinion, particularly from geologists, on the question of survival of plant and animal life after the 1883 eruption. Escher suggested that soon after the eruption ash would have cascaded down the steeper upper slopes, which would mean that some plants that he saw near the summit in 1919 could have been species that survived the eruption. He also suggested that deposits falling from very high eruption columns must have been cooled down considerably. Docters van Leeuwen responded by saying that a few years ago he had found a rather thick layer of ash and pumice on the summit. P. Marshall drew an analogy with the great 1886 eruption of Tarawera in New Zealand. At Tarawera, all vegetation close to the eruption center was destroyed, but trees were killed only where deposits were greater than 2 m thick; where deposits were thinner, they sprouted again. Marshall noted, however, that no comparison could be made between texture and temperature of the deposits in the two cases (Stephen Self has indicated to me that Tarawera's deposits would probably have been the hotter).

F. V. Colville, of Washington, pointed out that in the Katmai eruption in Alaska, on the immediate slopes where the layer of ash was deep and hot, all plants, including trees, were killed. J. Kuyper of Medan, Sumatra, contributed the practical observation that in tobacco cultivation experiments using cleared forest tracts, in soil heated to 90° C for 45 minutes by piping steam into it, all living plants were killed, and no weeds arose, whereas control tracts, not so treated, were full of weeds. As an example of likely soil-surface temperatures, Dammerman informed the meeting that Anak Krakatau's surface layer registered 94° C six weeks after its most recent eruption (this island was in process of emerging at the time of the congress).

E. van Everdingen, from Utrecht, responded to Escher by pointing out that pyroclastic clouds ascending to great heights would have been transported laterally by air currents to fall elsewhere. Those deposits falling back on to Rakata would have been the heaviest, and so could not have been carried very high and thereby greatly cooled. Had the overlying deposits been cleared in places by rain soon after the eruption, and had plants survived, one would have expected the vegetation to have spread from survival foci, but no such pattern was reported. Docters van Leeuwen confirmed this, saying that photographs taken in 1886 clearly showed that the resurgence of plant life was scattered. W. A. Setchell, of California, observing that rapidity of recovery is no argument against dispersal, cited vegetation found growing within three months of Falcon Island's emergence from several fathoms near the Tonga group in the southwest Pacific. (Backer had pointed out, however, that the two coconuts concerned had been planted by a Tongan chief.) A. C. de Jong, of Bandung, observed that since the Krakataus' forest trees were not evident after 29 years, they must have been eradicated. In 1912 he saw only coconuts and two coastal casuarina woods, one at "Krakatau South" and one on Sertung's spit. There were no other trees, Rakata's southern slopes being covered with *Saccharum*. Docters van Leeuwen noted that this description tallied with a plate in a publication by Haeckel describing a voyage passing close to the island in 1921.

Observing that there was no direct evidence for or against survival because details of the conditions at the time of the eruption were not and would not become available, E. J. Goddard, of Brisbane, Australia, believed that the only sensible course to adopt was to consider indirect evidence, such as floral records of subsequent surveys, the order of appearance of components of the flora, and its present constitution. From his experience of successions on islands of Australia's Great Barrier Reef, he was convinced that the observed succession of Rakata's flora and the increasing number of species recorded at each succeeding expedition supported the view that the present flora of Rakata represented recent, new populations. L. F. de Beaufort, of Amsterdam, reminded the meeting that for animals to have survived they must have not only survived the eruption itself, but also found food in the aftermath (a point to be taken up later by Dammerman).

Reactions to Backer's monograph

One cannot help feeling that it is a pity that Backer did not attend the Jakarta discussion; he would probably have relished the odds and enjoyed the fray! He did have a couple of supporters, but they were not present at the congress. Wing Easton suggested that the ash layers laid down in May and early August 1883 may have formed an insulating layer, protecting many seeds and other propagules from the effects of the later, massive, hot ash of August 27 (Wing Easton 1929). Griggs (1930), reviewing Backer's book in *Science*, stated (p. 133) that three years after the eruption of Katmai volcano in Alaska, "abundant plants of many species" had emerged on both upper and lower slopes. These had been buried beneath "several feet of ash for three years and were later cleared by flood waters." He agreed with Backer that complete sterilization cannot be assumed without careful investigation and thought it "likely that plants survived here and there on Krakatau." (It should be noted, however, that the thickness of Krakatau's ash was several tens of meters rather than "several feet.")

Sir Arthur Hill's review of Backer's book in *Nature* in the same year has some flavor of the sharpness of the debate that must have been raging in international scientific circles at the time. Sir Arthur wrote (Hill 1930, p. 628) that "Mr Backer himself had the opportunity of visiting Krakatao in April 1906, . . . but as he had not properly equipped himself for the expedition, he failed to climb higher than some 400 metres up the mountain. Had he reached the upper slopes, the botanical information he could have given would have been of the utmost value. As, however, he did not himself fully explore the mountain or make the detailed investigation which, as he rightly points out, was so much needed, his criticisms of others, and more especially of Dr. Docters van Leeuwen, . . . seem singularly out of place."

Sir Arthur agreed with Backer that the eleven species of ferns that Treub found at Zwarte Hoek in 1886 may have survived "either by means of rhizomes or spores lodged in clefts of these basaltic cliffs, since neither pumice nor ashes could have lain there to any depth, and would quickly have been washed away by the rain." He concluded that, except for littoral plants, the mode of origin of Krakatau's new vegetation must "be regarded as unsolved, since we know so little of what was actually growing on the Island in the first few years after the eruption, and it was not until twenty-three years had elapsed that any attention was paid to the vegetation on the eastern and south-eastern sides" (Hill 1930, p. 628). This, of course, was precisely Backer's point.

The eminent plant systematist and phytogeographer van Steenis placed great weight on Stehn's discovery of charred wood at the bottom of a pumice layer 5 m thick covering Rakata's summit, and concluded that although the sterilization theory was not yet proved, the probability was between 90 and 99 percent (van Steenis 1938). Backer's book was reviewed in the journal *Ecology* by Cooper. This favorable review ended: "Backer has apparently proved his case. It would seem that he might have done so more effectively in fifty pages than in three hundred—and we cannot suppress the intense desire to know what Dr Docters van Leeuwen might say in reply" (Cooper 1931, p. 426).

Docters van Leeuwen, of course, had much to say. But before he did so, his champion, Ernst, wrote a scholarly, 187-page paper entitled "The Biological Krakatau Problem" (Ernst 1934), which was itself summarized and reviewed in a number of journals.

Docters van Leeuwen's Reply

Finally, in 1936, Docters van Leeuwen published his reply as part of a 507-page monograph reviewing the first fifty years of vegetational and floral development on the islands. The first 74 pages consisted of his response to Backer (and others).

Docters van Leeuwen (1936) began by agreeing that the post-eruption visits were too few and far between and that they covered only small parts of Rakata. He noted, however, that the early vegetation was extremely uniform and the small areas covered were likely to have been typical. The long intervals between surveys would have been more of a problem after the first decade, when changes were more rapid, which was why he and Dammerman had agreed in 1919 to make regular visits. As to the early investigators' relative ignorance of the Javanese flora, Docters van Leeuwen again partially conceded the point but noted that plant systematists did not take the opportunity to visit the islands. Backer himself failed to follow up his 1906 and 1908 visits, and his abandonment of the extirpation theory, like the critiques of Scharff and Lloyd Praeger, was not based on renewed field investigations but on theoretical grounds.

Backer's assertion that only a thin layer of ash covered the upper slopes was based on a statement by Verbeek, but Docters van Leeuwen referred to Stehn's finding (Stehn 1929) that the deposits were 5 m thick on the summit even in

1928, 45 years after they were laid down; in 1883 these must have been thicker and hot enough to produce the completely carbonized plant remains found beneath them.

Docters van Leeuwen also disputed the assertion that Treub's assumption of no survival was based on preconceived views. He observed that Treub had reported seeing a number of plants, on the shore, on high ground, and even toward the summit. He pointed out, however, that Verbeek (who was the only person to reach the highlands soon after the eruption) wrote that during his ascent of the south side of Rakata in October 1883 he did not find a single dead tree trunk; the whole slope was completely bare. Verbeek did find carbonized tree trunks, remnants of the former vegetation, at the boundary of the old volcano's rocks beneath the 1883 deposits. Docters van Leeuwen countered Scharff's suggestion that the tsunamis would have cooled the deposits by pointing out that two months after the eruption the surface was too hot to be walked on barefoot and steam was given off after rain. Moreover, the tsunamis would have mainly inundated deposits that had formed new coastal land and thus would have had no plant propagules beneath them. (In Chapter 3 we saw that at least some tsunamis occurred before the massive ignimbrite deposits were laid down, and that according to one modern school of thought the great tsunami could have been produced by emplacement of pyroclastics into the sea, which means that this tsunami could have swept over the deposits immediately after they were formed.) Docters van Leeuwen also observed, as Ernst had done, that the constant accretion of the coast in places (such as "Krakatau South") would explain why the latest plants to appear in the coastal vegetation were found on its seaward rather than its landward edge.

According to Docters van Leeuwen, the trees seen by Escher in 1919 were *Macaranga tanarius* and one or two figs, possibly *Ficus fulva*. These are small, early-successional, heliophilous trees, not trees of mature forest, and Docters van Leeuwen doubted that they would have been components of the pre-1883 forest. Seeds of *Macaranga* planted in Bogor grew into trees 8 m high in two years; had they been pre-eruption survivors they would therefore have been more than 10 m tall in 1886 and could hardly have been missed by Treub on the island's otherwise relatively bare surface. On the point raised by Lotsy concerning the cycad, Docters van Leeuwen noted that the growth rate of these plants had since been found to be much faster than previously supposed, so the plant's size in 1905 was not evidence of its survival.

A recent find is relevant here. In the bottom of an eroded gully on Rakata, at an altitude of 80 m and some 250 m from the coast, John Flenley's group in

1979 found cylindrical, carbonized timber, 10 cm in diameter, of *Macaranga tanarius* (Baas 1982). This lay beneath an ash layer 4 m thick, which rested on shallow clay layers covering basalt (Newsome, Richards, and Flenley 1982). Although a radiocarbon assay dated the sample as at least a century before 1883 (Switsur 1982), posing interesting problems about its origin, the find appears to counter Docters van Leeuwen's assertion that this species was unlikely to have been present before 1883—it may have had a role as a gap-filler.

Docters van Leeuwen dismissed Wing Easton's theory that earlier deposits of ash could have insulated propagules from the huge deposits of August 27, by referring to Stehn's finding (Stehn 1929) that the earlier eruptions deposited material only in the immediate vicinity of the craters, an area which disappeared either into the atmosphere or into the caldera on August 27. Moreover, he continued, no traces of deposits from these eruptions had been found, the massive deposits of the twenty-seventh being emplaced directly on Krakatau's soil.

It must be appreciated that in attempting to justify the extirpation theory, Docters van Leeuwen was in the difficult position of having to prove a negative. Backer had maintained that it had not been proved that there was no survival. Such proof, of course, would have required a complete survey of the whole surface of the island at regular, short intervals after the eruption, as well as monitoring the arrival and subsequent fate of each propagule so that survivors could be distinguished from colonists. Twenty years before the Krakatau eruption, Charles Kingsley wrote the words heading this chapter, indicating that absence of evidence is not evidence of absence; proof of a negative is difficult and often impossible. Docters van Leeuwen did, however, advance several lines of indirect evidence.

First, he expanded on Goddard's argument. He pointed to the differences, at each survey, in the number of species in the beach flora generally, the flora at Zwarte Hoek, and the total flora of all the islands, and showed that there was a steady and substantial increase over the years, too large to be accounted for by the inadequacy of early surveys. He suggested that it would be unreasonable to assume that all the species newly discovered at each survey were present, but missed, at the time of Treub's 1886 visit. Here Docters van Leeuwen seems to have been demolishing a straw man, forgetting that Backer did not claim that no colonists reached the islands, merely that some survival may have occurred in the higher parts of Rakata. His data do indicate, however, that the number of surviving species, if any, must have been only a very small proportion of the flora in 1934.

Docters van Leeuwen next argued that both the timing of the appearance of plant species and their distribution provide indirect evidence for the extirpation theory. No plants could be seen anywhere, on Rakata or on the other two islands, three months (Verbeek) or nine months (Cotteau) after the eruption. He pointed out that in other cases (Kanagatake in Japan, Katmai in Alaska, and Sebesi), plants that had not been killed by the covering of ash reappeared within months. In fact Cotteau had remarked upon the clearly evident recovery of vegetation on Sebesi in contrast to the absence of growth on the Krakataus. In the 1911 eruption of Volcano Island, in the middle of the crater lake of Taal volcano in the Philippines, the ash deposits were neither very deep nor very hot, and about three years after the eruption Gates (1914) found grasses, shrubs, and young trees. These included rather dense thickets of *Saccharum*, which, like some of the trees, were believed to have arisen from surviving plants. In contrast, on Rakata three years after 1883, Treub had found neither trees nor clumps of *Saccharum*.

Perhaps Docters van Leeuwen's strongest indirect evidence, reinforced by later studies of the plant succession (Whittaker, Bush, and Richards 1989), concerned the dispersal spectra of the floras encountered in successive surveys. In 1886 only sea-dispersed coastal species and wind-dispersed species were found. A few species dispersed by birds or bats were first encountered in 1896, and the animal-dispersed proportion gradually increased in later surveys. As Docters van Leeuwen cogently argued, "It is hardly to be expected that the anemochorous plants would be more resistant than the zoochorous plants [and thus survive and appear earlier]" (Docters van Leeuwen 1936, p. 78). The most resistant plants, he believed, would be those with rhizomes, bulbs, tubers, resistant seeds or fruits, and lianes. Plants with rhizomes, however, except for a few ferns, were rare in Rakata's new flora, yet 13 species were present in that of Sebesi. Lianes, too, differed in their representation on the two islands; 4 species were present on Rakata, these and 19 others on Sebesi, many in the uplands. Moreover, the development of Sebesi's vegetation was very different from that seen on Rakata. The forest on Sebesi's slopes in 1921 consisted of 49 species, being more than twice as rich as Rakata's forest at that time, and above 400 m there were 12 species of shrubs, all of which were lacking on Rakata.

Backer had cited Handl's discovery of non-carbonized plant remains when digging a well near the southeast coast of Rakata in about 1917, and van Steenis (1931) commented that these remains probably represented "new" vegetation that had been covered as a result of floods and landslides. They were certainly not found beneath the 1883 deposits, which were some 30 m thick at the location of Handl's house, and Docters van Leeuwen reiterated that in any case before 1883 this area was sea, the coast having been extended to this point as a result of deposits from the eruption.

Dammerman's Response

Dammerman (1929) put forward his argument for the high probability that no animal life survived the eruption chiefly as a response to Scharff—Backer's monograph had appeared almost at the same time that Dammerman's reply was published. Backer's work is not even mentioned in Dammerman's 1948 volume, perhaps also because by then Docters van Leeuwen had made the case to Dammerman's satisfaction.

Dammerman noted that the deposits were so hot that beneath them tree trunks 60 cm in diameter were completely charred, and that even on Rakata's summit a layer of pumice 5 m thick was found as late as July 1929. Although acknowledging that nematodes can survive a temperature of 80° C and tardigrades and rotifers 150° C, he pointed out that animals depending on plant food or on other herbivorous animals could not have survived the long plantless period that ensued, even if they had survived the eruption itself. He disagreed with Michaelsen's view that the previously unknown species of earthworm found on Rakata was probably a surviving endemic species, believing it more likely to have been a colonist, as the fauna of the surrounding region was poorly known. Dammerman recalled that Jacobson had made an effort to search for earthworms in 1908 but found none, and argued that it was unlikely that a surviving species would be so rare after 25 years as to escape detection and then suddenly appear in large numbers eleven years later.

After observing that the only two reptiles present by 1908, the monitor and house gecko, were species with very good dispersal powers, Dammerman argued that the later, sudden, and localized appearance of the two skinks in large numbers indicated successful colonization events rather than a build-up of numbers from eruption survivors. (The recent finding on the Krakataus of several animal species, such as the carnivorous gecko, *Gekko gecko,* in large numbers in restricted areas has also been interpreted as evidence of recent colonization.)

The mollusc fauna provided further evidence for extirpation in 1883. Of the five species recorded before the eruption, in 1867 (see Chapter 5), none had been found on the island since (and they have still not been found). Six years after the eruption, in 1889, Strubell could find only marine or semi-marine species. True land molluscs (two snail species) were not found until 1908. One of these, the tree-living *Amphidromus porcellanus,* was abundant in ravines on Rakata and found as high as Jacobson's party reached (370 m). On Dammerman's own visits, six further species of ground-dwelling land snails were found. Dammerman thought it unlikely that Strubell, whose specific aim was to collect molluscs, would not have discovered any of these in 1889, particularly those occurring under dead leaves and the large and conspicuous *A. porcellanus,* had they been present.

Dammerman stressed that it was particularly those insects with subterranean larvae that Scharff believed could have survived the eruption deep underground which were absent or were discovered only on later surveys. For example, in 1921 six species of tiger beetles (Cicidelinae) were found on Sebesi, but none had been recorded on the Krakataus by 1929; nor were cockchafer beetle larvae (Melolonthinae) present, although they too were present on Sebesi. The larvae of both these insect groups inhabit soil. The burrowing blind snake, *Ramphotyphlops braminus,* was absent during the whole of Dammerman's survey period (it was first found in the 1980s). Jacobson did not record the large green cicada, *Dundubia rufivena,* in 1908, the larvae of which, like cockchafer larvae, live underground feeding on plant roots. Had the cicada been present Jacobson was unlikely to have missed it, for it stridently announces its presence every morning and evening and flies frequently to lights. Dammerman questioned where the adults emerging from the supposedly surviving pupae of ground-pupating moths would have oviposited, and on what would the resulting caterpillars have fed during the plantless months following the eruption?

Sebesi was much richer than Krakatau in Orthoptera, gall-producing flies, thrips, myriapods, and molluscs, and its fauna included many animal groups, including two species of snakes, a number of bugs, Lepidoptera, stick insects, leaf insects, tiger beetles, and slugs, that were absent from the Krakataus (Dammerman 1922). He believed that this contrast indicated survival of these groups on Sebesi but not on the Krakataus (although Sebesi's proximity to the mainland may also be at least a partial explanation).

Later Views

Sir Arthur Hill was convinced by Docters van Leeuwen's detailed response. He made his position clear in 1937: "I fully agree therefore with Dr. Docters van Leeuwen, after carefully reading his able review of all the evidence, that not a single valid argument has been given that any part of the original vegetation of Krakatau was saved, and that the present plant covering of the islands must consist wholly of immigrants" (Hill 1937, p. 137).

In 1951, however, van Steenis revived the controversy. He compared reports of the development of Anak Krakatau's vegetation between 1939 and 1949, during which time there was thought to have been little volcanic activity on the young island, with that of Rakata (van Steenis 1951). The development was slower on Anak Krakatau than it had been on Rakata after 1883. He also noted the finding, in 1951, of charred, uncharred, and only superficially charred bases of tree trunks that had been partly buried in their upright position in the basal layer of Rakata's ash cover. He was thus led to accept the possibility that diaspores of some plants survived the 1883 eruption. In a firsthand account of the 1951 Anak Krakatau expedition, van Borssum Waalkes (1960, p. 34) referred to "eruptions during the war, which probably devastated the existing vegetation" and observed that "whether all life [on Anak Krakatau] was killed by these eruptions is not known with certainty." If such eruptions did occur, of course, this would account for the slow rate of development on Anak Krakatau, without the need to postulate post-1883 survival on Rakata.

The tropical forest botanist Richards regarded Krakatau as "the most spectacular example of a primary plant succession [i.e., starting on a substrate not containing any living plant material] of which there is any record" (Richards 1952, p. 269). Clearly he came down on the side of complete eradication. He considered that the observed succession was typical of almost all primary successions in regions where the natural climax is forest and summed up his views (p. 270) as follows: "The arguments for survival, especially those of Backer, cannot be lightly dismissed. From the nature of the evidence a rigid proof of the 'sterilization hypothesis' is unattainable; as in most questions of plant geography, we must be content with high probability, without complete certainty. In the light of the counter-arguments put forward by Ernst (1934) and Docters van Leeuwen (1936) one is forced to admit that the probability of complete sterilization is very high indeed."

All of the recent group of Krakatau biologists, so far as I know, support the extirpation theory, except the nematode specialists. At least 77 genera of nematodes occurred on the archipelago by 1985. These include a terrestrial genus, *Geomonhystera*, which is known to survive extreme environmental conditions. Had it survived the 1883 event, this nematode would have been able to feed on pioneer organisms such as cyanobacteria after the eruption. *Eudorylaimus,* another Krakatau genus, has a diverse diet and has been found feeding on protozoans and cyanobacteria. In the absence of data on the combinations of temperature and exposure time that are lethal to nematodes and their resistant stages, it was concluded that the survival of soil nematodes through the eruption and its aftermath is a possibility (Winoto Suatmadji et al. 1988). Stephen Self has also pointed out to me that in the 1991 Mount Pinatubo eruption in the Philippines vegetation was also buried under several tens of meters of ignimbrite but, following rapid erosion and thinning of deposits in a tropical climate similar to that of Krakatau, there was regrowth within a year or less. He thinks it probable that some plants, perhaps as seeds, also survived the Krakatau eruption.

Verbeek does not appear to have taken the temperature of the pyroclastic layers on his first visit after 1883. Sudradjat and Siswowidjoyo (1987) reported, however, that during the 1984 eruption of Mount Merapi in central Java the temperature of a pyroclastic flow was 225° C 6.5 km (about 4 miles) from its source. Fisher and Schmincke (1984) cited emplacement temperatures of pyroclastic flow deposits at Mount St. Helens in 1980 of above 500° C, and stated that in a flow at St. Augustine volcano in Alaska, temperatures ranged from 500 to 600° C at depths of 3–5 m a few weeks after emplacement. They noted that high temperatures may persist for several years in deposits that are several meters thick; fumarole temperatures in poorly welded ash deposits in the Valley of Ten Thousand Smokes, Alaska, were as high as 645° C seven years after deposition.

In a recent pilot study on the Krakataus, seeds of 11 species of spermatophytes, recovered from beneath deposits of Anak Krakatau ash that were emplaced probably in 1930–1933, germinated after more than 60 years. Five of these species first appeared on the islands in 1897, 2 or 3 in 1908, 2 in 1920–1922, 1 in 1979, and 0–1 in 1992. Whittaker, Partomihardjo, and Riswan (1995) concluded that it would have been possible for a few species to have survived burial in 1883 until subsequently exposed by erosion, but they saw no reason to challenge the consensus view of near-complete sterilization. They pointed out that the impacts of Anak Krakatau's eruptions have been far less than those of 1883, and that all the species that germinated in their trials have effective dispersal mechanisms, 9 of the 11 having colonized Anak Krakatau, implying that their recolonization was just as likely, perhaps even more likely, than their survival.

Perhaps the strongest argument in favor of near-complete destruction of the biota in 1883 is one that was unavailable to Docters van Leeuwen and Dammerman—the development of a biota on Anak Krakatau. This island, because it emerged from the sea, unquestionably originated without land organisms. The general pattern of community assembly on this virgin land, following its emergence in 1930 and since its undoubted self-sterilizing eruptions of 1952/53, has been remarkably similar to that seen on the lowlands of Rakata after 1883 (Chapter 12).

It seems that if there was any survival at all after 1883 it was by only a handful of species, and rather than negating the value of the "Krakatau experiment," as Backer had argued, it simply added a little extra "noise" (Whittaker, Partomihardjo, and Riswan 1995).

7 ARRIVAL

Remember how recently you and others thought that salt
water would soon kill seeds. Remember that no one knew
that seeds would remain for many hours in the crops of
birds and retain their vitality . . . Remember that every year
many birds are blown to Madeira and to the Bermudas.
Remember that dust is blown 1,000 miles over the Atlantic.
Now, bearing all this in mind, would it not be a prodigy if
an unstocked island did not in the course of ages receive
colonists from coasts whence the currents flow, trees are
drifted, and birds are driven by gales?

Charles Darwin to J. D. Hooker, August 3, 1866 (from Darwin
and Seward 1903)

Getting there. That is the problem. Or, more precisely, accounting for the ways
land organisms manage to arrive on an island in viable condition is, as we shall
see later, the first part of the problem, if we are to explain how biologically sterile islands acquire communities of land plants and animals.

One obvious way by which land organisms may reach an island is by being
dispersed in the air (anemochory). Some organisms accomplish this passively, by being borne on steady air currents or storms or carried on some airborne vehicle, such as a leaf or twig. Others, of course, may fly. Organisms
may also arrive on the sea surface (thallassochory), again either passively, by
floating with surface currents or rafting on floating objects, or actively, by
swimming. Plants, and a few animals, may be carried by animals (zoochory).
Propagules may be carried on the outside of animals in various ways. Some
molluscs adhere by means of a sticky secretion that they produce; some seeds,

such as those of *Dodonaea viscosa,* have a sticky outer coating or are contained in a sticky fruit; other propagules, like the seeds of many grasses, become attached to fur or feathers by means of hooks, barbs, bristles, and the like; and very small seeds may be dispersed simply by being enclosed in mud adhering to the animal disperser. Plant propagules (and a few animals) may also be eaten and thus carried in the gut of the disperser (endozoochory, or endochory). After their carrier's landfall, these may be deposited in a viable condition, usually in the animal's feces, sometimes by regurgitation, and occasionally by the death of the carrier. A special category of zoochory, and usually classed as a fourth, is transport due to human activity (anthropochory), which will be discussed in Chapter 14. Many organisms can be dispersed by more than one of the above means, but the likelihood of a particular one is usually characteristic for a given group of organisms. Bordage (1916), Docters van Leeuwen (1936), and Dammerman (1948) all considered this topic with reference to the Krakataus.

Proving how an animal or plant is dispersed is usually difficult. The evidence is almost always indirect, and often a decision as to the mode of dispersal rests on intelligent assumptions. Simple experiments or lucky observations have demonstrated that a particular mode operates in a few cases. Charles Darwin in England, Docters van Leeuwen in Indonesia, Sturla Fridriksson's team on Surtsey, and our own group, as examples, have shown the means of dispersal in certain instances.

Dispersal by Sea

The stretches of sea surrounding the Krakataus, although a barrier to the dispersal of many land organisms, may serve as highways for species that are well adapted for dispersal on the sea surface. Many of Krakatau's beaches are littered with stranded flotsam—logs, pumice, the floating fruits and seeds of plants, marine animal carcasses (and in 1986 and in 1995 human skeletons)—as well as a variety of jetsam. Most has obviously originated outside the archipelago. Pumice and logs occasionally are capable of carrying organisms over long distances, and although it became fashionable some decades ago for biogeographers to discount and even to deride transport on floating logs as "anecdotal" and of no significance over long distances, in the funnel-shaped Straits of Sunda it is certainly significant. Very soon after the 1883 eruption, from a large floating tree trunk in the straits, a macaque monkey was picked up, fur-charred and miserable but alive and otherwise in good shape. Monkeys have not yet colonized the Krakataus, but given time and a reliable food supply on the islands they probably will. In August 1986, on our arrival at Anak Krakatau's southern beach we found a mass of vegetation some 20 m^2 in area, which included complete palms some 3 to 4 m tall with green foliage.

On a clear day some of the possible sources of this natural flotsam ring the horizon. Not only the mountain of Sebesi, some 14 km (9 miles) distant (Figure 20), but the massive Gunung Rajabasa on Sumatra can be seen over 40 km (25 miles) to the north, the mountains of Java over 50 km (31 miles) away to the east are visible, and from tree-fall gaps high on Rakata's southern slopes the hills of Panaitan Island and Ujung Kulon may be discerned some 60 km (40 miles) to the south. Ernst (1908) convincingly argued that flotation from the mainland coasts to the Krakataus could be accomplished in a day.

Taken over the year, the prevailing current in Sunda Strait is from northeast to southwest, from the Java Sea to the Indian Ocean. The level of the Java Sea is actually a few centimeters (from 10 in the west monsoon to 35 in the east monsoon) higher than that of the Indian Ocean, and there is a slight fall down the strait (Wyrtki 1961). There is a clear annual variation in current, however. From about March to November, which includes the east monsoon, or "dry" season (April to October), the flow tends to be from northeast to southwest,

Figure 20. Sebesi, with Gunung Rajabasa in the distance, viewed from the tip of Sertung's spit, 1991. Note the stranded flotsam in the foreground.

from the Java Sea to the Indian Ocean, and surface-current speeds are from 38 to 65 cm/s (1.4–2.3 km/h). The Krakataus could be reached from Sebesi in from 5 to 10 hours. At this time southeasterly winds prevail and flotsam accumulates on the southern and southeastern coasts of the Krakataus. The dangers of using local craft, which are prone to engine failure, are thus greater in this period, since a drifting vessel would be carried toward the wide southern opening of the strait and out to the Indian Ocean, with the quite high probability of the next stop being Africa. During our 1986 visit two young American women and their two Indonesian crew drifted out in this way. After a vain air search they were given up for lost, but they came ashore several weeks later on Sumatra's southwest coast, having survived on toothpaste and rainwater toward the end of their ordeal.

In November the flow reverses for three or four months, with a weak stream to the northeast, rarely in excess of 25 cm/s (0.9 km/h). Despite the greater width of the opening of the strait in the southwest (it is over 100 km wide, while the opening to the north is 24 km), evidence from salinity and surface-current measurements suggests that Indian Ocean water has a relatively weak influence on the movement of surface water because of the north-south gradient.

With the exception of the brief, early, fern phase on Rakata, the first vegetation of the Krakataus appeared on or immediately behind the beach. The *Ipomoea pes-caprae* and *Terminalia* forest associations, the majority of this early flora, almost entirely comprise seaborne plants with ranges encompassing the shores of the Indonesian islands and the western Pacific. The beach creepers *I. pes-caprae* and *Canavalia rosea* and the halophytic herb *Wedelia biflora* possess floating seeds, and the beach shrub *Tournefortia argentea* has seeds that are capable of germination after many months afloat and are often found in the drift zone. Guppy (1906) found the dried fruits of *Scaevola taccada* on beaches of the Cocos-Keeling islands, 1,100 km west of Java, the nearest continental landmass. Littoral trees, such as *Calophyllum inophyllum*, *Cerbera manghas*, and *Hernandia peltata*, have floating fruits, and the pods of *Erythrina orientalis* float, the seeds remaining viable for a long time. The buoy-like fruits of *Barringtonia asiatica* and the almond-like ones of *Terminalia catappa* are also dispersed over water. Apart from these species, other ecologically important members of the sea-dispersed component of the flora are the trees *Casuarina equisetifolia*, *Albizia retusa*, and *Pithecellobium umbellatum*, shrubs such as *Pandanus tectorius*, *Hibiscus tiliaceus*, *Clerodendrum inerme*, and *Colubrina asiatica*, and a few grasses, such as *Ischaemum muticum*, *Spinifex littoreus*, and

Thuarea involuta (Treub 1886; Ernst 1908, 1934; Backer 1929; Docters van Leeuwen 1936). Treub found fruits of *Barringtonia, Terminalia,* and *Pandanus* in the drift zone three years after the great eruption.

During our 1990 and 1991 expeditions, made in different monsoon seasons, a detailed study was made of plant propagules stranded on the beaches of Anak Krakatau by Partomihardjo (Partomihardjo et al. 1993). No fewer than 66 species of 38 families of plants were found as fruit, seeds, or seedlings, and disseminules of 36 species were considered to have originated from outside the archipelago, the most common of these being the coconut. Seventeen of the species, mostly cultivated plants common in both Java and Sumatra, did not even occur on the Krakataus. Seedlings of 21 species were found in the drift zone, most of them maritime plants but including some that grow just inland of the strand, such as *Morinda citrifolia* and *Dodonaea viscosa,* and two typically inland species, *Entada phaseoloides* and *Timonius compressicaulis.* Rather more species were represented in the east monsoon (53 species) than in the west monsoon (44). The species with by far the most disseminules was *Casuarina equisetifolia,* and these occurred commonly in both seasons; wind, as well as sea, plays a part in the secondary spread of its light, winged seeds, often to disturbed sites some distance inland. *Terminalia catappa*'s seeds were the most widespread, and there was little difference in their numbers between seasons. Both these species are important trees of strand and coastal vegetation. Thirty-eight of the species found by Partomihardjo's group were already established on the island, and the diversity of their collection demonstrates the importance of this mode of dispersal in quickly building up coastal plant communities.

Sea-dispersed plants formed a large component of the archipelago's flora in the first few decades after 1883 and in 1908 still accounted for more than half. It was not until 36 years after the eruption that the increasing numbers of plant species dispersed by wind and animals outstripped the sea-dispersed component. At about this time the seaborne component began to level off, fewer and fewer new immigrants being sea-dispersed. Docters van Leeuwen (1936) showed that although between 1920 and 1934 this component increased only from 68 to 76 vascular plant species, numbers of wind-dispersed and animal-dispersed species about doubled (from 64 to 112 and from 35 to 68 species, respectively). He believed that the seaborne component was close to its maximum in 1936 and would not increase greatly. Botanical surveys in the 1980s confirmed this: the beach assemblage, virtually comprising the sea-dispersed component, was shown to be a stable community with few losses or gains over the years (Whittaker, Bush, and Richards 1989).

Dammerman (1948) mentioned a number of invertebrates that could have rafted in trees or parts of trees, including certain earthworms, millipedes, oribatid mites, and the eggs, larvae, and pupae of insects found under bark or in fresh or decayed wood, such as longicorn, buprestid, and bark beetles and some chafers, cossid moths, flies, several ant and termite species, earwigs, cockroaches, and bugs. The first oligochaete worm to be found, in the first faunal survey in 1908, was a species living not in soil but in decayed tree trunks. True earthworms were not discovered until 13 years later. Leaf-mining larvae of flies, moths, and small beetles, scale insects, moth cocoons, bagworms, spider egg cases, and gall insects may be carried on leafy branches or even detached leaves, and whole nests of the leaf-sewing ant genus, *Oecophylla,* may occasionally be washed ashore on a tree branch.

Docters van Leeuwen showed that gall midges could be dispersed by sea as pupae while still within their plant gall. He immersed twigs of *Clerodendrum inerme,* densely covered with midge galls, in Sertung's brackish lagoon for a week. The insects inhabiting the galls survived well. He also placed mite-galled branches of nine Krakatau plant species, four of them beach plants, in artificial seawater for a week, and the mites survived. Noting also that galls occur on many beach plants on coral islands that have never been connected to the mainland, he concluded that floating with sea currents was important in the spread of gall producers, particularly those occurring on beach plants (Docters van Leeuwen 1936).

Sometimes the likelihood of a particular means of dispersal having operated may be assessed by comparing the characteristics of species that have colonized with those of the same group that have not.

Smith and Djajasasmita (1988) noted that molluscs inhabiting natural ground litter have been more successful colonists within the archipelago than have those that are usually found on trees. Nine of the 12 litter species found on the archipelago occurred on more than one island, 5 of them on three, whereas of the 7 arboreal species only 2 occurred on two islands and 5 on only one. This led them to suggest that rafting was a more effective means of inter-island dispersal of molluscs than air currents, arboreal species being assumed to be the more likely to be wind-borne. The first land snail to be found on the archipelago after 1883, however, *Amphidromus porcellanus,* although the only tree snail known there until the 1980s, is well adapted for rafting. It is of a group that produces a hardening secretion at the mouth of the shell so that the animal can adhere extremely firmly to the surface of bark. Darwin (1859) investigated dispersal powers in land snails and reported that species of *Helix* could withstand 3 weeks in seawater if the operculum were intact, and 25 of 100 individuals of 10 different species were able to survive submergence for two weeks. To assess the relative efficacy of air and sea dispersal of molluscs to the Krakataus properly, one would need to know whether the ratio of litter-inhabiting to tree-inhabiting species on the Krakataus (12:7) differs from that of the mainland mollusc fauna. This information is not yet available for mainland molluscs, so the comparison cannot be made. The comparison is possible in an insect order containing both ground-nesting as well as tree-nesting species—the termites.

Although at least 5 of the 24 species of termites in coastal areas of Java and Sumatra, each side of the strait, nest either under (subterranean) or on the surface (epigeous) of the soil, of the 9 species on the Krakataus not one is known to be subterranean or epigeous. Seven nest in fallen wood, one is a tree nester, and the nesting site of one is unknown (Table 7.1). Moreover, five of them are hardwood nesters (lower termites), which on the nearest mainlands are outnumbered by four to one (6 of 24 species). The winged reproductives of termites have a maximum flight range of no more than a few kilometers, much less than the distance even from the island of Sebesi to the Krakataus, let alone that from Java or Sumatra. Although subterranean termites feed on wood, they nest in soil, which means that any natural wood rafts would usually contain only their soldiers and workers; in contrast, in wood-nesting

Table 7.1. Distribution of termite species, by life-style, on the Krakataus and on Java and Sumatra. In comparison with the mainland termite faunas, that of the Krakataus is characterized by the preponderance of hardwood termites, the paucity of softwood termites, and the absence of soil nesters.

Species	H	S	Trees (S)	Soil	?	Total
On Java/Sumatra	6	6	7	5	0	24
On the Krakataus	5	2	1	0	1	9

Note: H = species that nest and feed in fallen hard, dry wood; S = species that nest and feed in soft, wet wood; Trees (S) = species that nest in dead part of trunks of living trees, feed on soft wood; Soil = species that make mounds or subterranean nests, feed on soft wood, fallen leaves, or humus; ? = life-style unknown.

Sources: Abe (1984); Thakur and Thakur (1992); Yamane, Abe, and Yukawa (1992); Abe (pers. com., January 19, 1995).

species reproductives may also be present in a potential raft, which would also serve as food and shelter for the population on arrival. Abe considers hardwood termites, those which both nest in and feed on hard, dry wood, to be preadapted for sea dispersal. They are coastal and thus prone to have their nests swept into the sea, their nest wood is relatively resistant to seawater, the insects themselves are relatively tolerant of seawater and of the coastal conditions they would encounter on landfall, and their caste differentiation is flexible. It came as no surprise, therefore, to find them in beached driftwood on the coasts of Sertung and Anak Krakatau in 1982. Hardwood termites have been the most successful colonizing group and now dominate Krakatau's termite fauna (Abe 1984; Yamane, Abe, and Yukawa 1992; Thakur and Thakur 1992).

This "disharmonic" lack of soil and litter inhabitants, compared to the mainlands, is parallelled in Krakatau pseudoscorpions. Of the 11 species present in 1984, 7 are subcortical in habit (living under bark) and 4 are litter inhabitants. One species, *Goryphus maldivensis*, was found under the bark of a dead tree on the shore. In contrast, of the 9 species collected on the same expedition at Ujung Kulon (on the western tip of Java) and in southern Sumatra, only one was found under bark, the remaining 8 being from litter (Harvey 1988).

Muir (1930) thought that nymphs of the homopteran bug family Derbidae may also have been carried to Krakatau in old logs. Derbids have poor flight powers, but 6 species were present by 1920. Springtails (Collembola), which are wingless, are another sea-dispersed invertebrate group. One species, *Mesira calolepis*, was found on Panjang in 1931 in beach drift, and Womersley (1932) regarded it as having been carried there by drift material. Hôgm Bödvarsson believed that 3 of the 4 species of springtails found on Surtsey in 1978, 15 years after its emergence, had arrived on the young island by drifting on the sea surface (Bödvarsson 1982).

Many of Krakatau's reptiles are also likely to have rafted. The beach skink, *Emoia atrocostata*, the first skink to be found on the Krakataus, in 1919, and not seen since 1933, was always confined to the coast of Sertung's spit. It is known to enter the sea without hesitation and it takes refuge there if pursued. It is closely associated with driftwood in the supralittoral zone, and Dammerman believed that it was dispersed to the Krakataus in this way. In the Bismarck Archipelago, off New Guinea, it occurs on offshore reefs. Its confinement to the Sertung spit would be explained if this were its only landfall; the configuration of the islands and the pattern of surface currents are such that the spit's beaches are a terminus for flotsam on the Krakataus, rather than a source (Rosengren 1985).

With their well-known ability to secure a firm purchase on almost any surface, geckos are prime candidates for raft dispersal, and six species have reached the Krakataus. The eggs of the commonest house gecko of the region, the *chechak, Hemidactylus frenatus,* are often laid under bark or in tree hollows, and the animal is common among debris of the supralittoral zone, a habitat predisposing it for dispersal in driftwood and flotsam. Dammerman found a decayed tree trunk on the shore of Rakata with the eggs of this species deep within it. The *chechak* was present on two of the islands by 1908 and became one of the archipelago's most successful colonists. Its eggs, and those of another Krakatau gecko that is also common in habitations, *Cosymbotus platyurus,* were shown to be resistant to salt water for at least 11 days by Brown and Alcala (1957). The *chechak* is often seen on boats and in cargo, however, and both species are probably just as likely to have been accidentally introduced. *C. platyurus* did not arrive until 1928 and then failed to become established. *Lepidodactylus lugubris* is another house gecko present on the Krakataus. It has a very wide coastal distribution, usually being found in buildings and on rocks and the trunks of mangroves, palms, and other coastal vegetation. It has been seen entering beached canoes and laying its eggs in them, and it is readily transported by boats. The very small *Hemiphyllodactylus typus* is most commonly found in the leaf axils of *Pandanus,* tree ferns, or coconut palms and under the bark of dead tree stumps, and is most likely to have reached the islands on natural rafts. The large, carnivorous *tokay, Gekko gecko,* which occurs in natural situations as well as in houses, and the king gecko, *Gekko monarchus,* probably also rafted, the former most likely on boats, the latter, which is less likely to be found in houses, probably on natural rafts.

Amy and T. H. Schoener have shown that floating is a possible dispersal mechanism for small lizards. They tested 39 individuals of the small Caribbean lizard, *Anolis sagrei,* in seawater tanks with waves simulating moderately choppy seas. All floated for one hour, 12 of them for 24 hours. The Schoeners suggested that surface phenomena were probably involved. Although the lizards' heads were held out of the water with the forelimbs hanging vertically below the body, surface tension was noticeable around the central body region. Moreover, when detergent was added to the seawater to reduce surface tension, the lizards began to sink immediately. The Schoeners noted that free flotation may have exposed these lizards to predation, presumably from both above and below the sea surface, but concluded that it may well have contributed, along with rafting, to this lizard's notable success in colonizing Caribbean islands (Schoener and Schoener 1984). Free flotation cannot therefore be entirely ruled out as a possible dispersal mechanism for the smaller Krakatau geckos and skinks.

The large five-banded monitor, *Varanus salvator,* called the *biawak* (Figure 21), was first seen only six years after the 1883 eruption and was soon found on all the islands. An excellent swimmer, it is the most aquatic of all Krakatau's land vertebrates. Jacobson caught one swimming in the sea off Panjang, and individuals often run into the sea to escape capture. Individuals of this species have been seen swimming in the Gulf of Thailand between islands that were well over a mile apart (Smith 1932), and Raven (1946) remarked that unidentified monitors, possibly of this species, were common on the beaches of an islet 30 miles off the coast of Borneo. On our 1992 expedition, Caroline Gross, walking along Anak Krakatau's deserted eastern beach one afternoon, watched a meter-long monitor emerge from the sea in front of her and walk up the beach into the coastal vegetation. Moreover, Simon Cook and Natasha Schedrin saw a large monitor at sea between Rakata and Panjang, swimming toward Rakata about a kilometer from shore. No scientists were on either island at the time. In Luzon, in the Philippines, these monitors were observed to

Figure 21. The monitor was a successful early colonist; Anak Krakatau, 1992.

dive into the water from overhanging tree trunks when approached (Neill 1958). My colleague Dennis Black has seen them lying up on tree branches over water also in New Guinea, and at Ujung Kulon I have also seen them apparently lying up for the night on the boughs of trees overhanging streams. So it is possible that occasionally young monitors could have rafted to the islands.

A male estuarine crocodile, *Crocodylus porosus*, was shot near Sertung's brackish lagoon in 1924 (Dammerman 1948), and Mrs. R. W. van Bemmelen is reported to have been "surrounded by snapping alligators" on Panjang, while her geologist husband was away from camp working for the day on Rakata, presumably in the 1950s (Furneaux 1965, p. 190). It is almost certain that the van Bemmelens could distinguish monitors from crocodiles. This largest of all crocodilians has been reported frequently in the open sea by Malayan fishermen operating off the coast of New Guinea. Its range is from south China to northern Australia, but Wood-Jones (1909) reported that two specimens had reached Cocos-Keeling Atoll from a nearest possible source 600 miles (about 950 km) distant, and the species has also reached Palau and Fiji. In the 1980s members of our expeditions saw crocodiles on three occasions in the sea covering Rakata's caldera. The species has not colonized the archipelago, nor is it likely to do so, there being no streams or rivers.

Dammerman believed that the reticulated python, which also occurs on Sebesi, reached the Krakataus by swimming. It is reputed to be the world's second largest snake, and like many other large snakes it is an excellent swimmer. Guppy (1906) mentioned a large female boa constrictor bearing many young being washed ashore in good condition on St. Vincent island in the Caribbean on a large cedar, which must have traveled at least 100 miles (160 km), probably 300 miles (480 km), on the Orinoco drift. The paradise tree snake, *Chrysopelea paradisi* (Figure 22), which can flatten and hollow the whole length of the body so that it acts as a parachute, facilitating gliding and retarding falls from tree to tree, is abundant in the Javan teak forests. It is a fast-moving, essentially arboreal hunter, and a very efficient swimmer. The species inhabits Sebesi and could have been carried to the Krakataus on floating vegetation, or may even have made the journey of 12 km or so (about 7 miles) by swimming, assisted by the current. The worm-sized blind snake, *Ramphotyphlops braminus*, is widespread in the Pacific, occurring on several remote islands. It burrows in the upper 15 cm (about 9 inches) of forest soils and feeds predominantly on termites. It is thought to be widely distributed through agricultural trading, but the almost complete lack of such activity on the

Figure 22. The paradise tree snake, evidently a fairly late colonist, now occurs on all four islands.

Krakataus, together with the snake's association with termites, suggests dispersal on driftwood, possibly in a termite nest, as eggs or adults. In 1984 when we found the first two individuals in a rotten log on Sertung, one promptly laid three eggs.

Large mammals of the mainlands, such as deer, banteng (wild ox), Javan rhinoceros, leopard, and tiger, are clearly unsuited for rafting, and in any case they would find inadequate habitat on the Krakataus. Smaller non-flying mammals, such as mouse deer, squirrels, gliding "lemurs," tree shrews, civets, small cats, monkeys, and almost all rodents, however, are also absent. Although some terrestrial mammals are capable of traveling over sea for considerable distances, Heaney's studies of the distribution of non-volant mammals in the Philippines indicate that seawater gaps of even 5–15 km (3–9 miles) have been major barriers to most mammals that are not human commensals (Heaney 1986). Rats are probably the best oversea colonists of all terrestrial mammals and, although they usually travel as stowaways on boats, there is some evidence that they may occasionally make their own way across stretches of water. In an experiment which presumably would be illegal now, laboratory rats were forced to swim in glass jars in water of different temperatures until death. In water at or near body temperature they swam on average for 50 hours, one individual for 72 hours, non-stop (Tan, Hanson, and Richter 1954). Dammerman (1948) reported two instances in Thailand of rats emigrating *en masse* over a front of several kilometers when short of food, and swimming away from land for long periods. The country rat, *Rattus tiomanicus*, which occurs on the Krakataus, covered a distance of about 35 km. The only other rat on the islands is the house rat, *Rattus rattus*, which undoubtedly was introduced (Chapter 14).

Pigs were first recorded on Panjang in 1982 but have not been reported from any other Krakatau island. Adults have been seen from time to time, and their tracks and rooting areas are evident in the forest. Two species occur in both Java and Sumatra, *Sus scrofa* and *Sus verrucosus*, and a third, *Sus barbatus*, in Sumatra. Unlike the Gadarene swine of the Bible, normal pigs, not possessed of devils, are good swimmers. Wallace (1880) recorded them covering over 6 miles (about 10 kilometers), almost the distance from Panjang to Sebesi, where *Sus scrofa* occurs. Sebesi fishermen frequently complain of pigs breaking their nets far out to sea, and Dammerman forecast the arrival of pigs on the Krakataus from Sebesi.

On our 1990 and 1991 expeditions signs of otters were reported on all islands, and twice a "furred mammal larger than a rat" was glimpsed moving into vegetation from the shore. Early one morning in April 1991, Peter Rawlinson and I photographed otter footprints in the wet sand of the upper beach of Anak Krakatau's east foreland. Heaney (1986) found that otters are almost invariably absent from all but the largest islands of the Philippines, presumably because of the high probability of extinction on small islands. He listed three species from Sumatra, the web-footed sea otter, *Lutra sumatrana*, the smaller clawless otter, *Lutra cinerea*, and the Eurasian river otter, *Lutra lutra* (also clawless). The sea otter and the clawless otter were recorded by Hoogerwerf (1970) from Ujung Kulon; the former fishes out at sea and the latter also takes freshwater crabs. From the facts that the prints were of a single individual (the sea and clawless otters both usually occur in groups) and that the tracks we saw lacked claw marks, Rawlinson tentatively concluded that the species was the Eurasian otter, *Lutra lutra* (Rawlinson et al. 1992). If this is correct, then colonization is unlikely since this species' freshwater habitat is not available on the Krakataus.

Dammerman was of the opinion that the sea was much less important than the air in the dispersal of animals to the islands. He estimated that in 1908 92 percent of the total fauna had been dispersed by air, and that this proportion was maintained in the 1921 and 1933 faunas. Docters van Leeuwen believed that the opposite was the case with regard to Krakatau's plants.

Dispersal by Air

During the wet, west monsoon (roughly November to April), the prevailing winds on the Krakataus are from the direction of Sumatra and travel at an average velocity of almost 20 km/h (12 mph). In the dry season (about May to October), the east monsoon prevails, winds chiefly coming from west Java, with an average velocity of over 22 km/h (about 14 mph). Thus small animals or plant propagules carried in the air may reach the Krakataus from either source area in about two hours, and at times of strong winds in even less time.

Plants dispersed by air evidently played an early, minor, but crucial role in the colonization of Rakata. The arrival on air currents of trichomes (the propagules of some cyanobacteria) and the abundant small, light spores of ferns led to Rakata's sparse interior plant cover consisting predominantly of these groups only three years after the eruption. Eleven of the 16 wind-borne vascular plants then present were pteridophytes, but few of the early flowering plants, apart from grasses such as *Imperata cylindrica* and *Saccharum spontaneum*, and a few heliophilous orchids, were wind-dispersed species. Three species of orchids were found, all having self-fertilizing flowers, and *Neyraudia madagascariensis*, a wind-dispersed grass, and three species of composites that have seeds with a parachute-like pappus were already established. This early wind-borne flora almost certainly played an important part in altering the conditions of soil and microclimate in favor of the establishment of other plant species.

The "fern phase" was short; it was seen on Rakata only in 1886 and was no longer evident at the next survey in 1897. Docters van Leeuwen estimated that in 1897 there were 28 wind-dispersed plant species, 30 sea-dispersed, and 6 animal-dispersed. By 1908 pteridophytes made up only 37 percent of the wind-borne component of the flora of 32 species, and now 60 vascular plant species were sea-dispersed and 23 animal-dispersed. As the grasslands declined and mixed woodlands began to take their place over a few decades, however, the wind-borne plant immigrants consisted almost exclusively of forest ferns that require shade and moisture or are epiphytes (growing on higher plants). Wind-borne flowering plants colonized this newly forming forest habitat about a decade later. The vast majority of wind-dispersed spermatophytes that have colonized are non-arboreal species and include many epiphytes. Just over half are orchids, 17 percent are composites, and 13 percent Asclepiadaceae (Whittaker, Bush, and Richards 1989).

The minute seeds of orchids are powdery fine and possess air sacs that increase their buoyancy. Gandawijaja and Arditti (1983) found that in 10 orchid species the percentage of the seed's volume taken up by air space ranged from 75 to 96 percent, and that this percentage was correlated with the seeds' flotation time in air. In one species with 96 percent of its seed volume taken up by air spaces, the seeds took over 10 seconds to fall through 1.5 m. Orchid seeds are, so to speak, minute balloons and are dispersed widely, even by very weak breezes. They have reached the Galápagos, Hawaii, the Azores, and New Zealand, over distances of 900, 1,000, 1,350, and 2,000 km, respectively. The small seeds have no reserve of food, however, and root fungi (mycorrhiza) must be present at the point of landfall to provide the nutrients needed for germination and growth, which is usually slow. Should the seeds be brought in mud on the legs of birds, as has been one suggestion, the necessary mycorrhizal fungi could, of course, be brought along with the seeds.

Only a few tree species are thought to have been brought to the islands by wind. The most important of these is *Neonauclea calycina*, which has been the dominant canopy tree on Rakata since the 1920s. Well before 1983 (when it was first discovered), *Crypteronia paniculata*, a large canopy tree with winged seeds, and more recently *Pterospermum javanicum*, also a large tree, with somewhat larger winged seeds, arrived. Other wind-dispersed tree colonists are *Radermachera glandulosa* (first found in 1919), *Vernonia arborea* (1922), and *Bombax ceiba* (1983). All but the *Bombax* and the *Crypteronia* are known from Sebesi (Whittaker and Jones 1994).

The first living thing to be found on the devastated islands after the 1883 eruption was an animal that had almost certainly arrived by air. It was a spider. Immature spiders of many families are commonly reported in temperate countries as "ballooning," that is, being carried in the air on a length of gossamer thread, parachute-like. In tropical regions this method of dispersal has been recorded less frequently, and Robinson (1982) commented that this may mean either that the phenomenon is rare in the tropics or simply that it is a sporadic occurrence that is likely to escape general notice. The Javanese, however, have a phrase, *hujan lawa* ("spider rain"), for unusual falls of spider gossamer.

The family of lacewings (Neuroptera) that is best represented on the Krakataus is the Coniopterygidae, which comprises small, vagile insects. Ng and Lee (1982) found these to be the most frequently caught neuropterans in Malaise traps (which intercept flying insects) in a Malaysian primary forest.

Flight, probably aided by wind, was also the likely means of arrival of many butterflies, moths, and aculeate Hymenoptera (stinging wasps, bees, and ants). Docters van Leeuwen (1924) reported a remarkable mass flight of at least 16 species of Lepidoptera, mostly owlet moths, and including large numbers of hawk moths and a skipper, flying on to his ship for an hour some 20 km from the shore of the Malay peninsula. The wind was slight but, like the insects, came from the land. Dammerman (1948) cited instances of mass flights of 18 species of Lepidoptera, mostly butterflies, elsewhere in the Indonesian region. By 1934, 11 of these had been found on the Krakataus, and 3 more were found in the 1980s. All the pierid butterfly species now present on the islands are known to make mass flights in Java. A monsoon storm in northwestern India was believed to have diverted a mixed flight of the nymphalid butterflies *Danaus genutia, Tirumala limniace,* and the migrant *Hypolimnas bolina,* and carried the insects almost 2,000 km (1,250 miles) over the sea to Arabia (Larsen and Pedgley 1985). Much less powerful assistance would suffice to bring butterflies to Krakatau from the adjacent mainlands. In fact, *D. genutia* was present on Rakata by the first survey, in 1908. *H. bolina* was seen coloniz-ing Canton Island, a mid-Pacific atoll 1,100 km (700 miles) from Samoa, in 1941 (van Zwaluwenberg 1942), and has been found in all surveys of the Krakataus since 1919. Many of the larger Hymenoptera are also quite good fliers. Yamane (1988) has commented that for many of the non-ant aculeate Hymenoptera the distance between the Krakatau islands themselves is a negli-gible barrier. Both he and members of our teams have often seen the large, strongly flying carpenter bees, *Xylocopa latipes* and *Xylocopa confusa,* flying from one island toward another, and it is likely that occasionally the longer dis-tances from Sebesi or even from the Javan or Sumatran mainlands might be successfully traversed. Yamane also noted that even the wingless females of mutillid wasps, known as "velvet ants," may make oversea crossings because the males are strong fliers and pairs are frequently captured in the air *in copula.*

Dragonflies are strong fliers and can cover long distances. The almost cos-mopolitan *Pantala flavescens,* together with *Diplacodes bipunctata,* became es-tablished in rain pools on Canton atoll (van Zwaluwenberg 1942), and Dammerman (1948) reported that *P. flavescens* was seen in mass flight in the Indian Ocean 460 km (about 290 miles) from land. A species of *Diplacodes, D. trivialis,* was one of the first two dragonfly species seen on the Krakataus, in 1908, and *P. flavescens* has been present since 1921. Because of the scarcity of fresh water, it is thought that most dragonflies are adventive, short-term visitors to the Krakataus, rather than residents.

Spiders and a host of small, light insects—such as thrips, chloropid flies, bark lice, the plant-sucking jassids, aphids, and psyllids, parasitic wasps, small beetles, ants, and springtails—are well-known components of the "aerial plankton," the population of organisms carried in the upper air. Our simple experiments on barren areas of Anak Krakatau (see Chapter 11) showed that aerial fallout on to the island is considerable. Docters van Leeuwen, who was a specialist on the insects of plant galls, noted that most gall producers are weak fliers, many, such as the gall mites, being wingless. He concluded, nevertheless, that most gall insects are dispersed passively by wind. Both gall mites and the larvae of gall midges are capable of surviving desiccation; both will rehydrate and revive from a shriveled state when they have access to water. The gall midges of the 14 species that he found on the Krakataus are small, and in spite of constant vigilance he could not find the gall of the *alang alang* grass *(Imperata cylindrica)* midge, a gall which is conspicuous and common on Java. The midge producing this gall is a poor flier and relatively heavy (Docters van Leeuwen 1920).

Comparisons within animal groups between successful and unsuccessful colonizers also indicates that air transport has been the predominant mode of dispersal of animals. The most extreme case of differential dispersal ability is of course that in which certain species in the theoretical pool of available colonists are such poor dispersers that their chances of reaching the Krakataus are virtually nil. They are simply not part of the *effective* available pool. An extreme example is the Javan rhinoceros. A rather less extreme example is the bird family Phasianidae, which includes open-country species, such as quails and peacocks, as well as forest birds such as jungle fowl and argus pheasants. On purely ecological grounds the family may thus have been expected to contribute some of its species, present in areas of West Java and southern Sumatra, to the developing Krakatau fauna. All the species, however, are poor fliers, and there are no representatives of the family on the Krakataus. Similarly, buttonquails (*Turnix* species), birds of grassland and open habitats, did not colonize the Krakataus when the habitat was suitable; they are poor, reluctant fliers and are also absent from the islands of Simeulue and Siberut about 100 km (63 miles) off the southwest coast of Sumatra. The families of pittas (Pittidae) and babblers (Timalidae) are important components of mainland forest communities, and some 18 species of babblers occur in the forests of Ujung Kulon (Java) and Barisan Selatan (Sumatra). Like buttonquails, they are poor dispersers, and no representatives of either family have colonized the Krakataus, nor Simeulue and Siberut, although Mees (1986) recorded a hooded pitta, *Pitta sordida,* reaching Anak Krakatau from Sumatra in 1955.

The archipelago's early representation of bats is also informative. Many fruit bats are nomadic and habitually forage over considerable distances, being more wide-ranging than insectivorous bats. They arrived on the Krakataus before insectivores, between 1908 and 1919, at about the time of forest development, and by 1933 three species were established, in spite of the fact that food became available for them later than for the insectivores. The single female horseshoe bat (an insectivore) found at 600 m on Rakata in 1928 (Hill 1983) was probably a chance stray, or perhaps could have represented a failed colonization attempt (although the species has not been recorded on the adjacent mainland). Insectivorous bats were otherwise not recorded in any of the dozen Dammerman surveys from 1919 to 1934. Several species were detected in the 1980s (Tidemann et al. 1990), and they must have colonized when forests were well established, long after their insect food became available. Insectivores make up 75 percent of the Sumatran bat fauna and 85 percent of that of Java (Dammerman 1948), yet on the Krakataus the proportion was still only about 60 percent by 1992. Their lower vagility, perhaps with their generally more specialized roosting requirements, probably accounts for their later colonization.

Similar examples may be found in invertebrates. Dammerman (1948) noted that of the 59 species of ants known to have arrived by 1933, there was not one in which the female is wingless, such as those of the doryline subfamily, which were nevertheless present on Sebesi. Nor were there any wingless tenebrionid beetles, although apterous species are quite common in the subfamily. Many groups of spiders do not balloon. Most of the primitive mygalomorph group, for example, are ground dispersers. They occur on continental islands off Australia. On the Houtman Abrolhos group, Western Australia, which include both continental and reef-formed islands, they are present only on those that are continental. They are absent from islands of the Great Barrier Reef, although 7 other spider families are represented (Main 1976, 1982). Although non-ballooning families of spiders are well represented on Java and Sumatra, only 18 of the 100 spider species present on the Krakataus in 1933 were thought to be non-ballooners (Bristowe 1931, 1934).

Large differences in dispersal ability between animals, even within a single group of related forms, may derive from aspects of their biology other than the simple capacity to make the journey. For example, it is well known that some birds are reluctant to cross water barriers although perfectly capable of doing so. For them the barrier is "psychological" rather than physical. Behavioral differences not directly related to dispersal may also sometimes affect the chances of colonization of islands by other animals. For example, as discussed above, differences in the oversea dispersal ability of termites appears to depend on nest-founding behavior, the selection of wood rather than soil as a nest site, and habitat selection in land molluscs probably affects their dispersal probabilities.

On the Krakataus there is a dearth of social wasps and bees that found colonies by swarming, rather than by individuals. Swarming is a complex and easily disrupted behavior, and in most cases it is unlikely that swarms would remain coherent over a long sea journey. *Apis dorsata*, however, a species of swarm-founding honeybee which occurs commonly on the coast of Java together with two other species of *Apis*, is evidently an exception. It has particularly coherent swarms that may endure for three or four times the distance from the mainlands to the Krakataus. *A. dorsata* became established on the Krakataus by 1919, but the other two species have not colonized. Similarly, at least three species of the wasp genus *Ropalidia* that found colonies by means of independent individuals have colonized the Krakataus, two of them by 1908, but no swarm-founding species of the genus has done so. Many social Hymenoptera with non-swarming colony-founding behavior, including ants, have also colonized the islands (Yamane, Abe, and Yukawa 1992).

Behaviorally induced differences in dispersal ability may also be responsible, at least partly, for differential colonization success in parasitoid insects of the hymenopteran family Braconidae. Parasitoids lay their eggs either on the surface cuticle or within the tissues of other insects, usually when the latter are larvae. On hatching, the parasitoid larvae eat and destroy the internal tissues of their host insect, and parasitoids can thus be regarded as specialized internal predators. One group of braconids comprises what are known as koinobiont parasitoids. These usually attack insect larvae, such as army worms or looper caterpillars feeding in an exposed, open situation, and the developing parasitoid feeds and grows within its caterpillar host without killing it. The caterpillar continues to feed and move about until it is mature, when the parasitoid finally emerges from its living, walking food supply. Thus, although koinobionts cannot immigrate to an island with the adult stage of their host, there is a chance that they may be carried within dispersing host larvae. In contrast, other parasitoids, known as idiobionts, kill or paralyze their host when they oviposit, which means that for the host-parasitoid system to reach an island, separate immigrations of host and parasite would always be necessary. K. Maeto and I thought that this might give koinobionts a slight edge over idiobionts as early colonizers, so we checked the relative representation of these groups on the Krakataus. Dammerman listed 9 species of braconids as having been collected on Rakata, Sertung, and Panjang between 1908 and 1933, and 7 were probably koinobiont parasitoids of Lepidoptera. On the biologically young Anak Krakatau the proportion is, coincidentally, the same (14 of 18 species, 78 percent). In contrast, in the period 1984–1986, after over a century of colonization, there were at least 92 species of braconids on the three older islands but only 26 (28 percent) were koinobiont parasitoids of Lepidoptera. These comparisons support our hunch that koinobiont braconid parasitoids are better (earlier) colonizers than are idiobionts (Maeto and Thornton 1993).

Dispersal of Plants by Animals

Gut parasites of animals arrive along with their hosts, of course, and if their hosts become established so will they, unless they require an intermediate host that is absent from the islands in order to complete their life cycle. In a few cases it is possible for non-parasitic animals to be dispersed within the alimentary tract of birds. Docters van Leeuwen believed that some gall insects could be spread endozoically when birds eat galls that are formed on the plant in response to the insect. He thought that the inner cavity of the galls, containing the larvae of the gall insect, may pass through the bird's alimentary canal undigested (Docters van Leeuwen 1920, 1936). It is as a dispersal mechanism of plant seeds, however, that endozoochory is of major ecological significance.

Darwin (1859) grew seven grass plants of two species from seeds obtained from locust excretory pellets sent to him in England from South Africa, and large short-horned grasshoppers *(Valanga nigricornis)* have been on the Krakataus for many decades. Fruit-eating birds and bats, though, are probably the main agents of this kind of dispersal.

Animal-dispersed plants (6 species) were first found on the Krakataus in 1896 (Penzig 1902). All were probably brought by birds, and by 1905 23 species were thought to have arrived in this way (Docters van Leeuwen 1936), although there are no precise records of land birds—Muller (1897) mentioned seeing "birds" in 1896 and the Selenkas (1905) a year later—until zoologists first surveyed the islands in 1908. Docters van Leeuwen noted that shortly after 1929 a number of endozoically dispersed plants arrived at all or almost all islands simultaneously, and suggested that there may be a connection between the arrival of the animal agents of dispersal and the changes in vegetation. Whittaker, Bush, and Richards (1989) have established that this component of the flora has increased steadily and shows the highest rate of increase of any dispersal category over the past 60 years. This, of course, is the only plant-dispersal mode with positive feedback, as fruiting plants brought by animal dispersers attract more dispersers, which bring more fruiting plants, and so on (Thornton 1984). The animal-dispersed component has been, and no doubt will continue to be, the most important one ecologically, since it includes the vast majority of the tree species of the islands' inland forests. Whittaker and Jones (1994) consider that birds have had a greater role than bats, in the sense of contributing the greater number of colonists; birds also appear to have colonized earlier and thus provided the earliest set of dispersers to initiate the positive feedback.

For dispersal to an island to be successfully achieved the plant propagules must be eaten in a source area and deposited on the island—where they grow to maturity. So in assessing the effectiveness of an animal as a dispersal agent a number of questions must be considered. First, does the animal actually ingest propagules of the plant, such as seeds, as opposed to merely removing them from the plant in order to obtain some associated part, such as the aril, which could not act as a propagule? Second, if the animal ingests the seeds, does it digest them, obtaining nutriment from them, thus being in effect a seed predator of the plant? If so, do any seeds at all survive and pass through the animal in a viable condition? Or does the animal obtain nutriment from, say, the pulp of the fruit, the seeds being unaffected by the mechanical and chemical processes involved, virtually all of them being passed in a viable condition? For what length of time are seeds retained in the animal's gut before being voided, and what are the animal's normal cruising speed and flight range? Is the animal capable of reaching the island under normal or abnormal circumstances and, if so, could it make a landfall before the seeds were voided? And finally, if the seeds arrive in a viable condition, are they likely to be deposited in an environment favorable for their germination and subsequent development and growth? In order to answer these questions fully one would need information on the animal's feeding habits, gut physiology and retention time, and speed and range of movement. One or more of these is usually unknown, which means that an assessment of the dispersal ability of an animal often rests on assumptions of similarity between the animal concerned and other, related animals for which the data are available.

Birds as dispersers

The small sunbirds—which resemble American hummingbirds in their habit of hovering in front of the trumpet-shaped flowers from which they extract nectar and in the brilliant metallic colors of the male plumage—feed predominantly on nectar, insects, and spiders. The plain-throated sunbird, *Anthreptes malacensis* (which colonized by 1919), occasionally takes soft fruit, and the Javan sunbird, *Aethopyga mystacalis* (arrived by 1952), sometimes takes seeds. Another bird to arrive by 1919, the white-breasted waterhen, *Amaurornis phoenicurus*, feeds on invertebrates and grass seeds. It is the frugivorous birds, however, that have been ecologically important plant-dispersers.

Although only 10 percent of lowland forest tree species in Sarawak, for example, bear fruit that is edible by birds (Fogden 1972), dispersal by birds is nevertheless important in the natural afforestation of islands because frugivorous birds disperse the seeds of important canopy tree species of secondary forest. On the Krakataus the important dominant trees of Sertung and Panjang, *Timonius compressicaulis* and *Dysoxylum gaudichaudianum*, and the summit shrub *Schefflera polybotrya* are bird-dispersed. The trees *Buchanania arborescens, Bridelia monoica, Ardisia humilis, Leea sambucina, Trema orientalis*, and the recently discovered (1989) large tree species *Knema cinerea*, understorey components *Leucosyke capitellata* and *Villebrunea rubescens*, and the early small-tree cover *Macaranga tanarius* are other ecologically important bird-dispersed species. Whittaker and Jones (1994) suggested that birds disperse a more balanced range of plant growth forms than do bats, and they estimated that some 80 Krakatau plant species have been dispersed by birds.

Some birds, such as the large imperial pigeons (species of *Ducula*), are obligate frugivores, feeding only on fruit, and some specialize, to varying degrees, on particular fruits (such as the pink-necked green pigeon, *Treron vernans*, on figs). The red cuckoo dove takes seeds as well as some fruit, bulbuls and orioles include insects, leaves, and flowers in their diets, the emerald dove also takes seeds and insects, and the large-billed crow is omnivorous. These are facultative frugivores, capable of surviving without fruit. Wells (1988) found that in the lowland forests of the Malay peninsula frugivores that included arthropods in their mixed diet outnumbered all other generalists. Generalist, facultative frugivores predominated among the early fruit-eating colonists of

Krakatau; one specialist and five facultative frugivores colonized within 25 years of the great eruption.

Wells observed that in Malaysian lowland forests bulbuls and flowerpeckers (frugivore-insectivores) concentrated on fruitings of trees of the pioneer and building phases of plant succession at disturbed sites, typically taking small fruits, many of which could be harvested for a low expenditure of energy. Two of the facultative frugivores to colonize Krakatau within the first 25 years were bulbuls, the sooty-headed *(Pycnonotus aurigaster)*, a bird of open woodland and forest edges, and the yellow-vented *(P. goiavier)*, which usually inhabits open scrub. The sooty-headed bulbul was not seen again and Dammerman regarded it as an unsuccessful colonist. Dr. R. Corlett has told me that in Singapore the yellow-vented bulbul seems to be the most important plant-dispersal agent in open country, taking fruit of up to about 12 mm in diameter. He has seen sooty-headed bulbuls taking fruits of the same size range and believes that bulbuls regurgitate rather than defecate seeds with a diameter of more than about 8 mm.

Lambert and Marshall (1992) suggested that small frugivorous birds with small gapes, such as *Pycnonotus* bulbuls and flowerpeckers (*Dicaeum* species), may not be able to eat many large forest fruits unless they are sufficiently succulent (like most figs) to allow the birds to break them up while still attached to the tree. Docters van Leeuwen (1936) recorded seeing yellow-vented bulbuls on the Krakataus (the only bulbul present during his visits) eating fruits of *Leukosyke capitellata, Pipturus argenteus,* and *Cayratia trifolia,* and Whittaker and Turner (1994) reported it taking seeds of *Dysoxylum gaudichaudianum,* an important canopy tree, on Rakata. On Anak Krakatau it has been seen eating the fruits of *Cassytha filiformis*—a sea-dispersed trailing parasite of beach creepers of the *I. pes-caprae* zone—and Zann, Male, and Darjono (1990) flushed the seeds from the birds' crops. In Docters van Leeuwen's Bogor aviary *Pycnonotus* species ate the black berries of *Musseanda frondosa* and the fruits of *Scaevola taccada,* both Krakatau species (Docters van Leeuwen 1936). Ridley (1930) stated that *Melastoma affine,* one of the first shrubs to appear in *alang alang* grassland and which colonized the Krakataus between 1886 and 1897, is largely dispersed by yellow-vented bulbuls, and this bird also took fruit of *Clidemia hirta.* In Singapore and peninsular Malaysia this species has been recorded as dispersing seeds of many other Krakatau plants, including *Ficus benjamina, Olea maritima, Lantana, Passiflora foetida,* and *Anacardium occidentale.* Lambert (1989a) discovered that in Malaysian lowland rain forest, a species of *Pycnonotus* retained seeds for periods of from 5 to 41 minutes before voiding them. These times indicate that bulbuls may occasionally carry seeds to the Krakataus from Java or Sumatra, but that they are more likely to bring them over the shorter distance (about 12 km) from Sebesi, where the bulbul also occurs. They certainly would be capable of dispersing seeds between the Krakatau islands.

Another facultative frugivore that arrived within 25 years of the eruption was the black-naped oriole, *Oriolus chinensis* (Jacobson 1909). This species frequents the tops of tall trees and feeds on figs, leaves, flowers, seeds, and insects. Zann, Male, and Darjono flushed 29 food items from the crops of two orioles on Anak Krakatau: a seed, a flower, 11 casuarina leaves, 14 fragments of figs of *Ficus septica,* a cockroach, and a cricket. Species of *Oriolus* have been recorded as feeding on figs (Ridley 1930; Lambert 1989a) and on Melastomaceae (Ridley 1930) in Malaysia. It is likely that the black-naped oriole was one of the first land birds to colonize the Krakataus, and since Lambert found seed retention times in this species to vary from 21 minutes to 4 hours, it could have been an important agent of the dispersal of figs, for example, to the archipelago.

The omnivorous large-billed crow, *Corvus macrorhynchos,* also present at the first survey, prefers open country and feeds opportunistically on the ground and in trees or bushes. It is present on Sebesi. Ridley reported that it feeds on figs and a species of *Digitaria* (*D. rhopalitricha* occurs on the Krakataus). On the Krakataus, Bartels (1919) found its gut to be full of the pulp and seeds of the papaya, *Carica papaya,* and Docters van Leeuwen (1936) frequently found seeds and seedlings in the crow's droppings and thought that it was responsible for the spread of papaya in Sertung's casuarina woodland. Seed retention times of 8–17 hours have been found in another species of *Corvus,* the raven (Proctor 1968), and the large-billed crow could have played an important role in the early colonization of fruiting trees.

The small emerald dove, *Chalcophaps indica,* was also already established on the Krakataus by 1908. It is a ground feeder in dense undergrowth, exploiting fallen fruit and seed as well as taking ants and flies and probing for other insects. Frith (1982) reported that in Australia it sometimes walks along tree branches taking fruit that is within easy reach from species of *Calamus, Euphorbia,* and *Lantana,* all genera present on the Krakataus, as well as the Krakatau species *Melia azaderach.*

Species of doves and pigeons of the genera *Chalcophaps, Macropygia,* and *Treron* have thick-walled muscular gizzards containing grit, and ingested seeds are mechanically abraded and digested. Thus although these birds are fruit eaters they are in fact seed predators and may only occasionally act as seed dispersers. Although Crome (1975) showed that the gizzard of *C. indica* is of the abrading type, facilitating the digestion of small seeds such as those of figs, Lambert (1989b) was able to collect some intact seeds from birds fed in captivity, and they germinated within a few days. Seed retention time was almost 4 hours and in one case over 12 hours. Wells reported that the species has crossed from the Malay peninsula to Sumatra and covered 800 km (500 miles) from the place where it was ringed, indicating that it could occasionally act as a long-distance disperser.

The only obligate frugivore to arrive within the first 25 years was *Treron vernans,* the pink-necked green pigeon. Species of *Treron* were the most abundant pigeons in Lambert's study area in peninsular Malaysia. He found that they were fig specialists, depending on figs entirely, and that they were responsible for a third of all figs taken from one strangler fig in a day. Ridley reported that *T. vernans* also ate the fruit of *Oncosperma tigillarium, Macaranga robiginosa,* and *Melastoma affine* in the Malay peninsula, as well as those of *Ficus benjamina* and *F. retusa.* Species of *Macaranga* and *Melastoma* are important pioneer invaders of grassland communities on the Krakataus. Like *C. indica,* the pink-necked green pigeon has a muscular, grit-containing gizzard and a long, coiled, narrow intestine (Cowles and Goodwin 1959). Lambert (1989b) again found, however, that about a third of the birds lacked gizzard grit and passed intact viable seeds, and intact seeds were found in feces in the field. Cowles and Goodwin (1959) dissected wild-caught museum specimens of *Treron* species. Two individuals had pieces of figs and intact seeds in the crop, and both grit and whole and partially-ground seeds in the gizzard. Lambert found that *Treron* individuals ranged over vast areas and he recorded seed retention times of from 1 to over 8 hours. Wells reported that this pigeon (like *C. indica*) disperses at night in search of localized fruit abundances. Another species of *Treron* was timed by stopwatch in Aden flying at a speed of 34 mph (54 km/hr) over 400 yards (Meinertzhagen 1955), and if speeds rather below this were sustained, the pink-necked pigeon could conceivably reach the Krakataus from Java or Sumatra before any recently ingested seeds that had escaped digestion were voided.

By 1919 and 1920 five more frugivorous birds had colonized the islands, one obligate (an imperial pigeon) and four more facultative fruit feeders (a starling, koel, flowerpecker, and whistler). The Asian glossy starling *(Aplonis panayensis)* usually feeds on fruit high in the treetops but will also take ants, beetles, and insect larvae. It is the most effective disperser of the early-successional *Macaranga tanarius*. Seeds of this plant recovered from the gut of a starling shot on the Krakataus germinated and grew into young plants. In Bogor Docters van Leeuwen noted that the starling nested and roosted in the crown of an *Attalea* palm. Several plants were growing in the crown, including *Ficus ampelas, Claoxylon polot, Homalanthus populneus, Clidemia hirta, Allophyllus cobbe, Leea sambucina, Lantana camara,* and *Morinda citrifolia,* all Krakatau plants. On the Krakataus the starling was seen feeding on fruit of *Leucosyke capitellata, Pipturus argenteus,* the bright red fruits of *Tinospora glabra,* and the black, shining berries of *Cayratia trifolia,* which Docters van Leeuwen believed was spread into the interior in this way. In captivity the starling ate fruits of two Krakatau plants, *Scaevola taccada* and *Tinospora glabra,* and seeds of the latter that were recovered from the feces germinated on sowing (Docters van Leeuwen 1936). Whittaker and Turner (1994) saw the starling taking seeds from fruits of *Dysoxylum gaudichaudianum* on Rakata. Although there are no data on flight speed or seed retention times, this species may have played an important role in the dissemination of a number of plant species to the Krakataus and thus the diversification of the archipelago's forests (even though half of the 14 species mentioned above were present by 1908, before the starling was first recorded).

The Asian koel *(Eudynamys scolopacea)* and the mangrove whistler *(Pachycephala grisola),* both facultative frugivores, were also present in 1919. The koel, a large cuckoo, feeds on figs, berries, insects, and small invertebrates; Ridley reported finding fig seeds in its crop in Java. The whistler is mainly an insectivore of tree branches but also takes some fruit (Zann and Darjono 1992).

The small flowerpeckers, like bulbuls, tend to take small figs. In peninsular Malaysia orange-bellied flowerpeckers *(Dicaeum trigonostigma)* accounted for 3 percent of the figs taken from a strangler fig in one day (Lambert 1989a). This species, however, is of particular interest because of its relationship with plants of the mistletoe family, Loranthaceae, the seeds of which are spread almost exclusively by flowerpeckers (Docters van Leeuwen 1954). Loranthaceae were represented on the Krakataus in 1860 but have not recolonized since the 1883 eruption, although the flowerpecker was found on Rakata and Sertung from 1919 to 1933 and now occurs on Rakata from an altitude of 150 m to the summit. The flowerpecker's main food is spiders, but it also eats insects, worms, and flower nectar, as well as small fruits (including mistletoe berries). The species can obviously survive without access to mistletoe berries. The intriguing question is, why has the flowerpecker not brought representatives of the Loranthaceae to the Krakataus? Docters van Leeuwen discovered that seeds pass through the flowerpecker quite quickly, in from 12 to 20 minutes and, moreover, the birds are fairly inactive for a while after feeding. He observed that the flight from Java or Sumatra would take a bird of this size about 2 hours, too long for the seeds to be retained until reaching the Krakataus. From Sebesi the flight would still take over a half-hour, and in any case Loranthaceae were not recorded on Sebesi in 1921. It is possible that Loranthaceae will reach the Krakataus, however, because after leaving the cloaca of the bird the sticky seeds often adhere to the surrounding feathers, and the bird then tends to rub its cloaca on branches to free the seeds, often unsuccessfully. The seeds could thus arrive by being carried part of the way internally and the rest of the journey (perhaps following a stopover on Sebesi) externally, attached to feathers.

Bartels recorded the pied imperial pigeon *(Ducula bicolor)*, an obligate frugivore, on Rakata in 1919, and it was found on Sertung in the following year. This large, nomadic pigeon inhabits coasts and small islands of the region (Wells found that it nests exclusively on islands in peninsular Malaysia), large flocks occurring for a time in an area, then departing. The gizzard of *Ducula* species is thin-walled, with a weak musculature, and the intestine relatively short (Gadow 1933). Quite large fruit is swallowed whole, and horny ridges and nodules on the inner wall of the gizzard abrade only the fleshy coating of pulp on the fruit, and the seeds or stones pass out unaffected by the digestive processes. Specialist fruit-eating birds that do not digest seeds, such as large imperial pigeons (species of *Ducula*) and fruit-doves (*Ptilinopus* species), are well known as important dispersers of many forest trees. Imperial pigeons are quite powerful fliers and can cover long distances. A Torres Strait pigeon (a species of *Ducula*) was taken on Lord Howe Island in 1913, some 1,300 km (800 miles) from the coast of its native Australia (Frith 1982), and Wells (1988) recorded a distance of 800 km (500 miles) being covered by an individual of another *Ducula* species, from its banding site in peninsular Malaysia to Sumatra. Homing pigeons average 48–56 km/h (30–35 mph) against the wind and 96 km/h (60 mph) with a tail wind, several speeds of more than 230 km/h (128 mph) having been recorded over courses of 128 km (90 miles) (Meinertzhagen 1955). Ridley found seed retention times in an unspecified species of *Ducula* of from 3 to almost 8 hours, Proctor (1968) recorded 3–24 hours in pigeons, and Lambert (1989b) found times of several hours in some of the

Krakatau columbids. So with flight speeds as low as 48 km per hour, imperial pigeons could easily carry a variety of viable seeds to the Krakataus from the mainlands. In Australia, Frith (1982) noted that pied imperial pigeons voided whole seeds of the following genera that have reached the Krakataus: *Canarium, Ficus, Calamus, Solanum, Terminalia, Antidesma, Pandanus, Piper,* and *Freycinetia*. Docters van Leeuwen remarked that this pigeon could be responsible for the arrival on the Krakataus of the palm *Corypha utan.*

Another columbid, the red cuckoo-dove, *Macropygia phasianella*, was the tenth facultative frugivore to arrive, by 1932. It takes berries, fruits, and seeds in trees and bushes of undergrowth or the middle storey of shaded forest edges, and it is usually seen in pairs or small parties. This cuckoo-dove, however, has a muscular gizzard, with gravel or sand, that facilitates digestion of seeds. In Australia it has been recorded as feeding on the following plant genera (and species) that occur on the Krakataus: *Hibbertia (scandens), Glochidion, Elaeocarpus, Dysoxylum, Cissus, Freycinetia, Trema,* and *Calamus* (Frith 1982).

A third bulbul, the olive-winged bulbul, *Pycnonotus plumosus*, was recorded in 1951 (Hoogerwerf 1953), and the chestnut-capped thrush, *Zoothera interpres*, was first noted in 1984 (Zann et al. 1990). The thrush is a shy forest bird of thick undergrowth and could have colonized long before it was discovered. Lambert (1989a) recorded it taking fallen figs in lowland rainforest in peninsular Malaysia, and it may thus be capable of dispersing seeds, at least between islands of the Krakataus. A stray individual of this species unaccountably turned up on Anak Krakatau during our 1986 expedition.

The most important plant dispersers to arrive in the past fifty years were two more efficient seed dispersers, the black-naped fruit-dove, *Ptilinopus melanospila* (by 1951), and a second species of *Ducula,* the green imperial pigeon, *Ducula aena* (by 1983) (Thornton, Zann, and van Balen 1993). Proctor (1968) found seed retention times of from 19 minutes to almost 9 hours in species of *Ptilinopus* and, like the pied imperial pigeon and perhaps also the crow and oriole, both these species would be capable of bringing seeds to the Krakataus from Sebesi some 11 km away, or even from the Javan or Sumatran coasts, each some 44 km distant. We have evidence that *D. aena* may disperse *Oncosperma tigillarium* on the Krakataus (Chapter 10).

Thus six frugivorous birds, one a fig specialist, arrived within 25 years of the 1883 eruption and five more, one an obligate fruit-eater and effective, long-distance disperser, became established in the second quarter-century. In the next 25 years two more facultative frugivores and another obligate frugivore and excellent long-range disperser arrived. Finally, by 1984, the twelfth facultative and the fourth obligate frugivore had arrived.

Bats as dispersers

It has long been appreciated that fruiting plants that attract bats differ from those attracting birds; in fact, characteristic "bat-plant" and "bird-plant" syndromes have been distinguished. Workers such as Ridley (1930), Docters van Leeuwen (1935), and van der Pijl (1957, 1982) have recognized different suites of adaptations promoting the dispersal of seeds by bats and birds, and their views have been supported in general by later investigators such as Lambert and Marshall (1992). "Bird-fruits," fruits eaten by birds, are often small, juicy, without scent, usually borne within the foliage of the crown, and of a color (red, blue, black, or orange) that, when the fruit is ripe, contrasts conspicuously with the foliage. In contrast, "bat-fruits," fruits usually eaten by bats, are generally dull-colored (brownish, yellowish, or greenish) when ripe. Bat-fruits are also usually large, fleshy, and have a characteristically rancid or musty odor when mature, and often ripen at night. Like bird-fruits, they are generally carried on the tree until after they ripen, but are borne on a part of the plant that is easily accessible to fruit bats, which are less maneuverable than birds. Bat-fruits are usually found next to a clear space, such as on the surface of the crown, on the branches of a "pagoda tree" (*Terminalia* is a good example), hanging outside or under the crown, or arising directly from the trunk or main branches beneath the crown, as in many fig species. These categories are not absolutely rigid, however. Dr. R. Corlett has remarked to me that in Hong Kong, fruit bats seem to eat many bird-fruits, and he thinks this may be related to the absence there of any bat-fruits other than figs. Moreover, Lambert (1989a) found that 14 percent of bird-dispersed fig species in his peninsular Malaysian study site had dull-colored fruit, and four attracted some of the largest concentrations of frugivorous birds. Following a survey of the literature, Whittaker and Jones (1994) estimated that some 43 Krakatau plant species could have been dispersed by either birds or bats.

Frugivorous bats are not seed predators. Large fruits are usually consumed *in situ* but smaller ones are bitten off and carried to a feeding roost, which may be from 30 to 100 m or more away. If the seeds are large they are separated from the soft parts of the fruit before it is ingested. Smaller seeds are separated within the mouth by the action of the tongue on ridges of the palate as gentle suction is applied to the mass of fruit material; the soft pulp is swallowed, the seeds and solid matter being spat out, sometimes as a pellet. Some small seeds, including those of figs, may be swallowed and voided as feces, usually at the feeding roost but possibly much further away from the parent tree. Food can remain in the gut of some fruit bats for quite a long time. Utzurrum and Heideman (1991) obtained times of from 20 to 45 minutes for seeds to pass through the gut of two genera of Philippine fruit bats.

Dog-faced fruit bats (*Cynopterus* species) were the first bats to be seen on the archipelago. They are fairly small, usually with a wing span of about 14 cm (about 6 in). They bite fruits from the tree and carry them away to feeding roosts, where they swallow the fluid and small seeds, spitting out skin, fiber, or large seeds, which are undamaged, then fly back for more. Near the West Javan fishing port of Pelabuhan Ratu I have seen *Cynopterus* bats drop *Terminalia* seeds 100 m from the nearest possible parent tree. Their feces are semi-fluid and small seeds voided in flight disperse, falling separately. Seeds of *Ficus racemosa* from the gut of a *Cynopterus* bat on Pulau Peucang, West Java, and others found in droppings beneath feeding roosts germinated when sown (van der Pijl 1936). As their food plants, Ridley (1930) listed figs and the fruits of *Adinandra, Eugenia, Elaeocarpus, Calophyllum,* and *Melia,* of the plant genera present on the Krakataus, as well as palms. Marshall (1985) listed the additional Krakatau genera *Cocos, Mangifera, Canarium, Terminalia, Cyrtandra, Artocarpus, Musa, Pandanus, Piper,* and *Morinda.*

Cynopterus sphinx was found on the Krakataus in 1919, when there was an extensive stand of *Cyrtandra sulcata* (whose oblong white berries are eaten by both bats and birds) on Rakata above 400 m. On Timor, Goodwin (1979) reported that this fruit bat preferred the fruit of *Eugenia grandis* when available, otherwise figs were the preferred food, as they are on Java, Hong Kong (information from Dr. R. T. Corlett), and on the Krakataus. The larger dog-faced fruit bat, *Cynopterus horsfieldi,* was present on the Krakataus from 1920 to 1930, and L. Shilton found it again in 1996. Dammerman (1948) recorded this species as also taking figs. *Cynopterus titthaecheilus* was first recorded on Rakata in 1974, and in 1986 we found the floor of its roosting cave on Panjang littered with the seeds of *Terminalia catappa.* It, too, takes figs (Dammerman 1948).

Cynopterus brachyotis was first recorded on the Krakataus in 1985. The species takes fruits of coastal plants such as *T. catappa, Calophyllum inophyllum,* and *Morinda citrifolia,* and for decades before its discovery it was probably responsible for the spread of these species inland to Rakata's upland slopes. Seeds of *Ficus* species, *Piper aduncum,* and *Cyrtandra sulcata* have been found in its droppings and many seedlings are often found under its feeding roosts. Boon and Corlett (1989), studying this species in Singapore, observed that fruits weighing up to 20 g were carried to feeding roosts, usually within 100 m of the parent tree. Heavier fruits were eaten while still attached to the tree. The largest fruit taken in the study was that of *T. catappa* (20–25 g) (although according to van der Pijl loads of up to about 75 g can be carried by *Cynopterus* species). Seeds larger than 5 mm were spat out or dropped undamaged, singly, those of intermediate size in clumps, under feeding roosts. Boon and Corlett discovered that only seeds less than 2.5 mm long and 1.5 mm wide were swallowed, and these were defecated both at the roosts and in flight. Fecal samples were collected on plastic sheets spread under mist-nets (large pieces of fine-mesh netting used for catching birds in flight) and under bat flight paths, to detect defecation of seeds in flight. Captured bats retained seeds for 11–15 minutes only, so that dispersal of small seeds from Sebesi would be a possibility, but seeds are unlikely to have been carried from the mainlands. *Cynopterus* bats certainly spread seeds from island to island within the Krakataus.

Geoffroy's rousette, *Rousettus amplexicaudatus,* which had arrived on the Krakataus by 1933, is larger than *Cynopterus* and capable of flying 30 to 40 miles (48 to 63 km) in a night. In 1984 we found many fig seeds in the guano on the floor of its roosting caves on Panjang, and Dr. S. G. Compton grew seeds, from its droppings, into plants of *Ficus fulva* and *F. septica.* This species could easily reach the Krakataus from the adjacent mainlands, and if seed retention times are close to those found by Utzurrum and Heidemann (1991) for Philippine fruit bats (20–45 minutes), this species could have dispersed seeds from the mainlands, and certainly from Sebesi. A second, rather larger species of rousette, *Rousettus leschenaultii,* was found in 1992 (Figure 23).

A camp of the Malay flying fox, *Pteropus vampyrus,* was on Sertung in 1985 and another on Rakata in 1989. The species was probably on the archipelago from time to time before those dates. The largest of the world's bats, with a wing span of a meter and a half (almost 5 feet), these bats may cover 70 km (about 46 miles) in a night of foraging for fruit and are wide-ranging nomads, following mass fruitings of trees. To a flying fox the distance from Sebesi to the Krakataus (about 13 km or 8 miles) or from the mainlands of Java or Sumatra (44 km, about 27 miles) presents no real barrier. In 1923 Dammerman (1948) witnessed a flight over a distance of at least 60 km (38 miles) on the east coast of Sumatra. An individual was recorded alive at sea 200 miles (about 300 km) from land, and another reached Cocos Island in the Indian Ocean from Java, a distance of 700 miles (over 1,100 km), although it expired on arrival (Ridley

Figure 23. Leschenault's rousette was the seventh fruit bat to colonize, by 1992.

1930). *P. vampyrus* passes excreta very rapidly when feeding voraciously, but Ridley believed that it could transport fig seeds (for example, those of *Ficus benjamina*) for at least a few kilometers.

When breeding *P. vampyrus* aggregates in groups of hundreds or thousands for some months. *Pteropus* species feed in tall trees, and, unlike *Cynopterus* species, do not fly through dense forest, although they will come lower for fruits such as banana or papaya. Boon and Corlett (1989) observed that in Singapore *P. vampyrus* seemed to be more arboreal than *C. brachyotis*. It would land in the canopy and clamber about eating fruit, being thus less likely to carry fruit away from the tree, although van der Pijl reported that it could carry loads of up to 200 g.

The food transit time of Australian species of *Pteropus* was found to be 40 minutes in the laboratory, and seeds were only slightly damaged (Nelson 1965), and Tedman and Hall (1985) reported mean minimum times of 12–34 minutes. Transport of seeds from Sebesi to the Krakataus may therefore be possible at a flight speed of about 18 km/h (about 11 mph). Ridley reported that fruit of the following Krakatau plant taxa was taken by *Pteropus* species: figs (including *Ficus racemosa*), *Calophyllum inophyllum*, *Syzygium*, *Elaeocarpus*, *Melia azaderach*, *Pandanus*, *Carica papaya*, *Spondias*, *Mangifera*, guava, banana, and *Terminalia catappa*. To these, Marshall was able to add *Glochid-*

ion, Artocarpus, Cocos, Citrus, Planchonella, Passiflora, and *Solanum*—all Krakatau genera. Pieces are bitten out of the fleshy side of *Terminalia* fruit and the rest is dropped far from the parent tree. Wodzicki and Felten (1975) stated that on Niue island in the south Pacific (near the eastern limit of the range of the genus), *Pteropus tonganus* showed a high preference for *Pandanus,* but that a New Zealand botanist, Mr. William Sykes, suggested to them that fruits of *Dysoxylum forsteri* may well be eaten. A species of *Dysoxylum* is an important late-successional forest tree on the Krakataus, and Docters van Leeuwen reported finding its seedlings growing under bat feeding roosts. The role of bats in the dispersal of this tree clearly warrants investigation.

Sebesi—an Eroding Steppingstone?

The extent to which the island of Sebesi acts as a steppingstone to the Krakataus for components of Sumatra's biota is difficult to assess. With its companion island, Sebuku, it considerably reduces the water gaps confronting potential colonists from the areas of Sumatra to the north. Of the two, Sebesi makes by far the better steppingstone, being about the same height (844 m) and size as Rakata (Figure 20). Both Sebesi and Sebuku were themselves severely affected by the events of 1883, however. Ash fell on Sebesi to a depth of more than a meter, but its biota was not eradicated, and recovery was probably enhanced by proximity to the mainland. In 1921 not only was Sebesi's forest recovery more advanced than that of Rakata, but its vegetation was of a quite different composition (Docters van Leeuwen 1922a, 1923). There were more than twice as many forest tree species than in Rakata's forest, and the uplands above about 400 m, instead of being clothed, as were those of Rakata at that time, in an almost pure stand of the shrub *Cyrtandra,* carried a woodland of tall trees with a diversity of shrubs, including a dozen that were not present on Rakata. The situation is further complicated by the fact that Sebesi's environment has itself been drastically changed over the period of recolonization of the Krakataus, and in ways different from the change on the Krakataus.

Sebesi has permanent streams, and thus has been inhabited and considerably disturbed by agricultural practices for many years. Much of the island's lowland area was cleared and planted by Haji Djamaludin and his workers in 1890, and in about 1900 cattle, goats, and horses were introduced. About ten years later some of these animals escaped and were feral in 1920–1922. The goats inhabited the highlands, one being shot in 1921 at an altitude of 700 m, and Dammerman estimated that there were about a thousand cattle running free on the island (Dammerman 1948). Coconut plantations were now extensive, and there were fruit trees and *ladangs* (rice fields in cleared forest). Although there were then no permanent inhabitants, short visits were made from Sumatra, and coconut gatherers stayed for longer periods. Further clearance for cultivation and coconut plantations has encroached far up the mountain slopes, natural vegetation now being confined to mangroves and the small summit area above an altitude of about 740 m.

Unfortunately, Sebesi was not surveyed by biologists until 1921, otherwise it may have been possible to identify components of its early flora that arrived on the Krakataus much later, as indirect evidence of its steppingstone role. Whittaker and colleagues (1992) list 18 species of plants that occurred on Sebesi in 1921 but were not found on Rakata before 1936, as possible examples of species reaching the Krakataus via Sebesi. They acknowledge that this is not many (less than 10 percent of the Krakatau flora), and also that there are many reverse cases—that is, of plant species shared by the two islands but first found on the Krakataus—which by the same logic would have to be taken as indicating a movement in the opposite direction. They have also identified a number of plant families that have been well represented on Sebesi but have been slow to colonize the Krakataus. Three particularly striking examples are the Acanthaceae, with 5 species on Sebesi and none on the Krakataus, Araceae (7 and 1), and Myrtaceae (9 and 2, the 2 species on the Krakataus not being recorded up to 1951). The representatives of these families on Sebesi are mostly herbs, shrubs, or small trees that are dispersed by birds or bats. A total of 143 of the 265 species of plants found on Sebesi in 1921 have also been recorded on the Krakataus. These of course include a core of highly dispersive species, many of which occur on several of the Krakatau islands. For many of the 110 species which have not yet been found on the Krakataus, however, there are no obvious habitat restrictions on the archipelago. Thus there is little evidence from data on plant distribution that Sebesi has been an intermediary in the colonization of the Krakataus by plants.

In a 5-day visit to Sebesi in April 1920, the ornithologist Siebers recorded 30 species of resident land birds (Dammerman 1948). Haag and Bush (1990) could find only 12 species in 6 hours surveying the lowlands in August 1989. Most of the species present in 1920 but not recorded in 1989 now occur on the Krakataus. Three of the four that do not are birds of low open country and rural habitats, and the fourth is conspicuous by its song. So if these were pre-

sent on Sebesi in 1989 they were probably uncommon, and thus are unlikely to
be among the next colonists of the Krakataus. A full survey of Sebesi's birds is
now needed.

During the 1989 Sebesi visit Bush, Bush, and Evans (1990) spent a morning
surveying butterflies, although again the survey was limited to the disturbed
lowlands. The mangroves and high uplands were not visited. No *Ipomoea pes-
caprae* association was encountered on the coast, and probably largely because
of this the butterfly fauna differed somewhat from that of the Krakataus.
About 30 species were recorded from Sebesi, as against 48 found on the
Krakataus during a much longer survey. About 20 species were noted in the
richest habitats of Rakata over a similar period of time at the same time of day.
There was considerable overlap between samples from the two areas, 22 of the
Sebesi species also being found on the Krakataus in that year or a few years
previously. Four identified Sebesi species had not been recorded from the
Krakataus, and four more that were found only on Sebesi were of doubtful
identity. Probably because disturbance by humans has been more extensive,
more species of the Sebesi butterfly fauna occurred well into the hinterland,
and the fauna thus lacked the predominantly coastal distribution typical of the
Krakataus. Perhaps the Sebesi butterfly fauna is richer than that of the
Krakataus. The results of a recent, more extensive survey by Dr. J. Yukawa and
Dr. T. Partomihardjo should help assess the extent that Sebesi has served as a
steppingstone in the colonization of the Krakataus by butterflies.

There is little doubt that Sebesi's potential as a steppingstone was not great
in the years immediately following 1883 and that as its biota recovered, its
growing importance in this regard was negated by progressively increasing human disturbance. But the change in the island's steppingstone status was probably not merely an overall decline. There must also have been a qualitative
change in the set of putative colonists to the Krakataus from Sebesi. For forest
species, for example, the importance of Sebesi as a steppingstone must have
been diminishing for many decades as suitable habitat on the island was progressively reduced, whereas the newly created habitats on Sebesi may have progressively facilitated the crossing from Sumatra for some species of open
country. The brahminy kite *(Haliastur indus)* and large-billed crow, as examples, are birds of open areas that occur on Sebesi and might be expected to colonize any similar habitat surviving or becoming available on the Krakataus.
Yet two residents of Sebesi, the reed warbler and the spider-hunter, both insectivorous birds, have never colonized the Krakataus in spite of the fact that suitable habitat for them appears to have been available for decades.

In sum, the question of a steppingstone role for Sebesi is confused by the effects of the 1883 eruption on that island, its possession of freshwater habitats, the early and progressively increasing human disturbance, and the fact that many species of its biota for which suitable habitats appear to exist on the Krakataus have not colonized the islands. Although Sebesi may have facilitated the passage of a few species over to the Krakataus, as yet there is little, if any, evidence that it has been important in this regard.

8   KRAKATAU'S CHILD

> In any renewed activity of the volcano it is to be expected
> that islands will arise in the middle of the seabasin that is
> surrounded by Rakata Peak, Verlaten [Sertung], and Lang
> [Panjang], just as the Kaimeni arose in the Santorini Group
> and just as formerly the craters Danan and Perbuatan
> formed in the sea within the old crater wall.
>
> R. D. M. Verbeek, *Krakatau* (1885)

ℰ∂

In the early decades of this century fishermen were familiar with upwellings of sulfurous gases in the sea between Krakatau and the coast of West Java at three quite widely separated areas, which were given names. De Nève (1985b) thought that these upwellings may have been related to the revival of Krakatau's activity.

In June 1927 Verbeek's prediction, quoted at the head of this chapter, began to be fulfilled. Fishermen reported gas bubbles surfacing at a particular place in the sea covering Krakatau's caldera, and on December 29 submarine explosions began, with typically Surtseyan cock's-tail clouds and base surges. Five days later there were six points of activity along a line 500 m long, about halfway between the sites of Krakatau's former craters, Danan and Perbuatan. On January 26, 1928, volcanic products broke the surface, and an ash island 175 m long and 3 m high was formed. It persisted for about a week before being destroyed by surf. The island was later named, appropriately, Anak Krakatau (Krakatau's Child), by Petroeschevsky.

Activity was renewed at the end of 1928 and continued into January 1929. During this period water fountains and eruptions of steam, lapilli (small fragments of lava), and fine ash were accompanied by detonations and lightning, and the eruption cloud glowed at night. On January 28 Vulcanian activity from two eruption points produced solid material, and a second island was formed, 250 m long and 38 m high, which persisted for six months. In December 1929 the focus of activity moved about 600 m to the southwest, but had returned to the original site by January 1930, and in June a third island, 30 m high, was formed. Gas and ash were emitted with great force, accompanied by thunderous detonations and at night a glow could be seen. After having reached a height of 50 m, in August 1930 this island, too, disappeared, probably, as suggested by Siswowidjojo (1983), as a result of explosions. Within a few days, on August 12, however, a fourth island emerged. Two days later a strong phreatic eruption produced a gas and water column 100 m in diameter that rose over a kilometer into the air. From that time on, accumulation of ash above water proceeded at a greater rate than its erosion by wave action, and Anak Krakatau IV, which began its life as a growing tuff ring, has survived.

Two trends are recognizable in the island volcano's development. The first, documented by Sudradjat (1982), is a gradual shifting of the eruption center some 350 m to the southwest between 1930 and 1988. Several workers have suggested that the morphological development of the young volcano—for example, the southwestern slope being much steeper than the eastern and northern slopes, the frequent breaching of the crater wall in the southwest early in development, and perhaps also the direction of the shift in the focus of activity— were strongly influenced by the volcano's position on a steep submarine slope facing southwest. The movement (until 1992) of the point of activity to the southwest also brought the new island more closely into the northwest–southeast alignment of the early Perbuatan-Danan-Rakata vents and the 1883 eruption points. The 1992–1995 eruption center was to the north of the older craters, however, and obliterated them.

The second evident trend is the change in silica content of eruption products, noted by de Nève (1982). After a very early initial rise in silica content to over 60 percent, there was a fall to about 51 percent until the period of lava emission began in 1961, after which silica content has risen fairly steadily to about 55 percent. Van Bemmelen (1942) advanced the theory that the composition of magma changes in cycles: its silica content rises from a lower value to a higher one of over 65 percent, and its viscosity increases, making it more difficult for its contained gases to escape. When they do so there is a cataclysmic eruption. Silica content then falls, to begin the cycle again. Such a cycle may be seen in the activity of Ancient Krakatau, Krakatau, and Anak Krakatau. Since the silica content of Anak Krakatau's products before the 1992 eruptions was 55 percent, and during the present eruptive episode it has risen from 52.5 to about 54 percent (Sutawidjaja 1993 and personal communication), another cataclysmic eruption is generally thought to be several centuries away. Zen and Sudradjat (1983) and Camus, Gourgaud, and Vincent (1987), however, cautioned that the larger part of the volcano is still under water and that any disturbance along the fracture zone in which it is situated could trigger a hydrothermal or phreatomagmatic (explosive interaction of magma and water) eruption and (Camus, Gourgaud, and Vincent 1987) a potentially dangerous tsunami.

Three developmental stages are evident in the volcano's short life. During the first, Surtseyan phase, phreatic activity dominated, and developing young biotas were destroyed in 1933 and June 1939. A crater lake was formed in February 1932 and its rim destroyed in November. The lake was re-formed in 1933 and filled with sand in 1936, to reappear a year later.

In 1952 the second phase of the island's development began, a phase of increased activity and growth of the island ash-cone. The opening of the vent rose above sea level and a new inner cone, isolated from the sea, was formed, the old crater wall now becoming an "outer cone." A crater lake 440 m in diameter became closed off from the sea during the Strombolian eruptions of 1952. From 0.5 to 1.5 m of ash was emplaced on Sertung and Panjang, destroying part of their forests, and another strong eruption in September 1953 damaged them further. The cone was now 104 m high, and the crater lake 40 m above sea level. In 1956 a new cinder cone was formed inside the crater, but in January 1960 the composite cone was obliterated by strong Vulcanian activity, the lake began to dry, and another new inner cone was formed. A series of increasingly frequent eruptions ensued (there were 4,000 in April alone).

The first lava flow, some time between April 1960 and March 1961, which reached the sea, heralded the third phase, characterized by lava production during Vulcanian to Strombolian episodes. Further flows to the west and south occurred in 1967–1968, 1970, 1972, 1979, 1980, and in 1988, when two new craters were formed just to the south of the inner cone. There were also periods of Vulcanian to Strombolian activity of varying length and intensity in 1977, 1978, 1981, and 1982. The present episode (Figure 24) began in November 1992, and by August 1994 there were lava flows to the north and one small flow to the east. This third phase in the "life" of the island, one of substantial lava production, has virtually guaranteed the island's future existence by pro-

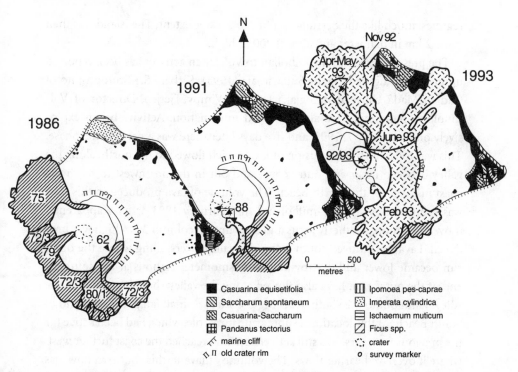

Figure 24. Anak Krakatau in 1986, 1991, and 1993, showing summit craters, major changes in vegetation types, and lava flows *(areas within thick, bold lines).* 1991 and 1993 flows sketched from oblique helicopter photographs; dating and outlines of earlier flows follow Sutawidjaja (1993).

viding a protective rampart against coastal erosion from southwesterly and northern swells.

Anak Krakatau's non-laval coast, however, has been subjected to considerable marine erosion. Remnants of erosional cliffs some 15 m high may be seen between the east foreland and northeast foreshore. These have now been eroded themselves by rain runoff and a black sand beach has formed below them. Behind the high-energy beaches in the south and (until covered with lava in 1993) in the north, however, the cliffs were continuous and increasing in height as the shore line was rapidly cut back. Rosengren measured the recession in the early 1980s and estimated its annual rate as about 5–7 m per year (Bird and Rosengren 1984). As a result of such strong erosion on only one of their flanks, the east and particularly the north forelands were gradually reshaped as they were subjected to a combination of erosional and accretional regimes, much like the Sertung spit but to a lesser extent. The island was then some 2 km in diameter and almost 200 m high.

The present period of Strombolian to Vulcanian activity has been reported upon, up to August 1993, by Sutawidjaja (1993). Kirbani Sri Brotopuspito of Gadjah Mada University, Yogjakarta, and Wimpy Tjetjep, Director of Volcanology, Bandung, have also provided information. Activity began explosively on November 7, 1992, and five days later there was a paroxysmal phase. This was followed by an effusion of lava, which flowed to the north, along the valley between inner and outer cones and out in the northwest to reach the coast just west of the North Beach. The volume of lava produced in this flow was estimated as about 2 million m³. In February 1993 activity apparently moved to a more southerly focus and an eruption column 2 km high was produced. Lava now moved south along the valley, overspilling where the outer rim became lower and destroying two seismometer stations at the southern end of the outer rim. Lava also flowed along the valley (here known as "Bone Alley") and out to the south, reaching the sea as a small fan at the west end of South Beach. At this location a submarine fumarole, which had been active for the previous few years, was still active. Lava also reached the coast further west, where it overlayed earlier flows. The volume of lava in this southern flow was estimated as 2.6 million m³. In April a third extrusion of lava flowed to the northeast. This filled the valley, embayed, and overspilled the outer cone on its northern flank at two points, and also on the northeast. Lava flowed down the outer cone's northeast flank, through developing mixed woodland on the northeast foreshore, to form a coastal fan extending out to sea for about 100 m on a front about 150–200 m wide.

The two northern flows merged and reached the sea on a broad front (Figure 25). They obliterated an area of casuarina woodland and covered almost the whole of North Beach, from which the lava extended about 300 m out to sea as a fan some 500 m wide and 10 m thick. This formed a substantial shield from northern swells, protecting a sector of the coast that had been subjected to rapid marine erosion for decades.

During my visits for ten days in early July 1993 and three days in early November 1994, eruptions occurred generally at intervals of from 2 to 15 minutes, with ballistic bombs from 5 to 150 cm in diameter being thrown mainly to the north and west. Some, however, fell to the south and destroyed a (third) replacement seismic station, which had been installed near the west end of South Beach in April 1993. Accumulation of ash was considerable by July

Figure 25. Lava entering the sea on Anak Krakatau's north coast in December 1992. Rakata is at right, in the background.

1993, over 20 cm on the east foreland and probably greater in the north and west. The two 1988 craters, just outside the previous summit, were much smoothed and infilled by ash, but both they and the 1988 lava flow were clearly recognizable from their contours. Lava bombs were seen near the coast on both the east and north forelands. The island has increased in area significantly and taken on a new appearance with the creation of a new, conical, summit, which now reaches some 250 m. By September 1993 the volcano had evidently quieted somewhat, but the radio teleseismograph recorded many shallow volcanic earth tremors. Following a lull of about six months, activity recommenced and by August 1994 there were sometimes 300–400 eruptions per day. Activity continues as of this writing (March 1995). This period of activity has already been one of the most prolonged since the 1950 eruption that produced the first inner cone, and one of the most productive in terms of volume of lava produced (initial estimates up to July 1993 exceeded 4.6 million m^3).

The emergence and establishment of Anak Krakatau already may have affected the erosional pattern of the archipelago. On the northern part of Panjang's west coast, some 2–3 km east of Anak Krakatau, cliffs of 1883 ash about

50 m high are now 50 to 70 m from the shore. They are fronted by a low fore-shore area, what is now known as Bat Cave Beach. It might be that the beach and foreshore areas are recent, post–Anak Krakatau features, having formed only since this part of the coast became protected from western swells by the emergence and growth of the new island to its west.

9 COMMUNITY ENRICHMENT

Who eats figs? Everybody.
D. H. Janzen, "How to Be a Fig" (1979)

ॐ

Since the 1929–1934 period (Chapter 5), the Krakatau forests have continued to mature and increase in diversity. The process may be followed most clearly on Rakata, the island most thoroughly surveyed and the only one unaffected by ash from Anak Krakatau's fairly regular bursts of island-building activity. Only on Rakata has there been an uninterrupted primary succession with continuous enrichment. Fifteen families of spermatophytes and one of ferns were recorded as having first arrived between 1934 and 1989 (Whittaker et al. 1992), all but two (one of these the single fern) animal-dispersed. Ten of the families are represented by animal-dispersed inland trees, seven on Rakata, and this group has steadily increased in relative importance. The enrichment has not been uniform throughout the flora, however.

The sea-dispersed component of the flora increased from 53 species in 1924 by only 6 species in the next 65 years (Figure 26) and its turnover has been minimal (Whittaker et al. 1992). This is in spite of the fact that the strand-line habitat, subjected to continuous change locally through marine erosion and redeposition of poorly consolidated deposits, is particularly dynamic on the Krakataus. The apparent paradox is illusory, however. This habitat has maintained its dynamism throughout the period, and sea-dispersed coastal plants are particularly well adapted for survive in just such a dynamic situation of constant renewal. Moreover, most of the species turnover in inland plant commu-

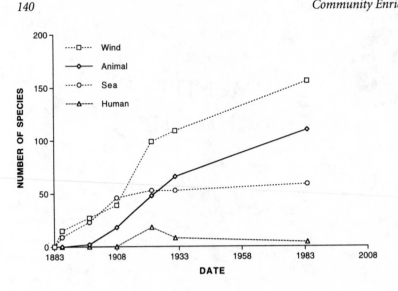

Figure 26. Dispersal-mode spectra of vascular plant floras present at successive survey periods. Surveys made in the periods 1920–1924, 1929–1934, and 1979–1989 are grouped and the data plotted at the midpoint of these periods. Data from Whittaker et al. (1992).

nities has been successional in nature, and the coastal vegetation has changed little in this respect for many decades (Whittaker, Bush, and Richards 1989).

In contrast, the wind-dispersed component on Rakata increased substantially in the same period, from 99 species (51 of them pteridophytes) to 156 in 1989 (81 pteridophytes), most of the increase being in ferns and orchids (Whittaker et al. 1992). This is a percentage increase five times that of sea-dispersed plants. The early-established ferns were pioneers of fairly open ground capable of withstanding considerable lack of moisture, and the orchids included heliophilous species. In both these groups subsequent increase has been through species adapted to the more shaded, mesic conditions following closure of the forest canopy and both include epiphytes restricted to the moist upland forests. There are still only 7 wind-dispersed tree species on the archipelago, however. The most important ecologically, the tree *Neonauclea calycina,* was first discovered, on Rakata, in 1908 and has been the dominant canopy tree there since the 1920s. It now occurs on all islands. Of 3 species first found only in the last 12 years, one, *Alstonia scholaris,* may have been brought attached externally to birds, and *Pterospermum javanicum,* found as a 5-m-tall specimen in 1992, has a sycamore-like, winged, wind-scattered seed that would be unlikely to have made the crossing from Sebesi (where it was recorded in 1921) under normal wind conditions. Several specimens of the third species found recently, *Crypteronia paniculata,* when encountered in 1983 were large enough to indicate a much earlier arrival (Whittaker and Jones 1994).

Although the animal-dispersed component colonized relatively late, it has shown the most significant increase, both as a percentage and in actual numbers (from 48 species in 1934 to 110 in 1989 and 134 in 1992), and animal dispersal has been of overwhelming importance in the development of the Krakatau forests. Only 10 species, mostly composites and grasses, are regarded as having been dispersed exochorously. Most of the endochorous colonists, relatively long-lived trees and shrubs of the interior forests, have suffered few losses since 1934. The roles in the community of many important heliophilous, early-successional, pioneer trees, such as the small figs *Ficus fulva* and *F. septica* and the early tree cover *Macaranga tanarius,* have changed, however. They are now opportunistic colonizers of small forest gaps, their early roles as canopy components having been gradually taken over, on Rakata, for example, by *Neonauclea* and several animal-dispersed trees, mostly figs and members of the Euphorbiaceae (Whittaker, Bush, and Richards 1989, Whittaker and Jones 1994).

The most important of the later-successional zoochorous trees are *Timonius compressicaulis, Dysoxylum gaudichaudianum,* and species of *Ficus* (which include large canopy dominants and strangler figs). In the Southeast Asian tropics many figs are stranglers, and at least six occur on the Krakataus *(Ficus elastica, F. virens, F. retusa, F. annulata, F. sumatrana,* and *F. tinctoria).* Stranglers usually begin life as epiphytes (for example, the last four), growing high on another tree rather than on the ground, obtaining water from rain, mist, or even the moisture in the air, and their nutritional requirements from humus derived from rotting organic detritus lodged on the host tree. Later, the strangler's roots grow over the host's boughs and trunk to the ground, anastomosing and embracing the host as they grow, often (but not always—for example, *Ficus tinctoria)* finally killing it and taking its place in the forest.

Fig species are important components of mixed forest in terms of both number of individuals and number of species, providing fruit practically year-round, and have a special importance in the diversification of Southeast Asian lowland tropical forests. On the Krakataus, almost two-thirds of the tree species present after four decades were fig species, and 17 of the total of 24 fig species that have colonized the islands had done so by that time. In some areas, such as western Rakata, figs are now a major constituent of the forest canopy, and they are significant canopy components of all the islands' forests. Their colonization is, of course, closely bound up with that of their animal dispersers, and incorporates positive feedback, more figs attracting more frugivores, which bring more seeds. Because the dispersers are seldom fig specialists, the positive feedback involves other fruit-bearing forest trees and has probably been largely responsible for forest diversification over the past six decades.

Figs and Forest Enrichment

Figs as a keystone component of lowland tropical forest

Keystone species are those that, like the keystone of an arch, are so important to the functioning and maintenance of the system that their removal would have a cascading effect, causing extensive changes to the interrelationships and representation of other organisms. Loss of a keystone species would alter the ecological networks so fundamentally that even if some type of community survived, the system would be drastically altered. One reason for considering *Ficus* species as a keystone group was summed up by Janzen's (1979) short comment appearing as the epigraph of this chapter.

From a platform 140 ft (about 45 m) high in peninsular Malaysia's tropical rain forest, McClure (1966) and others made regular observations of about 60 surrounding forest trees for five years, from 1960 to 1965. No fewer than 32 species of vertebrates were recorded as feeding during the day on the figs of a single strangler, *Ficus sumatrana*, growing on a tall canopy tree, *Shorea resina-negra*. In another intensive Malaysian study, of lowland dipterocarp rain forest at Kuala Lompat, Lambert and Marshall (1992) found that the 29 bird-dispersed fig species in their study area were exploited by at least 60 species of birds—about a quarter of the forest's total avifauna—and 17 species of mammals.

Several reasons have been suggested for the broad appeal of figs. They have a high proportion of edible flesh for their fresh weight; the seeds are non-toxic, even if ground up by the frugivore; crops are heavy; ripe figs of some kind are available almost throughout the year; they appear to lack secondary compounds that would restrict the range of consumers; and they come in a wide range of sizes (Janzen 1979). Several of the attributes of keystone species were evident to Lambert and Marshall. Fruiting in figs was usually synchronous within the crown of a tree but markedly asynchronous between trees of a species, and there was little or no recurrent annual pattern. Fig crops were large and, although relatively short-lived compared with the crops of other fruiting trees, one crop or another was available throughout the three-year study. Moreover, figs were most abundant just when other bird fruits were scarce, so that fruit-eaters could be sustained by figs during these generally lean periods. For wide-ranging fig-eaters, therefore, the living was, if not always easy, at least assured. Without figs, those frugivores depending on them at times of general fruit shortage would have had to leave the area, and other frugivore-dispersed plants would in turn be affected, as would animals dependent on them, and so on. This slowly acting cascade would have repercussions throughout the food web, and although the ecosystem would probably survive without the fig species component, it would be substantially changed. For these reasons Lambert and Marshall regarded bird-dispersed fig species as a keystone component of the forest community of their study area.

The increasing fig crop as a result of the spread of fig species within the Krakatau forests probably accelerated forest diversification by providing a staple diet for generalist fruit-eaters that in turn dispersed the seeds of other important forest trees, including, of course, other fig species. The flowering pattern and peculiar pollination system of fig species, however, would seem to make their colonization of islands particularly difficult.

The mutualism between figs and their pollinators

The fig-pollinator mutualism is one of the closest known between insects and plants. With very few exceptions, each one of the 800–900 fig species in the world can be pollinated (and can therefore set seed) only by the activity of females of its own particular species of very small wasps of the hymenopteran subfamily Agaoninae. Conversely, each species of pollinating wasp can complete its life cycle only if its eggs are laid in the developing flowers of its own particular species of fig (Wiebes 1986). Because the conflicting interests of the two partners, the fig to produce seeds and the wasp to produce offspring that feed on fig seeds, are balanced, the mutualism has been described as "reciprocal parasitism" (Bronstein 1992).

The fruit of a fig tree, the syconium (commonly termed a fig) is actually a swollen, recurved, vase-shaped receptacle for small flowers, usually several hundred of them. About half the world's fig species (including all the New World species and many strangler figs) are monoecious, meaning that on each tree all the syconia contain both female and male flowers: female flowers have a style for pollen reception and an ovary containing the ovule (the female sex cell that will become the seed if it is fertilized by pollen), and male flowers, fewer in number, produce pollen. Male and female flowers become mature at different times, so female flowers must receive their pollen from male flowers of another syconium.

Loaded with pollen, female pollinating wasps enter the fig through the narrow neck of the vase, which acts as a valve because of a series of interlocking, inwardly directed, stiff scales. Up to a dozen wasps may bite and push their way through, at considerable cost. Most of those successfully traversing the passage have usually lost wings or distal segments of the antennae. Once inside, the wasp readily oviposits (lays eggs) in flowers that have a style short enough for her ovipositor to penetrate to the ovary. The style lengths vary from flower to flower, and if the wasp penetrates one that is too long she withdraws her ovipositor and tries another. While engaged in this activity she transfers pollen received from another syconium to the stigmas of the flowers. Soon after oviposition she dies within the fig cavity.

Most of the longer-styled flowers, therefore, do not receive a wasp egg, and they develop normally, setting seed if they have been pollinated. It is mostly the shorter-styled flowers that have received wasp eggs and thus contain developing wasp larvae. These feed on the seeds' endosperm, deflecting development and often galling the flowers. Thus the price the fig species pays for the essential pollinating services of the wasp is that more than half of its female flowers produce wasps instead of seeds.

After about a month the wingless, eyeless male wasps of the next generation emerge from the flowers, cut a hole in the side of the seed coat of a flower containing a female, and mate with her. In some species they then cut the anthers free from the male flowers, which are often situated in a ring around the ostiole and which mature at this time, and bite tunnels through the syconium wall. They then usually die, never having left the dark chamber into which they emerged. The inseminated, winged females emerge from their flowers, acquire pollen from the male flowers, and leave the syconium either by forcing their way out of the ostiole, which may become more easily passable at this time, or, in some species, through a passage bitten through by the males. Adult females are thought to live outside the syconium for only a few days, and they immediately seek and enter young, receptive figs. Such figs signal their receptive state to female wasps by emitting chemicals (Hossaert-McKey, Gibernau, and Frey 1994; van Noort, Ware, and Compton 1989; Ware et al. 1993). The females enter, carrying pollen from the fig from which they emerged.

At least twelve of Krakatau's fig species are dioecious. Dioecious species have two types of trees; since female organs occur in flowers of both, the term for these species should strictly be "gynodioecious." In hermaphrodite trees, each syconium contains both male and female flowers, but the latter all have short, funnel-shaped styles down which the wasp can oviposit easily, preventing seed development. Such trees thus become functionally male, and their syconia, sometimes called gall figs, are a source of wasps, not of seeds. In female trees of the same species each fig contains only female flowers, all of which have narrow styles too long for the wasp's ovipositor to penetrate easily, although she may deposit pollen on the stigmas in her attempts. If they are pollinated, the flowers of these "seed figs" set seed. Seed figs of dioecious fig species are therefore wasp traps; female wasps entering them make a fatal mistake that dooms them to die there without offspring.

Fig dispersers on the Krakataus

Fig trees rely on vertebrate frugivores, particularly those that fly, for long-distance dispersal. Fig species that have large, dull-colored fruits, usually greenish when ripe, carried on the trunk or on main branches clear of the foliage (cauliflory), or hanging free of the foliage in a pendant group on a long, rope-like stalk (flagelliflory), are usually regarded as being bat-dispersed (van der Pijl 1941). Those with small, more brightly colored figs, often red or purplish when ripe, usually borne among the canopy foliage, are bird-dispersed, although there are several exceptions to the color criterion. Some species are suited to dispersal by both birds and bats. Van der Pijl (1957) listed the following Krakatau figs as being bat-dispersed: *Ficus ampelas, elastica, callosa, fistulosa, fulva, racemosa, hispida, ribes, retusa,* and *variegata.* To these may be added *Ficus septica, padana, hirta, virens, benjamina,* and *annulata. Ficus subulata, sumatrana, drupacea (= pilosa), virens, fulva, septica, ampelas, benjamina,* and *retusa* are known to be dispersed by birds, and the last six species, which occur in both lists, are dispersed by both birds and bats.

On the Krakataus, all the frugivorous birds mentioned in Chapter 7 are possible dispersers of figs. These include four obligate frugivores (one a seed predator) capable of dispersing figs from the mainlands to the Krakatus. At least seven of the facultative frugivores also retain seeds in their gut long enough to suggest that they might also do so. The bulbuls and whistler could act as dispersers from Sebesi, and the flowerpecker is more likely to be an intra-archipelago carrier. Midya and Brahmachary (1991) showed that the germination success of seeds of the Indian banyan, *Ficus bengalensis,* was significantly enhanced by being passed through the gut of captive mynahs, and Compton (1995) has obtained similar results with South African figs and their dispersers. We have seen in Chapter 7 that green pigeons (*Treron* species), although fig specialists, are fig-seed predators, but that occasionally they pass intact seeds. In view of the above findings on enhanced germination, the minority of *Treron* individuals that pass viable seeds could be important long-distance dispersers of fig species, although far less effective than the imperial pigeons (*Ducula* species) and fruit-doves *(Ptilinopus melanospila)*. The two rousette species may be involved in fig dispersal both to and within the Krakataus, but studies on gut retention time are needed. The flying fox may be capable of dispersing figs from Sebesi.

On the Krakataus small dog-faced fruit bats of the genus *Cynopterus*, as well as small birds such as flowerpeckers and bulbuls, probably disperse fig seeds only over short distances, both within and between islands. Ridley noted that *Cynopterus* species disperse the seeds of *Ficus benjamina*. Van der Pijl (1936) reported finding seeds of *Ficus racemosa* both in the gut of a *Cynopterus* species and in droppings beneath its feeding roosts on Pulau Peucang, off the Ujung Kulon peninsula. He was later able to germinate the seeds. In 1984 *C. sphinx*, caught in our mist-nets on Anak Krakatau, both spat out and defecated fig seeds, and seeds could be seen scattered over our tents and the island's bare volcanic ash, although at that time the young fig trees of that island were not fruiting. In Hong Kong *C. sphinx* and the rousette *R. leschenaultii* are the only fruit bats, and Dr. R. T. Corlett has suggested to me that they are the dispersal agents there of *Ficus fistulosa, variegata,* and *hispida,* and that they compete with birds for the fruits of other figs. In captured *Cynopterus* bats, passage times through the alimentary tract were found to vary from 11 to 15 minutes. Clearly, *Cynopterus* species are capable of dispersing fig seeds at least between islands but it is unlikely that they were responsible for transporting them to the Krakataus from the mainlands of Java and Sumatra, although transport from Sebesi is a possibility.

A fig species needs the animal dispersers to take its seeds in order to disperse them, and presumably this selective pressure has driven the evolution of bat- and bird-attracting figs. But it is also in the fig's interest that gall flowers be avoided by the fig-eater in order to conserve the supply of the all-important pollinating wasps in these flowers and, in dioecious species, to prevent dispersers wasting time on hermaphrodite plants when seed figs are available on female trees. Differences between seed figs and gall figs (or, in monoecious species, between seed flowers and gall flowers) that enable dispersers to discriminate between them would thus give a fig species a strong selective advantage.

Corner (1952) observed that in dioecious species gall figs are avoided by animals, perhaps because the insects that they contain make them distasteful, and he noted that in this way the pollinating wasps are conserved. In Singapore, however, Boon and Corlett (1989) found the partly eaten gall figs of the dioecious *Ficus fistulosa,* some with wasps still emerging, under feeding roosts of *Cynopterus brachyotis.* Corner's observation does appear to hold for bulbuls, however. Lambert (1992) demonstrated that a *Pycnonotus* species discriminated between the gall figs and seed figs of dioecious species. Hermaphrodite gall figs were avoided by bulbuls that fed avidly in female plants of the same species. Differences in color and texture (mainly in color) between figs of hermaphrodite and female plants were thought to be used as recognition signals by the birds. Thus, although bulbuls are but facultative frugivores, they may be more efficient dispersers of dioecious fig species than was thought. Lambert's finding will have considerable significance if it is found to apply to avian frugivores generally.

In Philippine primary rain forest, Utzurrum and Heideman (1991) made another remarkable discovery. Fruit bats appear to be able to discriminate between the viable and damaged seeds within the figs of monoecious species. Only about half the seeds from pieces of ripe fig dropped by bats and from ejecta (masses of chewed fig spat out) germinated, whereas over 90 percent of seeds collected from splats (fecal masses) did so. The differences in germination success of the three sample groups were correlated with the proportions of seeds that had tell-tale holes showing that they had been killed by developing agaonid larvae. There were no holed seeds in the splats, and their absence was thought to be due to differential ingestion of holed and intact seeds by the bats. Utzurrum and Heideman noticed that intact seeds almost always had a thin, slippery, gelatinous coating, which was absent from seeds damaged by developing fig wasps. They thought that as the bat applied gentle suction to the mass of chewed fig, it filtered out the inviable seeds and swallowed only the intact ones. Long-distance dispersal of plants by bats is largely through the seeds being deposited as splats, rather than as ejecta, and the splats would have a much larger proportion of intact, viable seeds than either the ejecta of the uneaten fig. Again, if this observation is confirmed more generally, then fruit bats, too, may be more efficient dispersers of figs than was previously realized.

The establishment of figs on islands

For figs to colonize an island, several conditions must be satisfied. Fig seeds must be carried there and deposited in a favorable habitat in viable condition. Having germinated, grown, and matured, however, a fig tree is nevertheless doomed to sterility unless its own pollinating fig wasp is present, and present, moreover, just when the figs are at the right stage for pollination. Wasps that may have arrived in the initial colonization, perhaps in a piece of fig fruit carried by a fruit bat, as found by Boon and Corlett (see above), would have lived for only a few days and could play no role in the establishment process. The right species of pollinators would have to arrive from outside the island much later, and at just the right time.

In most fig species, fruiting is synchronous within the crown of a tree, and if pollination rates are high any figs remaining unpollinated are aborted. When pollinators are in short supply, however, unpollinated figs remain attractive to wasps for much longer, disrupting the normal within-tree synchrony and allowing the figs to be pollinated over a longer period. Nevertheless, individual trees are usually incapable of maintaining their own pollinating wasp population because gaps in fig production may be weeks or months in duration, far too long for their wasps to survive. At the extreme, two trees of a fig species would be almost entirely dependent on the correct pollinator arriving repeatedly from outside the island. Figs usually colonize not as one or two plants but as a small group near a bird or bat roost. As Bronstein (1992) has pointed out, however, for a tree of a monoecious species either to have its pollen transferred or to receive pollen, it must flower three or four weeks out of synchrony with others of the same species, so that receptive syconia will be available for emerging female wasps carrying pollen from their natal fig. If flowering within the clump were precisely synchronous, no wasp would be able to find a receptive fig before she died, and the trees would be barren unless pollinators recolonized the island. In a very small colonizing group of figs, the chance of any figs of the right stage being available will be very low, and the fig population will be unlikely to be able to sustain its own wasp population. It will be dependent on immigrating pollinators. Only when the fig population has reached a critical size will the asynchrony of fruiting between trees ensure the availability of some figs at the right stage at the right time. The island fig population will then be self-sufficient as far as its pollinators are concerned, and its tenure secure.

How then do immigrating pollinating wasps arrive at just the right time to secure the foothold of the colonizing fig species and ensure its survival until the critical population size is reached? This is perhaps the least understood aspect of the puzzle of fig colonization. There is indirect evidence (next chapter) that some wasps are carried to the islands continuously in the air.

The assembly of a fig flora on the Krakataus

In 1897, 14 years after the 1883 eruption, four fig species were already present on Rakata: *Ficus fulva, septica, hispida,* and *padana* (= *toxicaria,* not found after 1932) (Figure 27). All are small, fast-growing, dioecious, bat-dispersed

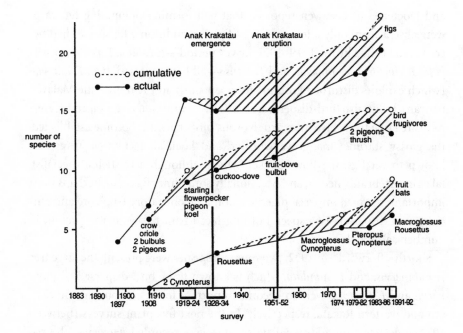

Figure 27. Colonization of the Krakataus by figs, avian frugivores (indicated at date of first record by common name), and fruit bats (by generic name), from 1883 to 1992. One bulbul became extinct between 1908 and 1919, one *Cynopterus* species between 1933 and 1984, and another bulbul and the crow were extinct in 1984. The nectarivorous bat *Macroglossus sobrinus* was recorded only in 1974; *Macroglossus minimus* was found in 1992. Boxes on the ordinate indicate grouped surveys.

trees, and *F. fulva* and *F. septica* are also dispersed by birds. In the first zoological survey of 1908, Jacobson (1909) recorded three unidentified species of chalcid wasps, one on each island. By then three more fig species, all small trees or shrubs, had arrived: *Ficus fistulosa,* another bat-dispersed, dioecious, early-successional species that exploits disturbed or open ground; *F. hirta,* also dioecious, which Corlett found to be dispersed by birds in Hong Kong; and *F. montana* (= *quercifolia*), the dispersers of which are unknown. Although Jacobson particularly noted the absence of fruit bats during his brief survey, six frugivorous birds were recorded, one of them the fig specialist *Treron vernans.* The others were two bulbuls (one of which did not persist), the black-naped oriole, large-billed crow, and emerald dove.

Dammerman (1948) found two pollinating fig wasp species (of the genera *Blastophaga* and *Ceratosolen)* in 1920 and 1933, together with two of their competing seed-predator fig wasps (species of *Philotrypesis* and *Sycoryctes*), and Docters van Leeuwen reported that gall-forming pollinating fig wasps were absent from only a few species of figs. Six additional fig species had become established by the 1919–1921 survey period—*F. callosa, F. ribes, F. variegata, F. tinctoria, F. ampelas,* and *F. retusa*—all bat-dispersed, two, *F. ampelas* (which exploits disturbed ground and open sites) and *F. retusa* (the Malayan banyan), being also bird-dispersed. *F. callosa, F. variegata, F. retusa,* and *F. tinctoria* are large trees. Five more frugivorous birds had also become established: the glossy starling, mangrove whistler, and koel, all facultative frugivores known to eat figs; the flowerpecker, which, although a mistletoe specialist, takes invertebrates, nectar, and occasionally sometimes figs; and, perhaps most important, the pied imperial pigeon, a nomadic obligatory frugivore and efficient seed disperser. Two species of dog-faced fruit bat were also present in numbers, although one did not persist.

Shortly afterwards, in 1922, two more fig species were present, the large tree *F. pubinervis,* and *F. subulata,* which is known to be bird-dispersed. *F. pubinervis* and *F. tinctoria* were to become important forest components on eastern and western Rakata, respectively. In the next five plant surveys (between 1922 and 1934), no additional fig species were recorded. However, the red cuckoo dove (a seed predator) had colonized by 1932, and the fruit bat, Geoffroy's rousette, appeared by 1933.

The next survey was not until 1951, and two more figs had arrived in the meantime, *F. annulata* and *F. elastica,* both monoecious stranglers and both bat-dispersed. There were now also two more fruit-eating birds, a facultatively frugivorous bulbul and a fruit specialist, the black-naped fruit-dove, which is also an efficient seed disperser. In 1974 a third species of dog-faced fruit bat was present, and in 1979 another fig, *F. heterophylla.* By the 1980s five more fig species had arrived: *F. racemosa* (bat-dispersed), *F. sumatrana* (bird-dispersed), and *F. lepicarpa* in 1982, and *F. virens* (bird-dispersed) and *F. drupacea* in 1989. Also during the 1980s and early 1990s three further fruit bats and two more frugivorous birds, one, the green imperial pigeon, an obligate frugivore and efficient seed disperser, were recorded. By 1992 *Ficus benjamina* (bat-dispersed) had arrived. Thus by the 1990s over a score of frugivorous flying vertebrate species had reached the islands, along with some 24 species of fig, 22 of which have persisted. Fifteen fig species colonized within four decades of the 1883 eruption, the remaining nine in the last four decades.

All seven pioneer fig species, which colonized in the first 25 years, are dioecious. In dioecious species about half the trees bear figs that contain only female flowers with styles too long and narrow for the wasps' successful oviposition. Having made the fatal error of entering one of these "seed figs" of a female tree, therefore, although they pollinate the flowers in their vain attempts to lay, the wasps die there without reproducing. Thus figs on female trees of a dioecious fig species are wasp traps or sinks. This means that in a small population of trees, maintenance of its own wasp population should be more difficult, other things being equal, for a dioecious fig species than for a monoecious one. For maintenance of its wasps, its critical population size would have to be double that of a monoecious species. Put another way, local extinctions of wasp populations could continue in a larger population of trees in a dioecious fig species, slowing its population growth and impeding its early establishment following the first foothold. So on theoretical grounds one might expect monoecious species to be the better colonizers of small islands. And yet, both on the post-1883 Krakataus and on Anak Krakatau, it is the dioecious species that are the successful early pioneers (Table 9.1).

Bronstein (1989) has suggested that in extreme, highly seasonal environments, dioecious species may in fact have the advantage. Through their female trees they may concentrate seed production at favorable times of year, whereas monoecious species generally produce seeds year-round. Indeed, this selective advantage was used as a partial explanation for the evolution of dioecy in figs (Kjellberg and Maurice 1989). Little is known of the seasonal fruiting pattern of dioecious figs, but Hugh Spencer, George Wieblen, and Brigitta Flick, in a five-year study of 198 trees of *F. variegata* in northern Australia, have data that show fruiting to be bimodal, fruit production in female (seed-producing) trees peaking with peak rainfall. Rainfall on the Krakataus is distinctly sea-

Table 9.1. Number of *Ficus* (fig species) recorded on the Krakataus, by date of arrival. Dioecious species tended to be early colonizers of the archipelago and predominate on Anak Krakatau. (The trends shown for dioecious and monoecious species in the first four columns of numbers are significant; Mann-Whitney $z = 2.73$, $p < 0.01$.)

Species	Period of arrival				Present on Anak Krakatau, 1992
	1883–1908	1909–1922	1923–1951	1952–1993	
Total	7	8	2	7	7
Dioecious	7	5	0	2	6
Monoecious	0	3	2	5	1

Sources: Data on individual *Ficus* species from Whittaker, Bush, and Richards (1989), Partomihardjo, Mirmanto, and Whittaker (1992), and unpublished information from R. J. Whittaker and T. Partomihardjo.

sonal in most years (Thornton and Rosengren 1988), and pioneer fig species usually germinate on the ground, where poor soil conditions may render seedlings particularly vulnerable to periods of low rainfall. If their fruiting pattern is similar to that found in *F. variegata* in Australia, then dioecious species on the Krakataus may well have an advantage over monoecious ones.

The Role of Animals in Plant Dispersal within the Krakataus

The effectiveness of dispersers over the distances within and between islands of the archipelago may differ, of course, from their effectiveness over the much greater distances between archipelago and mainland. Locally, ants, land crabs, rats, bats, or birds may be responsible for the gradual spread of plants that arrived by a different dispersal mode, by moving locally produced seed within the archipelago or within an island. The term *diplochory* is used to describe the involvement of two modes of dispersal in the establishment of a single species (van der Pijl 1936).

Figs and their seeds are consumed by a whole suite of insects, such as weevils and other beetles, lygaeid bugs, and pyralid moths, both while the fig is on the tree and after it has fallen to the ground. In Africa, lygaeids can destroy nearly all the seed crop that falls beneath the trees (Slater 1972). This is one reason (another is seedling competition) why dispersal of seeds away from the parent tree is so important to the fig. One of Slater's lygaeids is capable of carrying a seed in flight and could therefore act as a short-distance disperser.

Ants may be important secondary dispersers of fig seeds after the fig species has arrived through long-distance endochory. Kaufmann and her associates (1991) discovered that the fleshy outer seed coat of *Ficus microcarpa* is attractive to ants even after the seed has passed through the gut of a hill mynah. They believed that secondary dispersal by ants may reduce competition between seedlings and that some seeds of epiphytic figs may be carried into trees, to accumulations of humus, or to ant nests that provide anchorage, moisture, nutrients, and some protection from herbivores by the ant colony.

Curiously, land crabs are not included in Dammerman's systematic treatment of the Krakatau fauna, although three species of aquatic crabs in the Sertung lagoons are mentioned (Dammerman 1948). It was not until 1982 that four species, one in each of the genera *Coenobita*, *Metasesarma*, *Gecarcoidea*, and *Ocypode*, were recorded (Iwamoto 1986). About ten species of terrestrial crabs were present in 1986, including three species of *Ocypode*, three of *Coenobita*, *Birgus latro* (the robber crab or coconut crab)—a doubtful record, and one species each of *Grapsus, Metasesarma,* and *Gecarcoidea* (Morgan 1988). It is clear, however, that terrestrial crabs were present long before the 1980s. Dammerman somewhat casually mentioned seeing large land crabs—which "even if unprotected by their burrows . . . are a match for the rats" (Dammerman 1948, p. 317)—on Panjang in February 1928. Bristowe (1931) listed four unnamed species as present in 1931, and in a popular account of his 1931 visit to Panjang, he mentioned the "dark red land crabs" which "lived in holes all over the island and even up the hillside" (Bristowe 1969, p. 147). These were probably the purplish *Gecarcoidea lalandii,* not formally recorded from Panjang until 1982. Evidently land crabs colonized some time before 1928 and the first species to arrive was probably *G. lalandii.* Dammerman made no mention of terrestrial hermit crabs, but Bristowe (1969) wrote of dislodging quantities of hermit crabs from the branches of scrub on Panjang in 1931. These were almost certainly species of *Coenobita,* which are known to be good climbers and were first recorded from Sertung in 1982.

On Fanning Island in the central Pacific, Lee (1985) found that almost all ripe fallen fruit of the *pandan, Pandanus tectorius* (a sea-dispersed species present on the Krakataus), are spread locally over land by the Pacific land crab *Cardisoma carinifex.* Mortality in seedlings that had been carried up to 20 m to sheltered spots, to be consumed away from other crabs, was much less than in those growing from seeds close to the tree. Dispersal by crabs appears to be important for the successful reproduction of *Pandanus* on Fanning. Land crabs may also have a detrimental effect on plant dispersal. By a series of simple exclusion experiments on Eniwetok Atoll in Kiribati, Micronesia, Louda and Zedler (1985) showed that fruits of *Terminalia catappa* were destroyed by terrestrial hermit crabs (species of *Coenobita*) and those of *Guettarda speciosa* and *Scaevola taccada* suffered severe damage. Thus, elsewhere in the Pacific, as generalized predators at high densities, land crabs are known to have different influences on the survival of fruit and seeds of four common plant species that also occur on the Krakataus.

Alexander (1979) believed that on Aldabra Atoll, in the Indian Ocean, where the three Krakatau species of *Coenobita* and two of the *Ocypode* species also occur, the coenobites may prevent the establishment of some sea-dispersed plants. On Christmas Island, also in the Indian Ocean, land crabs feed on fruits, seeds, or seedlings from the foreshore (species of *Coenobita*) to the interior rain forest (*B. latro* and particularly the endemic red crab, *Gecarcoidea natalis*). The red crab, which is related to *Gecarcoidea lalandii* of the Krakataus, is abundant and at high density. Its ecological impact was investigated by O'Dowd and Lake (1989, 1991) and Green (1993). The crab was a predator of seeds or fruits of several species, including *Barringtonia racemosa, Hernandia ovigera, Macaranga tanarius, Scaevola taccada,* and *Terminalia catappa,* all either present on the Krakataus or having close relatives there. Seedling predation precluded the growth of several plants, particularly the canopy dominants *Planchonella* and *Syzygium* species, and the crabs removed 39–86 percent of annual leaf fall. Only in the cases of two species with no close relatives on the Krakataus were the crabs seed-dispersers (moving seeds to nutrient-enriched sites and thus increasing the plants' chances of successful recruitment when canopy gaps occur) rather than seed-predators. The red crab thus plays a pivotal role as a generalist consumer in the Christmas Island ecosystem and has almost keystone status.

As these examples show, the influence of land crabs on the establishment of plant propagules may be significant and complex; they may facilitate the establishment of some species but impede the establishment of others. Within the entrances of land crab burrows on Rakata I have seen aggregates of *Terminalia* seeds some 20–30 m from the nearest possible source tree. Crabs clearly are short-distance dispersers of *Terminalia* there, although bats are even more important in this respect (see above). Iwamoto (1986) also reported finding substantial amounts of fruit in the gut of *Ocypode kuhli* and *Gecarcoidea lalandii* on the Krakataus. In recent germination trials on soil taken from within burrows of *G. lalandii* on Rakata, 10 species of seed plants germinated, but only 6 of these were among the 11 species germinating from samples of nearby topsoils, and Whittaker, Partomidhardjo, and Riswan (1995) regarded such medium- to long-term seed stores as seed "sinks" (losses from the system). The Krakatau crabs are themselves preyed upon by the monitor lizard, *Varanus salvator.*

Van der Pijl (1936, 1957) and Whittaker and Jones (1994) have listed the following Krakatau plants as being sea-bat diplochorous (that is, with overseas, long-distance dispersal by sea currents and inland or inter-island spread of locally produced seed by bats): the large tree *Calophyllum inophyllum,* medium trees *Hernandia peltata, Terminalia catappa,* and two species of *Erythrina,* small trees *Cerbera manghas, Guettarda speciosa, Morinda citrifolia, Spondias mombin, Ochrosia oppositifolia,* and *Gluta renghas,* the shrubs *Pandanus tectorius* and *Ximena americana,* the cycad *Cycas rumphii,* and the liane *Mucuna acuminata.* Many of these were significant canopy trees of the early coastal forests. Sea colonists believed to be secondarily spread by birds include only two trees, the large *Adenanthera pavonia* and medium-sized *Sterculia foetida,* together with the shrub *Scaevola taccada,* the trailing parasite *Cassytha fili-formis,* a small liane, *Cayratia trifolia,* and an herb, *Tacca leontipetaloides.* Three shrubs and a climbing herb may be secondarily spread by birds or bats. Whittaker and Jones noted that there are no herbs in the secondarily bat-spread list, in which, in contrast to the bird-spread group, trees and shrubs predominate. This observation confirms earlier suggestions that the role of these two sets of dispersers, although overlapping, may be somewhat comple-mentary (van der Pijl 1957). They also tentatively suggested that bats appear to contribute more than birds in the seeding of open habitats, a suggestion sup-ported by observations of the wide dispersal of small seeds by bats defecating on the wing (e.g., Boon and Corlett 1989, Tidemann et al. 1990).

There are few examples of sea-wind diplochory, the most important being *Casuarina equisetifolia* and the shrub *Meloicha umbellata.* The predominance of sea-animal diplochory among the early sea-borne colonists of the coastal woodlands means that the delayed positive feedback associated with endo-chory, mentioned in an earlier chapter, could well have begun quite early in the colonizing process, with plant species that were primarily sea-dispersed.

Animals that are primary dispersers of plants may of course also act as sec-ondary dispersers, but some winged frugivores that are incapable of mediating primary dispersal, having short flying ranges and/or short gut retention times, may be important agents of secondary dispersal for plants that arrived through the agency of long-distance dispersers. For example, the seeds of figs, *Piper aduncum,* and *Cyrtandra sulcata* have been found in droppings of the bat *Cynopterus brachyotis* on the Krakataus, and many seedlings are often found under its feeding roosts, yet it is probably an unlikely agent of primary colo-nization from the mainland. *C. sulcata* seeds have also been found in bird droppings (Docters van Leeuwen 1936) and this may be a case of double zoo-chory, birds being the primary dispersers and bats the important agency of secondary spread over Rakata's uplands.

Volant frugivores may also spread 14 species of Krakatau plants that were introduced by humans (Whittaker and Jones 1994), 5 of which were discov-ered only in the last six years and have not yet had time to spread. Only 3 species have established breeding populations—the mango *Mangifera indica,* the presumably bird-dispersed *Capsicum frutescens,* and *Gnetum gnemon,* first found on Rakata in 1919 (it is eaten as a vegetable in Indonesia) and which has been spread (presumably by birds) to Sertung and Panjang.

Both species of rats on the islands are omnivorous and climb trees, and Iwamoto (1986) found pieces of fruits and seeds in their guts. It appears, then, that rats may also be involved in the overland dissemination of some of Krakatau's plants.

Diversification of the Fauna

The fauna of the Krakataus has continued to diversify since the 1930s, but the rate of increase in species number and the loss of species with changing habitat have not been the same for all components. In the analysis of serial surveys pseudoturnover (false records of turnover) is a continuing problem. Because of the mobility and cryptic habits of animals, as well as changing survey methods, the problem is likely to be greater in faunal than in floral studies. In even the most exhaustive animal survey some species that are present may not be recorded, thus providing spurious records of absence and, at subsequent surveys, perhaps, erroneous records of new colonization. Of the vertebrates, the problem is probably most acute in the bats. Survey methods for microbats, particularly, have changed from collection by shotgun, net, and tennis racket to the use of monofilament mist-nets and harp traps, and the recording and identification of ultrasonic calls. The present size of the fauna and the comments of earlier zoologists on the lack of bats before 1919 nevertheless suggest that a substantial increase has occurred in the past 50 years and that insectivores probably were later colonists. Since the first 4 species of bat (including one insectivorous individual) were recorded by the 1930s, an additional 16 species have been reported, 10 of them insectivorous microbats. There are now at least 17 species on the islands, comprising the Malay flying fox, two rousettes, 3 species of dog-faced fruit bats, a nectarivorous bat, a carnivorous/insectivorous false vampire, and 9 insectivorous species. In spite of modern techniques, however, species can be missed even by bat specialists. This was brought home to us on three occasions recently. In 1990 a tomb bat was caught for the first time, by chance, in a furled-up mist-net that had been used to trap birds, and bones of the hill long-tongued fruit bat, *Macroglossus minimus*, a nectarivorous bat then unrecorded on the archipelago, were found on Anak Krakatau in barn owl pellets. Not surprisingly, the owl collected the species before bat specialists had detected it. In 1992, a large, unrecorded rousette flew into camp after fruit.

Although the first bats to colonize the Krakataus, *Cynopterus* and *Rousettus* species, as well as *Pteropus vampyrus*, may act as pollinators, the specialist nectarivorous bats of Southeast Asia are the cave fruit bat, *Eonycteris spelaea,* and the long-tongued fruit bats, *Macroglossus minimus* and *Macroglossus sobrinus.*

These are small bats with elongate snouts and tongues with brush-like surfaces over the distal part, well adapted for licking nectar and pollen from flowers. Projecting scales on their hairs trap pollen in the fur of head and shoulders as the bats probe, spending a second or two at each flower, and, like bees, the bats groom pollen grains from their hairs for food. They are important pollinators of several trees and vines that bear so-called bat flowers, with special characteristics that may have evolved as adaptations to nectarivorous nonflying mammals (Sussman and Raven 1978), or moths which also pollinate them (for example, *Barringtonia asiatica*), before nectarivorous bats evolved. These characteristics, reviewed by Heithaus (1982), include a bell-like flower structure into which the visiting bat must insert its head, or a "shaving-brush" form, with stamens, pistil, and nectar fully exposed. The flowers are usually white or dull in color (flower-visiting bats are color-blind and such colors are unattractive to daytime nectar feeders), usually open at night, often have a strong, stale odor, attractive to bats over long distances, are easily accessible to bats (flagelliflorous, cauliflorous, or on layered branches), and the plant may be leafless when flowering.

Macroglossus minimus is particularly associated with species of the mangrove genus *Sonneratia;* on the Malay peninsula it has not been recorded outside mangrove areas. This small bat (weight 16–24 g) ranges over at least 2 km to feed but has a more coastal distribution than *Macroglossus sobrinus,* which specializes on *Musa* (wild banana) flowers (Start and Marshall 1976; Itino, Kato, and Hotta 1991). *M. minimus* was recorded on the coast of West Java in 1982 and 1985, and in 1992 three individuals were captured in mist-nets beneath *Barringtonia* trees on Panjang. Two years previously, skeletal remains of this species had been found in barn owl pellets on Anak Krakatau. *M. sobrinus* was found on Rakata in 1974 (Hill 1983) but has not been seen since. It is a little surprising that *M. minimus* appears to be the first to have become established on the Krakataus, which have no mangrove vegetation.

The rather larger, cave fruit bat, *Eonycteris spelaea,* also occurs in West Java. In its guano in the Batu Caves, Malaysia, Start (1974) found pollen that must have been eaten at least 38 km from the caves, about the distance of the Krakataus from the adjacent mainlands. In contrast to species of *Macroglossus,* *E. spelaea* uses a rather diverse range of flowers as nectar sources, including *Barringtonia* species. Moreover, it roosts communally in caves and forages in flocks of a score or so individuals and its flight is fast and direct. *M. minimus* forages singly or in small groups, roosts in trees, and has a much weaker flight than *E. spelaea* (Gould 1978). These differences suggest that *E. spelaea* would be the more likely to colonize the Krakataus, and in fact there is evidence that it has already done so.

The tree *Oroxylum indicum* is an exceptional case of a plant that is adapted for pollination by just one species of nectarivorous bat (Gould 1978). Its flagelliflorous inflorescences have flowers that both in their structure and the timing of their opening and nectar presentation are exclusively adapted to the morphology and behavior of the cave fruit bat, *Eonycteris spelaea*. These characteristics preclude pollination both by bats that are larger, lighter, or have earlier feeding times than *E. spelaea* and by day-feeding nectarivores. Dr. T. Partomihardjo found *O. indicum* near Handl's Bay on Rakata only in 1992 but he has informed me that the individual was a mature tree producing pods and there were seedlings in the vicinity. Pollination had been achieved, so it is possible that this is a case of the plant "recording" the presence of its pollinating bat before the zoologists have done so.

The three Krakatau sunbird species seek nectar from certain plants (often with red, purple, or orange tubular flowers), such as *Erythrina orientalis* and species of *Hibiscus, Morinda, Carica*, and species of *Musa* with erect inflorescences and diurnal nectar production. Although there is no confirmed case of effective pollination by birds of any Malaysian forest plant (Wells 1988), it is likely that the sunbirds play some role in the pollination of some of these plants on the Krakataus.

Birds have also diversified since the 1930s, and in this case pseudoturnover is likely to be minimal, although again some species have been sighted only at times other than during formal surveys. In October 1951, Hoogerwerf and van Borssum Waalkes made an expedition to Panaitan Island, off the Ujung Kulon peninsula. A shortage of good drinking water forced them to leave Panaitan two weeks earlier than planned and they decided to spend the remainder of their time on the Krakataus. From this setback arose the only animal survey between the Dammerman era and the first of the recent series in the 1980s (Hoogerwerf 1952, 1953), as well as a valuable plant survey. They spent ten days on Rakata, based at South Bay, whence they explored the southern coast, penetrated the interior from several places, and climbed to the summit. Brief visits were made to Sertung (two hours) and Panjang (half an hour). Hoogerwerf discovered five bird species additional to those recorded in the Dammerman surveys: the crested serpent-eagle, *Spilornis cheela*; silver-rumped swift, *Raphidura leucopygialis*; Javan sunbird, *Aethopyga mystacalis*; black-naped fruit-dove, *Ptilinopus melanospila*; and olive-winged bulbul, *Pycnonotus plumosus.* One raptor, the crested goshawk, *Accipiter trivirgatus,* had been lost. In the following year the oriental hobby, *Falco severus,* was seen on Sertung. In the 1980s surveys several more species—the chestnut-capped thrush, *Zoothera interpres;* little swift, *Apus affinis;* green imperial pigeon, *Ducula aena;* barn owl, *Tyto alba;* peregrine falcon, *Falco peregrinus;* and changeable hawk-eagle, *Spizaetus cirrhatus*—appeared to have become established. The black eagle, *Ictinaetus malayensis,* was also seen (Thornton, Zann, and Stephenson 1990).

The increase in the reptile fauna has also slowed markedly since the 1930s, only the paradise tree snake and the blind snake being additional records since that time (Rawlinson et al. 1992).

The butterflies, taken as a group, show a reduction in the rate of species increase rather like that of land birds, as do the cockroaches and dragonflies. The colonization curves for the Nymphalidae and Hesperiidae, butterfly families clearly affected by the change from grassland to forest that began from 1908, exemplify this pattern of change. The change is not uniform across animal groups, however. Lycaenid butterflies, for example, have increased from none in 1908, to 8 species by 1934, and to over 20 in the 1980s. Few of Krakatau's butterfly colonists are forest dwellers; perhaps 15–18 species (as much as 77 percent) are dependent for their food plants on the early-successional *Ipomoea pes-caprae* association on accreting beaches, a rather transient habitat.

In ants and other aculeate Hymenoptera (the stinging wasps and bees) as well as in braconid parasitoids, diversity has increased markedly since forest formation, as it has also in thrips, lacewings, and land molluscs. Comparisons of land-mollusc surveys, however, like those of bats, must be treated with caution, particularly when specialists were not involved in all of them. Dr. W. Nentwig found that between the 1921–1934 period and 1991, the spider faunas of Sertung, Panjang, and Rakata had doubled. Representatives of six additional families, Theraphosidae, Oonopidae, Mimetidae, Nesticidae, Hersiliidae, and Philodromidae, were found, which Nentwig believes would almost certainly have been recorded had they been present earlier.

Ash fall resulting from Anak Krakatau's periodic volcanic activity is thought to have been responsible for the extinction of the beach skink, *Emoia atro-costata,* and to have contributed, along with vegetational change, to the demise of the zebra dove, which feeds on the ground on the fallen seeds of grasses and sedges; both these species were confined to the Sertung spit. The skink was always confined to the spit, where it appeared in numbers in 1919 and was persisting when the Dammerman surveys ended in 1933. Anak Krakatau's 1952 eruption is known to have devastated this area, and in spite of intensive searches the skink has not been found in recent surveys (Rawlinson et al. 1990, 1992).

The mobility of animals often permits them to minimize damage by ash fall, but such damage is thought to have played an important role in deflecting the succession of the forests on two of the islands. As the island group's plant and animal community has diversified, its forests have diverged, and the development of different forest types on the various islands has been a subject of intense interest to plant geographers.

10 DIVERGENCE OF THE FORESTS

> It is not a question of one single Krakatau problem; there
> are many and various problems, and time and again new
> ones will appear, attention to which will be fully worth
> while.
>
> W. M. Docters van Leeuwen, "Krakatau 1883–1933" (1936)

In 1929 Docters van Leeuwen had noticed differences in the makeup of the
forests of the three older Krakatau islands, particularly that the tree *Timonius
compressicaulis* was abundant only on Panjang: "One species of tree, *Timonius
compressicaulis*, which in Krakatau [Rakata] is only found on the NW side, occurs everywhere [on Panjang]" (Docters van Leeuwen 1936, p. 181). He suggested that chance landfalls of colonizing birds may affect the sequence of arrival
of animal-dispersed plant species on the various islands, implying that differences in sequence may lead to divergence of the forests on the three islands.

Up to 1932 the lowlands of all three islands carried vegetation of a similar
floral composition and state of development, with Rakata having a slightly
greater development of closed-canopy mixed forest. The high ground of Rakata
carried vegetation types not found on the other islands: a *Neonauclea*-dominated
forest above about 400 m, and a closed *Cyrtandra* scrub near the summit.

The Development of Rakata's Forest, as Seen in 1951 and 1979

On Sertung in 1951 the northern spit was largely covered in *Casuarina* woodland,
which sharply demarcated the spit from the broad-leaved forest of the rest of

the island (as it still does; see Figure 14, p. 53). *Terminalia catappa* was recogniz-
ably prominent in the forest covering the higher ground. Only the western side of
Panjang (near what is now known as Bat Cave Beach) was inspected. *Casuarina*
was growing at an angle from the tuff cliff backing the beach area, and from the
sea *Terminalia* and scattered groups of *Casuarina* could be recognized as components of the inland forest (Van Borssum Waalkes 1952, 1953, 1954, 1960).

The western part of Rakata continued to develop more slowly than the rest
of the island. Krakatau South had been completely eroded away, and to the
east, between what are now known as Owl Bay and South Bay, a new, small spit
had formed, on which grew *Casuarina, Desmodium umbellatum, Premna
serratifolia, Hibiscus tiliaceus, Pipturus argenteus,* and several *Ficus* species. The
Barringtonia association was well developed in the South Bay area and, unusually for the Krakataus, was here actually dominated by *Barringtonia asiatica.* The southern and western grasslands had now become isolated patches
separated by trees and shrubs. *Saccharum* persisted in isolated spots on the
ridges, and on one ridge at about 300 m on the northeast was a small remnant
of *alang alang* grass *(Imperata cylindrica).* On Rakata's northern cliff, barren
patches could still be seen but vegetation was now fairly dense up to the summit, *Casuarina equisetifolia* growing at an angle of about 60 degrees away from
the cliff face. *Terminalia catappa* was also common, and clumps of *Saccharum*
occurred on the higher parts.

The most important botanical finding of this visit was the discovery that
most of Rakata's southern and eastern slopes, from the shore to about 750 m
altitude, was covered in forest dominated by *Neonauclea calycina*. The
Macaranga–Ficus fulva forest, which Docters van Leeuwen had noted below
this height, had been replaced by *Neonauclea*-dominated forest. Unusually for
Rakata, this dominant tree (which, it will be recalled, is an upland tree on Java)
had extended down the slope of the mountain.

Several rather imprecise zones were recognizable within the forest. Up to
about 50 m *Terminalia catappa* was very common, and *Macaranga tanarius,
Melochia umbellata, Tarenna fragrans,* and small-tree species of *Ficus (fulva,
fistulosa,* and *hispida)* were included. The undergrowth was dominated by the
fern *Nephrolepis biserrata.* Beyond, up to 200 m, the forest became almost pure
Neonauclea, and the trees here were larger than those near the coast. From 200
to 500 m, large figs *(Ficus variegata, retusa, tinctoria,* and *pubinervis)* and the
smaller epiphytic creeping shrub, *Ficus montana,* were components of the forest, and the ground vegetation consisted mainly of pteridophytes, particularly
Selaginella plana. Epiphytic mosses and ferns were much more common at

this altitude than in the lower forest; for example, the bird's nest fern, *Asplenium nidus,* was frequent. Above about 500 m a thick layer of mosses covered the ground and festooned the branches, stems, and trunks of shrubs and trees.

From 750 m to the summit was a dense thicket of shrubs, with scattered young *Neonauclea* and *Macaranga* and frequent dense clumps of *Saccharum.* By now the bat-dispersed *Cyrtandra sulcata,* which had earlier dominated the island's upper slopes had been shaded out and, though still common, no longer dominated. *Neonauclea* trees now occurred even on the summit. A newcomer, *Schefflera polybotrya,* was now the most common shrub. This species is not common on the adjacent mainland but occurs at Cibodas, for example, in the West Java uplands. *Schefflera* species are usually dispersed by birds and bats, and *S. polybotrya* bears succulent, fleshy berries.

It was 27 years before the next botanical study of Rakata. John Flenley's 1979 expedition (Flenley and Richards 1982), the first of what may be called the modern series, made a 16-day survey of Rakata (and 3 days on Anak Krakatau). The summit area of Rakata, although not the summit itself, was investigated, but Panjang and Sertung were not covered. On Rakata the *Neonauclea* forest was still present, and the *Casuarina* woodland that had been in decline at Handl's Bay had almost disappeared, as had the patches of grassland on the south side.

During this expedition the first attempt was made to quantify the altitudinal differences in environment on Rakata. Forster found that the incidence of Rakata's 19 species of bryophytes (mosses and liverworts) in 10 sample vegetation plots varied with altitude, relative humidity, and air temperature (which themselves were intercorrelated). Surprisingly, the soil's salt concentration was not negatively correlated with altitude. Neither number of tree species nor tree density was significantly correlated with the representation of bryophytes. Air temperature declined 0.8–0.9° C per 100 m altitude, and Forster thought that this rate of change probably extends throughout the year for it was also found to be true of soil temperature. He concluded that temperature, and therefore relative humidity, exerted an important influence on the bryophyte flora. The relative humidity differed by 38 percent between sites at 1 m elevation (R.H. 72 percent) and at 670 m (98 percent), and Forster suggested that in the dry season desiccation-prone species would be able to survive only in the uplands (at the 650 m plot, 100 percent humidity was recorded between 7:15 P.M. and 7:15 A.M., and cloud covered the site for 15 of the 24 hours) (Forster 1982).

There is also some indication of differences between the upland and lowland faunas of Rakata. Compton's group in 1984 found that the fauna of chalcid wasps at 250 m on the northwest ridge was quite the most distinctive of six sites sampled on the archipelago, all others of which were located in the lowlands (Compton et al. 1988). Also, one family of soil nematodes has been found only in the *Schefflera* vegetation near the summit (Winoto et al. 1988).

The Discovery of Forest Divergence

In July and August 1982 a largely Japanese team led by Hideo Tagawa was the first since the time of Docters van Leeuwen to visit all islands and to make extensive surveys of Sertung and Panjang. The group spent 12 days on Rakata, 11 on Sertung, 7 on Panjang, and a few days on Anak Krakatau. An Indonesian expedition in July 1982, led by Hasiana Ibkar-Kramadibrata, also spent 3 days on Panjang and some time on Anak Krakatau. The two expeditions, surprisingly, were unaware of one another's activities.

The Indonesian group reported the presence on Panjang of *Terminalia catappa, Neonauclea calycina,* and *Dysoxylum gaudichaudianum,* the last two of these, particularly the *Dysoxylum,* being important components of the forest. Surprisingly, they did not find *Timonius compressicaulis* although they made three transects inland, one from the west coast and two from the east (Ibkar-Kramadibrata et al. 1986). One of their transects must have passed just to the north of a plot later to be established as a research plot by Whittaker's group in young, even-aged *Timonius* forest (Whittaker, Bush, and Richards 1989).

On Rakata, Tagawa's group found that the forest was still dominated by *Neonauclea,* and the decline of *Cyrtandra sulcata,* first noted some 30 years before, was complete. *C. sulcata* was now largely confined to ravines, under *Neonauclea,* and the summit area was covered by *Schefflera polybotrya* scrub. The team made the important discovery that the forests on both Panjang and Sertung were quite different from Rakata's forest. There was no *Neonauclea* forest on either Panjang or Sertung. Instead, the forests of these islands were dominated by *Dysoxylum* and by *Timonius,* forest types that had never before been encountered on the Krakataus, nor even reported from Java or Sumatra (this also applies to *Neonauclea* forest). Although floristically (i.e., in component species) there was little difference, particularly in the lowlands, between the forests of the three islands, the different dominant patterns gave them quite different characters.

The divergence foreshadowed by Docters van Leeuwen had taken place. Did this come about in the way that he suggested? And why did it begin in the 1930s? Was it related in some way to colonization of the archipelago by the later-successional tree *Dysoxylum gaudichaudianum,* or to the emergence of Anak Krakatau, both of which also occurred at about this time? Or was it, perhaps, related to differences in habitat between the islands? How had these different, unusual types of forest developed on islands only a few kilometers apart?

Tagawa and his colleagues approached the last of these questions by considering the likely successional relationships between the dominant trees and their possible sequence of arrival on each island (Tagawa et al. 1985).

Timonius compressicaulis (Rubiaceae) was first found on the Krakataus in 1929. It is a small tree, reaching some 25 m in stature, and in Sumatra it is a component of the *Barringtonia* association. Its rapid growth and ability to flower when only a meter tall makes it well suited for the invasion of bare ground. The fleshy pericarp of its stone fruits (drupes) is eaten by fruit-eating birds. Docters van Leeuwen had found it to be widespread on Panjang in 1929, and on Rakata he saw groups of well-developed specimens in a large ravine southwest of Zwarte Hoek but found the species nowhere else on that island. *Timonius* was unknown on Sertung until Tagawa's group recorded it in 1982.

Timonius often occurs in association with *Terminalia catappa* (Combretaceae). Treub had found *Terminalia* fruits washed up on Zwarte Hoek's west beach in 1886, and it was first found growing, on Panjang, in 1896. In the following year Penzig found it on all three islands. The large almond-shaped fruits, which can float for months, are dispersed over long distances by sea, and the species ranges from Madagascar to Polynesia. The fruits are taken by *Cynopterus* fruit bats (and land crabs) and eaten at feeding roosts some distance from the parent tree. Some of our expeditioners have been showered with the fruit when they unknowingly camped beneath such a roost. Rats and land crabs may also assist in secondary, short-distance dispersal.

Dysoxylum gaudichaudianum (Meliaceae) is a tree of early or intermediate successional stages of primary and secondary forest, and of swamp forest, bamboo woodlands, hedges, and *alang alang (Imperata cylindrica)* fields. It has a wide ecological amplitude and may reach 40 m in stature. The small, 1-cm-long seeds fall after maturation, and germination occurs very soon after seed fall. The seeds are dispersed by birds (for example, imperial pigeons, *Ducula* species) and perhaps also by fruit bats. The species was first recorded on the Krakataus from Panjang in 1932. Van Borssum Waalkes did not find it in 1951 during nine days spent on Rakata, a couple of hours on Sertung, and half an hour on Panjang. In 1983 many trees on Panjang had girths of over 200 cm (one 254 cm) and heights of over 20 m (the tallest 33 m) (Whittaker, Bush, and Richards 1989), and those on Sertung in 1982 were of similar size (Tagawa et al., 1985). It was found on Rakata in 1979 and, in contrast to the Panjang and Sertung trees, in 1983 the specimens appeared to be immature, mostly saplings (Whittaker, Bush, and Richards 1989). The species was originally misidentified by the Flenley and Tagawa groups as *Dysoxylum parasiticum* (= *caulostachyum*), which, unlike *D. gaudichaudianum,* is a component of mature forest canopy.

Tagawa's group found a high density of *Dysoxylum* seedlings and saplings in all three forest types, *Neonauclea, Timonius,* and *Dysoxylum* forests. *Timonius* seedlings and saplings, however, were found in *Neonauclea* forest only in tree-fall gaps, and *Neonauclea* was never found growing in *Dysoxylum* forest. Of these three tree species, then, *Dysoxylum* was the only one which appeared to be able to regenerate under canopies of the other two. Because *Neonauclea* and *Dysoxylum* forests have a more closed canopy than *Timonius* forest, Tagawa's team suggested that *Neonauclea* could overtop and succeed *Timonius* but that *Dysoxylum,* the tallest of the three, would become the dominant over *Neonauclea.* They provided a successional scheme for the vegetation of the Krakataus, showing *Dysoxylum* as the eventual successional dominant on the islands.

Tagawa realized that a problem followed from his proposed scheme, what he was later to term "the problem of the formation of *Neonauclea* forest" (Tagawa 1992). This involved two related questions: Why doesn't *Dysoxylum* invade Rakata's *Neonauclea* forest, and why isn't *Dysoxylum* a leading forest type anyway, on the island's now rich soil? In short, why have the forests diverged?

Divergence Hypotheses

Arrival sequence and colonization interval of potential dominants

In addressing the above questions, Tagawa and his colleagues took Docters van Leeuwen's hint and considered the arrival sequences of the three dominants on the individual islands as suggested by first records:

	Rakata	Panjang
Neonauclea calycina	1905	1929
Timonius compressicaulis	1929	1929
Dysoxylum gaudichaudianum	1979	1932

(All records are from Docters van Leeuwen, 1936, except for the appearance of *D. gaudichaudianum* on Rakata, noted by Flenley and Richards, 1982.)

Neonauclea was first recorded on Rakata in 1905, and although not recorded on Sertung until 1920, the trees were then quite large, so they must have arrived much earlier. It then became quite frequent in the Sertung hills and on the northern spit. *Neonauclea* was not found on Panjang until 1929, after which it became widespread from the shore to the hills. Between 1897 and 1929, however, Panjang was the objective of only two separate day-visits, and *Neonauclea* may have been overlooked on these cursory surveys. This wind-dispersed tree probably arrived on all islands a couple of decades or so after 1883.

Timonius and *Dysoxylum* are more tolerant of shade than *Neonauclea,* and Tagawa's group believed that on Panjang they arrived at about the same time as *Neonauclea* or soon after, giving *Neonauclea* little time to develop on the island. This may also have occurred on Sertung, but little can be said about relative times of arrival there. During the 62 years following Docters van Leeuwen's 1920 visit, the developing forest on the main body of the island, as opposed to the atypical early-successional northern spit, was visited only for a day in 1929 by Docters van Leeuwen. *Timonius* was found on Panjang in the same year as *Neonauclea,* 1929, and *Dysoxylum* was first recorded there in 1932 as fruiting trees with seedlings around them. On the relatively well-surveyed Rakata, however, *Neonauclea* was established well before *Timonius* and *Dysoxylum* arrived and was thus able to expand and become dominant. *Timonius* was not found on Rakata until 1929, 24 years after *Neonauclea,* and *Dysoxylum,* first recorded there in 1979, was evidently absent from Rakata in 1952 when van Borssum Waalkes spent nine days there.

Thus, Tagawa's group proposed that both the *Neonauclea* forest of Rakata and the *Timonius* forests of Panjang (and perhaps Sertung) would in time become dominated by *Dysoxylum,* and they regarded *Dysoxylum* forest as the most "advanced" successional association on the islands, to which all the archipelago's forests would converge. They suggested that the differences between the islands' forests in 1982 were due to the differing intervals between the arrival of the three potentially dominant tree species on the various islands and to the dominance relationships expected in the various sequences. Presumably there is a critical interval between colonizations, during which the first colonist can become securely established and after which the second colonist is unable to dominate.

Deflection of succession by volcanic disturbance

In 1986 the Indonesian group's identification of the *Dysoxylum* species as *D. gaudichaudianum* was confirmed. The species was thus not a tree of mature primary forest canopy, and the prediction that it would become the forest dominant on the Krakataus no longer conformed to what was known about the species elsewhere. Whittaker's group concluded that the Sertung and Panjang forests, rather than being more mature than that of Rakata, were, in general, at a younger stage of development. In this model, Rakata's forests are the most mature of any on the islands, and they are predicted to become more patchy and to incorporate a number of forest tree species, including species of *Ficus* (already well represented in the uplands), rather than develop into a predominantly *Dysoxylum* forest.

In explanation of the divergence, Whittaker's group emphasized the role of disturbance, particularly the effects of ash falls resulting from Anak Krakatau's continuing activity. In the 1930s more ash had fallen on Sertung than on Panjang, prevailing winds being from the east in the rather longer dry season and from the west in the wet season. Rakata was relatively protected from such falls by its great northern cliff-scarp and by the general directions of prevailing winds. A study of Krakatau soils confirmed that there were indeed more complex sequences and greater depths of ash on Sertung than on Panjang (Shinagawa, Miyauchi, and Higashi 1992). These falls during and since Anak Krakatau's emergence were thought to have deflected the plant succession on Sertung and Panjang and to be the reason that their forests have not reached the stature of those on Rakata. The ash-fall disturbance was presumed to have enabled the colonization of Sertung and Panjang by *Timonius,* which acted as a nursery crop, modifying the environment for the later-successional *Dysoxylum.* Large areas of very young *Timonius* forest were believed to represent recovery from ash falls. The resulting patchy mosaic is evident in the vegetation maps of Richards and Whittaker (1990). The disturbance model nicely accommodated the much more rapid change in canopy architecture on Panjang and Sertung than on Rakata since 1983, and the smaller increase in floral richness on the first two islands since the 1920s. As Forster (1982) had shown, however, the more heterogeneous environment of the 800-m-high Rakata may also have been an important factor in the more rapid diversification of its flora.

In 1989 the post-1883 history of ash falls on Panjang and Sertung was investigated (Whittaker, Walden, and Hill 1992). Soil profiles were made on Panjang, Sertung, Rakata, and on Sebesi. Chemical, physical, and mineral-magnetic analyses confirmed that in most sites on Sertung and Panjang there had been a number of ash falls in the early 1930s, when Anak Krakatau was emerging, and also during its 1952–1953 eruptions. Whittaker's group erected "disturbance models" for both Sertung and Panjang, summaries of ash-fall horizons in six soils beneath different forest types. They argued that if *Timonius* forms the first (relatively open) canopy after disturbance, and *Dysoxylum* then grows through this to form a higher canopy that shades out *Timonius,* one would expect sites covered by *Timonius* to show evidence of more recent disturbance than sites covered by *Dysoxylum* canopy. The data did not show this unequivocally, however. Whittaker's team also attempted to establish whether the more complex soil sequences on Sertung resulted from the 1930s phase of activity or from more recent volcanic episodes. In general, the answer appeared to be—from both. The researchers suggested that the southern end of Sertung, where *Dysoxylum* forests do not occur, received most of its ash fall in the past three or four decades.

The problem with these very detailed studies is that there is no direct evidence of the ecological effect of the falls. The effects of the 1952–1953 volcanic episode, probably the most severe, were monitored soon after its occurrence. Although there was damage on the inner coasts (those facing Anak Krakatau), the outer coasts were little affected. Moreover, within the short time between the eruption and the monitoring visit there was evident recovery of the vegetation (van Borssum Waalkes 1954, 1960). As Whittaker and his colleagues pointed out, however, much damage can be done to vegetation without actually killing a high proportion of established trees, and the creation of light gaps by localized, sub-lethal canopy damage may have an important effect on forest dynamics. Such damage may not be indicated at all by soil-profile analysis. In general, in spite of the work of Whittaker's team, which was both extensive and intensive, it appears that the ecological impact of the ash falls received by Sertung and Panjang cannot now be clearly demonstrated.

Whittaker and his colleagues summarized what they considered to be the only unequivocal deductions that are possible from the dates of first record of the three tree species on the various islands as: (1) *Timonius* was widely established on Panjang earlier than it was on Rakata; (2) by the time *Timonius* was was first recorded on Rakata, *Neonauclea* was well established there (the difference in first records is 24 years); and (3) *Dysoxylum* was the last of the three dominants to be found on each of the three islands. In fairness to Tagawa's group, perhaps the following should be added: (4) *Dysoxylum* was first discovered on Panjang only 3 years after *Neonauclea,* whereas on Rakata it was not discovered until 74 years after *Neonauclea,* and the difference in arrival times was probably at least 46 years.

Points 3 and 4 above appear to support the hypothesis that *Dysoxylum*'s arrival on Panjang close on the heels of *Neonauclea* prevented the development of *Neonauclea* forest there, whereas on Rakata, where *Neonauclea* had been established for decades before *Dysoxylum* could even begin to compete, its development was unhampered. The situation on Sertung is equivocal; it is possible that *Dysoxylum* followed *Neonauclea* on Sertung also, but with a much shorter lag than on Rakata.

Colonization interval and differential disturbance

Tagawa (1992) has suggested that *Neonauclea* seeds may have survived the 1883 eruption in deep gullies and germinated when rain erosion had cut through to the old soil surface. He also developed an alternative hypothesis, that the wind-dispersed *Neonauclea* was established on all three islands at about the same time but volcanic ejecta, received only by Sertung and Panjang, limited its spread on these islands. He suggested that *Timonius* and *Dysoxylum* seeds, being larger, are more resistant to burial under volcanic ash than those of *Neonauclea*. Unlike many secondary forest trees, *Neonauclea* does not produce stem or root sprouts following damage, and seed germination is low under a canopy of mixed secondary forest. Disturbance would thus put *Neonauclea* at a disadvantage. Rakata was unaffected by ash fall and *Timonius* and *Dysoxylum* arrived there later, by which time *Neonauclea* had become very well established. Its dominance on Rakata was thus checked by neither ash fall nor competing species. Thus Tagawa has also now incorporated Anak Krakatau's activity into an explanation of the divergence of forest types on the archipelago.

Perhaps the strongest and most obvious argument in favor of an ash-fall disturbance explanation of the observed differences in dominance patterns is the timing of forest divergence. It is since Anak Krakatau's emergence and the renewal of volcanic activity that the developmental sequences of the vegetation of Sertung and Panjang, on the one hand, and of lowland Rakata, on the other, have diverged. The divergence, however, has not been in the form of a check or a reversion to an earlier successional stage on Sertung and Panjang. The divergence took the form of a deflection along new successional trajectories on these islands. Their new canopy dominants were two trees that first appeared on the archipelago at about the time that Anak Krakatau emerged from the sea.

Two notable observations by Docters van Leeuwen do not sit well with the disturbance theories, however. First, he had noticed the beginnings of forest divergence in 1929, when he found *Timonius* to be common only in the Panjang forests. Second, in May 1932 he noted that on Panjang *Timonius* was common in the hills and that *Dysoxylum gaudichaudianum* (= *amooroides*) trees grew "gregariously" (implying animal dispersal), were in fruit, and had many seedlings around them. These first signs of the development of *Timonius* and *Dysoxylum* forests on Panjang were noted well before Anak Krakatau's ash fall could have been responsible for them. Ash first fell on Panjang in January 1928, so the fall could hardly have been the cause of the unusual representation of *Timonius* 18 months later or for the presence of mature *Dysoxylum* trees that were producing many seedlings, after four years. This evidence strongly suggests that the divergence, at least between the forests of Panjang and Rakata, was in train well before disturbance effects from Anak Krakatau could have been its cause.

Neonauclea calycina, normally found in uplands, has achieved dominant status only on the high island, Rakata, and it is surely significant that its spread here was, unusually for the spread of tree cover on that island, from the heights to the lowlands. Moreover, it is the more recently covered lowland areas that are now (but why only now?) giving way, to some extent, to *Dysoxylum*. There was no *Dysoxylum* above 450 m in 1989, even as a gap-filler; the highest *Dysoxylum* specimens noted were at about 100 m elevation in the east and 250 m in the west. *Dysoxylum*'s slow extension of range on Rakata thus excludes those areas long colonized by *Neonauclea*. These observations support the proposal that sequence of colonization and the extent of the intercolonizing interval, even when the sequence is the same, are important determinants of forest type.

A considerable body of innovative work by American experimental community ecologists, discussed in more detail in Chapter 13, appears to have thrown some indirect light on this problem. Techniques were developed for assembling "micro-communities," usually of aquatic organisms, in the laboratory. Dickerson and Robinson (1985), for example, found it relatively easy to produce different end-point communities (with respect to relative abundance of species, species composition, and vulnerability or resistance of the resulting community to invasion) simply by altering the sequence of invasion of the same set of species. This was confirmed by simulation models (Drake 1990b). Changing the interval between introductions, even with the same sequence, could also affect the final outcome (Robinson and Edgemon 1988). Drake (1991) suggested that the length of the initial period of competition-free population growth enjoyed by an invader may be at least partly responsible for the effect of the timing between invasions. The parallel, in general terms, with the explanation of Tagawa's group is obvious. They suggested that the time since *Neonauclea*'s establishment affected the colonization success of *Dysoxylum* and *Timonius*, and thus their importance in the resulting community. In the Krakatau "experimental situation," however, it is difficult to separate the effects of volcanic disturbance and island altitude from these priority effects.

Drake also suggested that disturbance events—both the events themselves, if they are of such a nature and magnitude that they change the context of the process, and the frequency and timing of the events—may lead to entirely novel species assemblages. The possible parallel with the different, unusual forest types that have developed on Sertung and Panjang, having no equivalent on the mainlands in terms of floral dominance pattern, is evident. Perhaps the Krakatau-type forests have not been seen elsewhere simply because there are few situations comparable to the Krakataus at present, or perhaps on Krakatau we are seeing a natural example of Drake's finding. The magnitude and frequency of disturbance, often largely randomly determined, may thus be of great importance in community assembly, depending on the generation times of the species making up the community and their relative susceptibility to the particular type of disturbance.

The results of the experimental ecologists appear to provide some support for both Tagawa's hypothesis, which places importance on the sequence of colonization and the length of the interval between colonizations of potentially dominant trees, and that of Whittaker and his colleagues, which stresses the role of disturbance.

Will Divergence Continue or Will the Forests Converge?

Dysoxylum was not mentioned as an important component of the lowland *Neonauclea* forest, either when it was first found on Rakata in 1979 (Whittaker 1982) or in 1983 (Whittaker, Bush, and Richards 1989). Yet by 1989 it appeared to be taking over from *Neonauclea* in lowland areas by exploiting large forest gaps (Bush, Whittaker, and Partomihardjo 1992). On Sertung, in the same year, an extension of *Dysoxylum*'s dominance was seen over substantial areas of the north and east. Although *Timonius* remained dominant in the south and west, there were numerous scattered emergent *Dysoxylum* trees, which were not then sufficiently dense to form a canopy over the *Timonius*. On Panjang, also, there was a dramatic expansion of *Dysoxylum* forest from about 50 percent of the island's cover in 1983 to about 75 percent in 1989. Within the *Timonius* forests it was evident that in many places *Dysoxylum* would soon overtop the 20-m-high *Timonius* (Bush, Whittaker, and Partomihardjo 1992). Thus it appears that the prediction of Tagawa's group, that *Dysoxylum* will become the forest dominant of the Krakataus, is already approaching fulfillment, with the notable exception of upland Rakata. Whittaker's group suggested that *Dysoxylum* "may yet expand further in colonizing the interior of Rakata" (Bush, Whittaker, and Partomihardjo 1992, p. 197) but remained unconvinced that a *Dysoxylum* forest would eventuate. They suggested that the largely coastal distribution of seed-dispersing birds on Rakata may limit its inland spread there (see below) and predicted a continuing divergence of forest types.

On Rakata there are pied and green imperial pigeons, both of which are known dispersers of *Dysoxylum*. There are at least four, probably more, species of fruit bats (at least six species occur on the archipelago), three other specialist fruit-eaters (the black-naped fruit dove and Asian glossy starling—both efficient seed dispersers—and pink-necked green pigeon), and nine facultatively frugivorous bird species (Zann et al. 1990). It is through some of these agents that *Dysoxylum* may slowly infiltrate the forest by filling gaps and thus perhaps extend its range inland to the higher ground.

In 1992 Whittaker and Turner kept a mature, heavily fruiting *Dysoxylum* tree in Rakata's lowlands under observation for five days. Black-naped fruit doves were seen taking the seed "duplexes" (two seeds joined by an aril), and the yellow-vented bulbul and Asian glossy starling were both seen taking seeds from fruits. *Ducula aena* was also seen to take seeds from fruits and to take whole fruits (Whittaker and Turner 1994). It was not clear that the birds ate the seeds rather than the red aril that connects the seed pair—very many single seeds were found on the ground and gut contents were not investigated. Nor was it clear if the birds dispersed the seed (feces were not analyzed), but bird dispersal of *Dysoxylum* has long been accepted. As noted above, Whittaker and Turner believed that several of the presumed bird dispersers do not penetrate far into non-coastal areas of Rakata, although this needs checking by careful surveys of the higher slopes. Since *Dysoxylum*'s range expansion on Rakata is also confined to coastal and near-coastal areas, this was taken as circumstantial support for the view that it is spread predominantly by birds, which thus limit its expansion. Docters van Leeuwen, however, stated that *Dysoxylum* seeds "are eaten by birds and particularly by bats. Young plants are often found under the resting-places of the last-mentioned animal . . . The way in which they occur in Lang Island [Panjang], where they stood in groups, points to dispersal by bats" (Docters van Leeuwen 1936, p. 398). As noted in Chapter 7, it has also been suggested that bats disperse a species of *Dysoxylum* in the Tonga Islands. Moreover, *Dysoxylum* fruits hang on pendulous branches and when ripe have a carrion smell, both features of bat-dispersed fruits. Even if bats do take the quite large seeds of *Dysoxylum* (about 1 cm maximum diameter), they probably spit them out at feeding roosts up to a hundred meters from the parent tree; it would therefore take them many years to spread the species over large areas. Nevertheless, fruit bats have probably been partly responsible for the dispersal of *Cyrtandra, Schefflera,* and several fig species into the uplands of Rakata. It may be that, for *Dysoxylum* at least, animal dispersers are limiting in the high interior of Rakata to the extent that in these regions bird dispersal is less likely.

Disturbance, of varying levels of intensity, continues to affect Panjang and Sertung, and this factor, together with the greater height of Rakata, could lead to further divergence of the islands' forests. The effect of disturbance on community development was considered by Whittaker and Levin (1977). Disturbance of low intensity that is very frequent (in comparison with the time neeeded for the succession) may fragment a community into a mosaic of different developmental phases, in which progress of the succession is balanced by regression due to disturbance. The community may then never reach the mature state typical of undisturbed conditions. If the disturbance regimes of the two affected islands differ markedly, this difference in itself (as well as differences in the arrival sequences of potential dominants) may lead to the formation of alternative, self-maintaining states. Different communities may emerge on the two islands from what is basically the same biota. Differences between the communities on Sertung and Panjang are not yet marked, and this may reflect a generally similar impact of volcanic ash on both.

પ્ર

Of course, as with many ecological problems, it is most probable that a number of factors are involved in the divergence of the islands' forests: not only colonization sequence and intercolonization interval but also frequency and magnitude of disturbance, and perhaps also simply the differences in height and size between Rakata on the one hand and Sertung and Panjang on the other. Moreover, as Whittaker's group has pointed out, Rakata is the only island large enough to have a true hinterland; no part of the other two can be said to be uninfluenced by coastal factors. The lack of *Dysoxylum gaudichaudianum* in the uplands of Rakata, and the importance of this species in the forests of Sertung and Panjang, may thus be due to its relatively late arrival on Rakata, the fact that *Neonauclea* forest was then well established (in the uplands), the slowness of bat dispersal of seeds of this size over Rakata's larger area, and the coastal distribution of its bird dispersers. Clearly, an investigation of the relationship between *Dysoxylum* and fruit bats, both on Rakata and on the other two islands, is now needed.

11 LIFE ON AN ACTIVE VOLCANO

An examination of the fauna on Anak Krakatau, the new
volcano risen from the sea, is of greatest significance, as we
are here certain that we have an island originally entirely
devoid of animal and vegetable life, with even a completely
sterile soil. It is therefore of the utmost importance that the
fauna and flora of this island should be constantly
examined, and at regular intervals, and it is greatly to be
hoped that this unique opportunity will not be neglected,
as it was in the case of Krakatau itself.

K. W. Dammerman, "The Fauna of Krakatau, 1883–1933" (1948)

An emergent, virgin, volcanic island that is colonized by terrestrial plants and
animals, as was Anak Krakatau, is a very rare and unquestionable example of a
primary xerosere: a biotic succession starting on a "clean slate," a substrate initially devoid of life. Like the extirpation of life on an island, the emergence of
land from the sea is another precious opportunity to study the dispersal powers of terrestrial plants and animals, the processes involved in colonization,
and the way a functional ecosystem is gradually assembled from the very beginning. Few such opportunities arise and they should be grasped and fully exploited by those seeking to understand the biogeography and ecology of
islands.

The emergence and growth of Anak Krakatau in 1930 (Chapter 8) was just
such an opportunity. Many biological questions raised by the great eruption of
1883 were posed anew, in relation to the new island. "Here was an island cry-

ing out for regular study if only it survived," W. S. Bristowe (1969) later wrote of the six-month-old island he visited in 1931. The island has survived, but it has changed, and continues to change, a great deal. And it has been studied (although unfortunately not regularly) for what it can tell us about the questions, first raised in 1883, about the assembly of a new ecosystem. New questions have also arisen, and they will be discussed in the next chapter.

Beginnings: the First Biotas, 1929–1952

The first island to emerge from Krakatau's submarine caldera was dissipated by wave action and was never visited by biologists. The second one was the highlight of the excursion that followed the Fourth Pacific Science Congress in May 1929. By the time of the excursion it was reduced to a mere sand shelf, but biologists collected several insects, completely desiccated by the heat, including a large black cricket, *Brachytrypes portentosus,* and a female of the brown ant *Prenolepis taylori,* both of them species known on Rakata, 3 km away. Shortly after the excursion the island disappeared. The third island lasted only from June to August 1930 and, like the first, was not monitored by biologists. On August 11, 1930, the present island, Anak Krakatau IV, emerged (see Figure 28).

The British arachnologist Bristowe was the first biologist to land on Anak Krakatau IV, in February 1931. Although the island was but six months old and there was no plant life, he found an anthicid beetle, *Anthicus oceanicus,* on the beach, and large numbers of a small collembolan (springtail), *Mesira calolepis,* associated with drift debris. *Camponotus* ants, a small cosmopterygiid leaf-mining moth, a mosquito *(Aedes vigilax),* and three species of spider were also present (Bristowe 1931, 1969; Dammerman 1948).

Fifteen months later Docters van Leeuwen found the northern beach littered with flotsam (and jetsam), including tree trunks, bamboo stems and rhizomes, seeds, and fruits, many of which lay above the high-tide mark, partly buried. He collected seeds and fruits of 18 species, and seedlings of ten of them had struck root. All the identified species except a mangrove were typical members of the *Ipomoea pes-caprae* and *Terminalia* forest associations. There was no sign of the film of cyanobacteria, which Treub had found on Rakata in 1886 (Docters van Leeuwen 1933, 1936).

In November 1932 Boedijn, visiting the island with Dammerman, collected a further six plant species, four as seedlings (including two animal-dispersed species not previously known from the archipelago, *Cissus repens* and *Murraya paniculata)* and one as a stranded fruit. Dammerman sampled the coarse ash,

Figure 28. Both Surtsey (*above,* from the north in 1991), off Iceland in the North
Atlantic, and Anak Krakatau (*below,* from the east in 1985) emerged from the sea as
virgin volcanic islands (in 1966 and 1930, respectively). Surtsey is about 1.5 km from
east to west and 152 m high; Anak Krakatau is about 2 km from north to south and
in 1985 was 195 m (now about 250 m) high. Their physical environments and the
biological richness of their source areas differ greatly, however, and studies of
colonization and community assembly on these islands reveal both similarities and
contrasts.

which had a high acidity (pH 4.7), a very low content of organic matter (0.6 percent by weight) and a remarkably low phosphoric acid content. The ash samples contained two species of fungi, one of bacteria, three of cyanobacteria, and ten of green algae. Two moth species not previously known on the archipelago were flying, both cosmopolitan pests of legumes as larvae (neither was found since), and tenebrionid beetles were found in the husk of a washed-up coconut and came to light at night. The anthicid beetles found by Bristowe two years previously were still present, and workers of a myrmicid ant were found among flotsam. On the shore Dammerman noticed a beach thick-knee (*Burhinus giganteus*) and three migratory greater sand-plovers (*Charadrius leschenaultii*).

Shortly after this visit, ash from further eruptions covered and killed all the vegetation. Two months later, however, in January 1933, Dammerman saw a few small moths flying and a swarm of flies, comprising eleven species of at least five families, hovered over the island's highest point. Volcanic activity had ceased by June, and in October Boedijn, visiting again with Dammerman, found signs of the beginning of a second flora, seedlings of two sea-dispersed species, *Canavalia rosea* and *Cerbera manghas*, but no animals were seen apart from four thick-knees on the beach. On a visit in December, Dammerman found three specimens, with cast wings, of a termite species, and a small arctiid moth was flying which was new to the islands (and was never again recorded). In April 1934 many of the plant species previously reported were present and now there were three more. In the first two floras, at least 17 species had reached the seedling stage on the island, all but two being sea-dispersed. Dammerman's last visit to the Krakataus was in August 1939. Anak Krakatau had erupted earlier in the year and was again bare and completely covered in ash. The only specimens found were a dried-up winged ant and a hover fly that settled close enough to be captured. The island's second biota appeared to have been eradicated.

Van Borssum Waalkes believed that eruptions during World War II may have devastated a third biota on the island, and in August 1947 the geologist Petroeschevsky, making an aerial reconnaissance, noted isolated casuarinas only about 3 m tall. In June 1949 he revisited the island, and in August the biologists van der Pijl and Toxopeus spent a morning there. Both groups appear to have landed on the north shore, where they found early stages of what was perhaps the fourth sequence of colonization. Young specimens of several herbs and woody plants were growing among the casuarinas seen two years before, which were now 6–7 m tall. The fact that attention was paid to the north of the island on these visits perhaps indicates that this was the first (and perhaps at that time the only) lowland area to be vegetated. On the ash slopes were scattered groups of *Saccharum* and a single specimen of a pioneer fern, *Nephrolepis hirsutula.* Altogether, 13 species of vascular plants were recorded, but no algae or mosses. Seven species had been components of the earlier seedling floras: *Calophyllum inophyllum, Canavalia rosea, Cocos nucifera, Ipomoea pes-caprae, Pandanus tectorius, Barringtonia asiatica,* and *Erythrina orientalis,* all sea-dispersed species. The others were the sea- and then wind-dispersed *Casuarina equisetifolia,* the sea-dispersed *Cyperus javanicus, Scaevola taccada,* and *Spinifex littoreus,* and the wind-dispersed *N. hirsutula* and *S. spontaneum.* (van der Pijl 1949).

Several herbivores and some of their predators were already present. A cosmopterygiid moth was found on *Casuarina,* and a eucosmid moth and the lycaenid butterfly, *Zizina otis,* on *Canavalia.* The acridid grasshoppers, *Atractomorpha crenulata* and *Valanga nigricornis,* and a mantis of the genus *Hierodula* (all of which were still present on the island 40 years later) were also noted, as well as one ant and two jassid species and unidentified tachinid flies and microlepidopteran moths. The monitor *Varanus salvator* had already colonized, and a skink, a *Mabuya* species, probably *Mabuya multifasciata,* was found on driftwood. The pioneer creeping legume, *Canavalia rosea,* was already attended by the carpenter bee, *Xylocopa confusa,* and *I. pes-caprae* by a small bee of the genus *Halictus* (Toxopeus 1950).

By the time van Borssum Waalkes and Hoogerwerf visited the island for an hour in October 1951, there were 21 species of vascular plants, 10 of them new arrivals. These included the wind-dispersed pioneer fern, *Pityrogramma calomelanos,* and several sea-dispersed species—the grasses *Imperata cylindrica (alang alang)* and *Ischaemum muticum,* and *Desmodium umbellatum, Thuarea involuta, Hernandia peltata,* and the cycad *Cycas rumphii.* Coastal trees, *Terminalia catappa* and *Morinda citrifolia,* had become established, and the sea- and animal-dispersed plant parasite *Cassytha filiformis* trailed over the beach creepers. A total of 23 plant species were identified from the 1949 and 1951 collections, five being wind-dispersed, the rest sea-dispersed. The ash slopes were bare, save for the two ferns and *Saccharum spontaneum,* and van Borssum Waalkes (1960, pp. 35, 40) explained this by the nature of the substrate—unweathered, hard ash that he described as "a kind of bladder lava." Presumably this was ash with a brittle, indurated crust 1 or 2 mm thick, remnants of which were still present on the lower eastern slopes of the island from 1982 to 1992. On the flat coastal strip in the north, swarms of grasshoppers were feeding in a quite extensive *Imperata* grassland and the casuarinas were now about 10 m tall, whereas on the eastern side the vegetation seemed to be rather younger, with smaller *Imperata* fields, one near a casuarina.

Hoogerwerf visited the island again for a few hours in August 1952 and, in a report of both visits combined, recorded the white-bellied fish-eagle, yellow-vented bulbul, savannah nightjar (heard in August 1952), and beach thick-knee, as well as five migrant birds—four shorebirds and the barn swallow, *Hirundo rustica* (Hoogerwerf 1952, 1953). The vegetation, particularly on the southeastern slopes, had been damaged by ash several centimeters thick, and most of the ferns were killed. On the more northerly slopes *Saccharum* appeared to have spread: "solitary tussocks . . . which did not occur there last year, were noted everywhere" (letter quoted in van Borssum Waalkes 1960, p. 44). Extremely violent eruptions occurred about six weeks later (October 10–11) and Hoogerwerf and van Borssum Waalkes monitored their effects on November 16, on the first official voyage of the Bogor Botanical Gardens' new research vessel, *Samudera*. (Oddly enough, the new boat had already "met" Anak Krakatau. *Samudera* passed close enough, toward the end of her journey from Europe, for the wet pain, freshly applied for the Jakarta arrival, to be covered with black ash.) By the November voyage, Anak Krakatau's lowlands were covered by several meters of ash, and all the vegetation appeared to have been killed. Four species of migrant birds on the beach were the only living things seen. Further eruptions in July and November 1953 again devastated the island, thus ending what was possibly the island's fourth biota.

The Development of the Present Community

After the 1952–1953 eruptions, the only signs of a previous community were the skeletons of the tops of tall casuarinas projecting above the layer of black ash which covered the island to depths of from 3 to 5 m. Unfortunately, Dammerman's earnest plea, quoted at the beginning of this chapter, went unheeded. In the next 25 years there were no systematic investigations of the assembly of Anak Krakatau's latest ecosystem. The interesting early pioneering phase of colonization was thus not recorded. From later studies, however, we can make some intelligent guesses and, contrary to the usual textbook dogma, there is good reason to believe that animals may have colonized before plants.

The pre-vegetation fauna

The first animals to arrive were probably dependent on energy sources outside the island. Both on Surtsey, which emerged off Iceland in the North Atlantic in 1963, and on small bare cays of Australia's Great Barrier Reef, the pioneer fauna was found to be dependent on marine detritus (Lindroth et al. 1973; Heatwole 1971). The first birds seen on Anak Krakatau, the Pacific reef-egret, *Egretta sacra*, beach thick-knee, and migrants like the common sandpiper, grey wagtail, Pacific golden plover, Mongolian plover, whimbrel, and great knot all forage on the shore, and some of these were probably among the first birds to feed on the island. Indeed, Hoogerwerf (1953) saw the last four species on Anak Krakatau's beach only six weeks after the devastating October 1952 eruption. In the previous year he had also seen the white-bellied fish-eagle and the thick-knee, two species capable of finding food inland and on the shore. The fish-eagle *(Haliaeetus leucogaster)* is one of the Pacific species that Diamond (1974a,b, 1975) categorized as a D-tramp, meaning a specialist at colonizing small islands with small biotas, found only rarely as a component of more mature communities. Bristowe had found a number of insects associated with beach flotsam when the island was but six months old (see above), and the plantless island could therefore also have provided occasional food for insectivorous birds. Like the sea birds that were the first arrivals on Surtsey, these early animal colonists would have derived their sustenance from energy sources outside the island itself and acted as what Heatwole called "transfer organisms," conduits through which energy enters the island ecosystem.

Energy would also have reached the island by air. In 1985 we showed that a continuous rain of air-borne arthropods falls on to the island. We set up small seawater traps on stakes 1.5 m above the surface of ash-covered lava (to avoid catching any surface-dwelling animals) near the southwest coast in an area devoid of vegetation. The site was chosen to minimize the likelihood of trapping insects derived from the island's own vegetation; the prevailing wind was from the sea and the vegetated areas of the island were on the other side of the volcanic cone. In 10 days we trapped 219 individuals of scores of species of arthropods, including spiders, mirid and lygaeid bugs, aphids, a cicadellid, a psocopteran, a moth, weevils and nine other families of beetles, several families of flies, ants and other aculeate Hymenoptera, fig wasps, and parasitic Hymenoptera. We had also set up a similar series of traps resting on the surface of the lava in the same area, and our catch in these was quite different. It included a large proportion of small crickets of the genus *Speonemobius*, with ground-dwelling lycosid (wolf) spiders, ants, and earwigs. Pitfall traps set at the same time along the rim of the inner volcanic cone caught ants, earwigs, ground spiders, and again the cricket. The large *Hierodula* mantis, first seen on the archipelago in 1919, was also found to occur on the bare ash (Thornton et al. 1988).

In 1986 we followed up this pilot study by setting up an array of water traps and baited bottle traps on the surface of the southwestern lava, and a similar array close to the vegetation of the east foreland. Our catches on the barren lava confirmed that the dominant species of this unlikely habitat was the cricket, which, we discovered, was flightless and crepuscular. In contrast, the cricket was not trapped in the eastern area near vegetation. The guts of the crickets contained remains of the insect fallout which our earlier trapping had revealed (New and Thornton 1988). Some of this fallout was taken before it settled by aerial insectivores, two or three species of dragonflies, the Pacific swallow, edible-nest swiftlet, glossy swiftlet, little swift, and the migrant barn swallow. This guild of aerially feeding species also exploits the "hill-topping" insects often found at the edge of Anak Krakatau's outer rim and noted by Dammerman soon after the island's emergence. These aerial exploiters of fallout may in turn have been preyed upon, as they were later, by the small, fast-flying, oriental hobby, *Falco severus*. On reaching the surface of barren ash, the fallout was exploited by a guild of arthropods dominated by the cricket but also including the mantis, wolf spiders, and earwigs, as well as two migrant birds, the common sandpiper and grey wagtail.

Air-borne fallout can be a major source of energy in habitats where primary productivity is deficient, such as deserts, alpine snow fields, and oligotrophic lakes (Edwards 1987), and was shown to be an important nutrient source in the recolonization of Mount St. Helens following its 1980 eruption (Edwards and Sugg 1993).

Communities very similar to those on Anak Krakatau's barren substrates are found on laval areas of the volcanic Canary Islands in the east Atlantic, Mt. Vesuvius in Italy, and, particularly, the island of Hawaii, where Frank Howarth, at about the same time, had found a related cricket, also flightless and nocturnal (Howarth 1979, Thornton 1991; Figure 29). As on Anak Krakatau, the Hawaiian cricket was not found in adjacent areas with plant cover where, presumably, competitors and predators occur that do not inhabit the barren tracts. Communities specializing in the exploitation of aerial fallout of arthropods in barren areas are likely to have evolved in parts of the world with long histories of volcanic activity, such as Indonesia. The evolution of such habitat

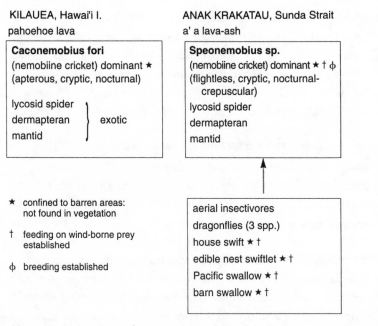

KILAUEA, Hawai'i I.
pahoehoe lava

Caconemobius fori
(nemobiine cricket) dominant ★
(apterous, cryptic, nocturnal)

lycosid spider ⎫
dermapteran ⎬ exotic
mantid ⎭

ANAK KRAKATAU, Sunda Strait
a' a lava-ash

Speonemobius sp.
(nemobiine cricket) dominant ★ † φ
(flightless, cryptic, nocturnal-
   crepuscular)

lycosid spider
dermapteran
mantid

★ confined to barren areas:
    not found in vegetation

† feeding on wind-borne prey
    established

φ breeding established

aerial insectivores
dragonflies (3 spp.)
house swift ★ †
edible nest swiftlet ★ †
Pacific swallow ★ †
barn swallow ★ †

Figure 29. Comparison of communities subsisting on airborne arthropod fallout in barren laval areas of the island of Hawaii and on ash-lava beds on Anak Krakatau. (Data from Howarth 1979; New and Thorton 1988; Zann et al. 1990.)

preferences has been called "adversity selection," although the term is rather inappropriate since it is clearly advantageous to the animals concerned. These communities are of unusual ecological interest for two reasons. First, they "short-circuit" the usual textbook energy route of plants-herbivores-carnivores, and second, they do so by directly exploiting an energy source coming from outside the system itself.

The monitor, or *biawak,* as it is known in Indonesia, was not seen on Anak Krakatau IV until 1982, when Dieter Plage was making the television documentary, *The Day That Shook the World* (from which I have taken the title of Chapter 2), but it was probably one of the first land vertebrates to become established. It was part of the island's short (?fourth) community in 1949 and has since been found on all the vegetated forelands as well as on ash beds and in canyons within the western lava flows. Semi-aquatic and a good swimmer, it is a carnivore and carrion feeder; for example, it digs up shore crabs as well as eggs of the green turtle, *Chelonia mydas,* a few individuals of which nest on Anak Krakatau's beaches. A monitor arriving on Anak Krakatau even at a very early stage of the island's biological development would thus have sustenance and would have acted as a transfer organism. Moreover, the species is capable of surviving considerable volcanic disturbance. After the 1930 eruptions had deposited heavy falls of ash on Sertung, Dammerman (1948, p. 25) found that on that island "where the denuded trees [around the lagoon on the northern spit] suggested a winter landscape . . . a monitor lizard walked along its banks."

Community development in the face of repeated volcanism

The first visitor to the island after the devastating 1952–1953 eruptions was the Pacific botanist Raymond Fosberg. On a brief visit in December 1958 he was unable to land because of continuous pulsating explosive eruptions, but through binoculars he saw a patch of tree-sized casuarinas at beach level in the "northeast," with a few other specimens scattered on the slopes above. He visited again in December 1963, after considerable further volcanic activity had waned, and this time he was able to land. Although much of the vegetation was dead, the small patch of casuarinas had survived, and he recorded some 25 species of plants in a grassy area around this patch, all well-known members of the *Terminalia* forest and *I. pes-caprae* associations (Fosberg 1985). Eighteen were part of the island's previous early floras, and six had been components of all of them.

Mildred Mathias and her colleagues were the next visiting biologists, in 1971, and again they were botanists. By now the vegetation had recovered considerably and there were four species of pteridophytes and 38 of spermatophytes (Fosberg 1985). Four of the 15 additional species recorded had been present in the pre-1953 floras. All the trees and shrubs then present (which included the island's first figs, *Ficus fulva* and *Ficus septica*) are dispersed by animals, at least over fairly short distances.

An eruption in 1972 was said to have severely affected the vegetation and reduced it in extent. In 1979, during an eruption in which the fallout drifted to the west, away from the island's vegetation, Barker and Richards made a vegetational and floristic survey of the east foreland. More wind-dispersed plants, including the tree *Neonauclea calycina*, three composites, two orchids, and a grass, were now present, and several sea-dispersed plants and the animal-dispersed *Antidesma ghaesembilla* had colonized (Barker and Richards 1982).

In August and December 1981 Tukirin Partomihardjo began what was to become his long-term study of the island's plants (see Partomihardjo, Mirmanto, and Whittaker 1992). On these visits four more sea-dispersed species were recorded, along with two grasses that probably were dispersed by humans. There were four more wind-dispersed colonists, including another grass and orchid, and important animal-dispersed species had colonized, including *Dysoxylum gaudichaudianum* and *Timonius compressicaulis,* the dominant canopy trees on Panjang and Sertung. A year later two more wind-dispersed, four sea-dispersed, and one animal-dispersed species had arrived.

Barker and Richards surveyed the island again in 1983, discovering the vine *Aristolochia tagala* (an important butterfly larval food plant, which is dispersed by sea and wind), four composites, *Carica papaya* (the well-known papaya, dispersed by the large-billed crow or by humans), two further species of *Ipomoea* (sea-dispersed), and *Cayratia trifolia* (also sea-dispersed). Two important new animal-dispersed species were *Pipturus argenteus* and *Macaranga tanarius.* The human-dispersed pigweed, *Mimosa pigra,* was found on this survey and is unknown elsewhere or in any other survey on the archipelago. Ninety species were recorded between 1979 and 1983, 46 sea-dispersed, 19 wind-dispersed, 13 animal-dispersed, and six by wind and by humans (Barker and Richards 1986).

Between 1989 and 1991, 138 plant species were recorded. The human-dispersed component increased considerably after 1983; nine such species were first recorded in 1989, including a palm of the genus *Phoenix,* three grasses, a composite, a *Capsicum,* and the tamarind, *Tamarindus indica.* A third fig species, *Ficus pubinervis,* was encountered, as was *Lantana camara,* which is bird-dispersed, probably by bulbuls. The first ferns since the 1953 eruptions had been found in 1971 (four species). By 1983 there were five, but in 1990 eight additional species were present. There were also five additional animal-dispersed spermatophytes, including three more figs, *Ficus hispida, F. fistulosa,* and *F. variegata,* and five more grasses, which were probably human-dispersed. By the following year a further grass species had arrived, the twelfth, and a species of *Mezoneuron,* a tree genus not yet recorded from the archipelago. In July 1992, in a lava lagoon on the northwest of the island, Partomihardjo found two species of mangrove—two individuals of *Lumnitzera racemosa,* which had been encountered as seedlings in 1991, and one of *Excoecaria agallocha.*

Initially, Anak Krakatau's vegetation was confined to the north and east forelands and consisted of grassland within which casuarinas grew. Photographic evidence of vegetational change on the island suggests that the eruptions of 1972 affected the north foreland only, or affected it to a much greater extent than the east foreland, allowing the latter to "overtake" it and putting the vegetation of the two "out of step" (Thornton and Walsh 1992). This successional asynchrony has persisted. Grassland, and then casuarina woodland,

developed on the northeast foreshore between the two forelands, and a continuous belt of vegetation around the northeast coast linked the lowland vegetated areas some time between 1986 and 1990 (Figure 25). By 1992 the east
foreland was in transition from casuarina woodland to mixed forest, with several animal-dispersed trees, notably the figs *Ficus hispida, F. pubinervis,* and
F. variegata, a *Glochidion* species and *Villebrunea rubescens,* and a number of
epiphytes. *Casuarina equisetifolia,* being intolerant of shade, is outcompeted
by early-successional plants such as *Ficus fulva, F. septica, Hernandia peltata,*
Hibiscus tiliaceus, and *Terminalia catappa* and, in contrast to these species,
does not thrive in rich organic soil. Casuarina woods therefore do not self-
regenerate. They can persist only where the substrate itself is constantly renewed and their colonists can extend to this new land, as was occurring on the
Sertung spit in the 1970s and 1980s and on Rakata's southern coasts in the
early decades after 1883. In 1992 the northeast foreshore of Anak Krakatau still
carried thickets of *Saccharum* and the casuarina woodland there had fewer
later-successional species. On the north foreland the change from grassland to
casuarina woodland did not occur until the early 1990s; fields of the grasses *Is-*
chaemum, Imperata, and *Saccharum* were still present in 1992, but the *Casua-*
rina cover was increasing rapidly.

By 1992 scattered clumps of *Saccharum* extended almost to the rim of the
outer cone, and occasional casuarinas also grew at this height and even within
the outer rim. The sharp demarcation between the 20–25 hectares of plant
cover on the coastal strip (of a total island area then of 235 hectares) and the
sparsely vegetated ash slopes is thought to result from the increasing depth of
the water table as the ground slopes away from the coastal plain (Partomihardjo, Mirmanto, and Whittaker 1992). On the ash-covered lava flows of the
south and west, the ferns *Pityrogramma calomelanos, Nephrolepis hirsutula,*
and *Nephrolepis biserrata* grew sporadically in shaded crevices and overhangs
of the *aa* lava. Species of both these fern genera are well-known pioneers of recent lava flows also in Hawaii and Japan. On Anak Krakatau's older flows there
were saplings of *Ficus fulva, Guettarda speciosa, Timonius compressicaulis,* and
Neonauclea calycina. Patches of the grass *Pogonatherum paniceum* occurred,
particularly on near-vertical surfaces, and on the southern lava apron, especially on older flows that had been covered by ash, occasional clumps of *Sac-*
charum grew. On the few ash-lava beaches in the south and west, the ground
creeper *Canavalia rosea* was colonizing in a few places, and in deep, shaded
crevices between lava ridges in the northwest there were occasional bushes of
Melastoma affine. A few isolated casuarinas grew on the lower lava fields.

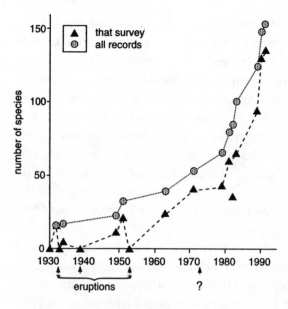

Figure 30. Increase in the number of vascular plant species on Anak Krakatau from its emergence in 1930 to 1991. The dotted line joins cumulative species totals from all records (circles, which are minimum estimates), and the dashed line joins totals for particular surveys (triangles). Arrows indicate known and presumed major eruptive events.

The increase in Anak Krakatau's floral richness is shown in Figure 30. At least three, and possibly four, earlier colonizing floras had been destroyed (in 1932–1933, 1939, and 1952–1953, and possibly one during World War II). Assuming a "clean slate" after the devastating eruptions of 1952–1953—an assumption about which serious doubts have recently been raised (Whittaker, Partomihardjo, and Riswan 1995)—45 species of vascular plants had become established on the island by 1971. This number rose to 90 by the time of the 1979–1983 surveys, and to about 140 by the 1989–1992 period. In 1990 about 25 hectares (0.25 km^2) were vegetated. Damage from volcanic activity since 1953 appears to have repeatedly set back the succession, the effect being greater in the north than in the east, but there is no evidence that any eruptive episode totally eradicated the vegetation. In 1988 the formation of two new large craters and a lava flow had little effect on the island's biota. The island's present biota may represent successful colonization since 1953 in the face of repeated constraints imposed by the volcano's activity.

Of the total of 37 plant species present in the presumed extirpated floras monitored in 1930–1934 and 1949–1951, 31 have since recolonized, 27 of which had been found in the earlier (1963 and 1971) surveys. Partomihardjo's team identified this group of 30 or more species as part of a deterministic core, largely comprising sea-dispersed plants of the strand line and coast and a few other wind-dispersed pioneers, present on all the islands. In surveys of Anak Krakatau's beaches during different monsoon seasons in 1991 and 1992, stranded propagules of 66 plant species were found; those of 36 species were thought to have originated outside the archipelago, mostly from the north. At this time the island's flora of about 140 species included 58 of the total of 63 sea-dispersed plant species growing on the archipelago, showing that the young island had already received almost the complete complement of this core component of the flora (Partomihardjo et al. 1993).

In contrast to the sea-dispersed component, the island's complements of wind-dispersed and animal-dispersed species in 1992 were only about a third and a quarter the size, respectively, of those of Rakata. Although still a minority of Anak Krakatau's 1992 flora, the animal-dispersed component has great importance because most of the flowering plant species, including trees, of the archipelago's interior forest communities are animal-dispersed. This component, which included a few important canopy trees such as *Canarium hirsutum, Dysoxylum gaudichaudianum* and *Timonius compressicaulis,* increased very rapidly on Anak Krakatau in the decade prior to 1992. The recent large increase in the human-dispersed component, which by 1992 was larger, proportionally, than on the other islands, is almost certainly because of the increased frequency of tourist visits to the island's then quiescent volcano.

The vegetation of Anak Krakatau in 1992 was therefore largely the result of a classic interrupted succession. Its flora was nevertheless almost as rich as that of Panjang, an island more than twice its age and with a vegetated area fifteen times more extensive.

Post-vegetation colonization by animals

Hoogerwerf had recorded the fish-eagle *(Haliaeetus leucogaster),* yellow-vented bulbul *(Pycnonotus goiavier),* and savannah nightjar *(Caprimulgus affinis)* in 1952. A hooded pitta, *Pitta sordida,* of the Sumatran subspecies, was found dead on the island in June 1955 (Mees 1986), and in 1976 Ben King recorded the migrant barn swallow, the golden-bellied gerygone, *Gerygone sulphurea,* and the peregrine falcon, *Falco peregrinus.* Early photographs and plant surveys show that the island's first vegetation was grassland. An *Ipomoea-Ischaemum* community developed on accreting beaches, *Imperata* fields *(alang alang)* a little further inland, and *Saccharum* clumps, later to become extensive patches, in the coastal area. The orthopterans, moths, skippers, beetles, and other insects associated with this grassland would have provided sustenance for the savannah nightjar, a nocturnal aerial insectivore. This ground nester could have become established before trees were present.

From knowledge of the birds associated with the north foreland's grassland areas in the 1980s, Zann and Darjono (1992) were able to make intelligent guesses as to other birds likely to have become established on the east foreland in the early 1970s, when it was largely covered by grasses. The lesser coucal, *Centropus bengalensis* (a large cuckoo), and white-breasted waterhen, *Amaurornis phoenicurus*, are normally associated with this habitat. Both were found in 1982 (Ibkar-Kramadibrata et al. 1986) and the coucal was still to be heard in the *alang alang* field on the north foreland in 1992, the waterhen inhabiting *Ischaemum* grassland on the east foreland. The collared kingfisher *(Todirhamphus chloris)* was characterized by Diamond (1975) as a Pacific island "supertramp," a species that is an extreme specialist at colonizing small islands and that is never a component of large, mature avifaunas. The kingfisher now nests within arboreal termite nests on Anak Krakatau, but it can nest in cliffs and gully sides and, like the nightjar, would not have needed trees in order to become established. Moreover, it is a generalist feeder taking a variety of foods including shore crabs, which would have been available from an early stage in the island's development.

The grassland was initially invaded by the pioneer tree *Casuarina equisetifolia,* which for some time was a component of an open parkland savannah, as represented on the north foreland in the late 1980s, the northeast foreshore in the early 1980s, and the east foreland in the 1970s. This vegetational stage would have provided habitat for the yellow-vented bulbul, a feeder on fruits and insects which prefers fairly open country and, with the nightjar, was part of the island's third biota in 1952. This bird probes and gleans insects from branches between 2 and 18 m above the ground in a variety of vegetation types. From the crops of 14 birds, Zann, Male, and Darjono (1990) flushed 34 food items, including 3 grass seeds, 11 seeds of *Cassytha filiformis,* a flower, and remains of beetles, wasps, crickets, and dragonflies. The large-billed crow, *Corvus macrorhynchos,* seen in 1983 by M. B. Bush and D. Newsome, is another generalist feeder (also a nest robber) that is an inhabitant of open country. Like the bulbul, its preferred habitat would have been declining on the other islands as the forests developed.

The change from savannah woodland to casuarina forest is rapid once it begins. Birds of casuarina forest include the white-breasted wood-swallow, *Artamus leucorhynchus,* found in 1982, which takes insects on the wing from a foraging perch on the top of a tall tree, and the magpie robin, *Copsychus saularis* (1982), mangrove whistler, *Pachycephala grisola* (1984), and golden-bellied gerygone (1984), which are casuarina specialists. In 1982 the barn owl, *Tyto alba,* and edible-nest swiftlet, *Collocalia fuciphagus,* were also present (Ibkar-Kramadibrata et al. 1986), and in 1983 Bush and Newsome saw a single migrant leaf-warbler (*Phylloscopus* species), which has not been recorded since. Additional species to arrive in 1984 were the little swift, *Apus affinis,* and Pacific swallow, *Hirundo tahitica* (Zann, Male, and Darjono 1990). Zann and Darjono (1992) believe that as the density of trees increased a second wave of bird colonists would have arrived, including the black-naped oriole, *Oriolus chinensis* (1982), mangrove blue flycatcher, *Cyornis rufigastra* (1982), plain-throated and olive-backed sunbirds, *Anthreptes malacensis* and *Nectarinia jugularis* (1984), emerald dove, *Chalcophaps indica* (1986), and pied triller, *Lalage nigra* (1989).

A dramatic increase in the number of bird species coincided with the fruiting of the island's first figs in 1985 (Figure 31). In the same year, although the crow had gone, the red cuckoo-dove *(Macropygia phasianella),* which feeds on fruit and seeds, and the Asian glossy starling *(Aplonis panayensis),* were first recorded. The starling is a known disperser of the early-successional tree *Macaranga tanarius,* which had been recorded in 1982. In 1986 there were four new frugivores: the pink-necked green pigeon, *Treron vernans,* a fig specalist; the chestnut-capped thrush, *Zoothera interpres,* which takes fruit, berries, and insects; the omnivorous house crow, *Corvus splendens;* and the emerald dove, a facultative frugivore that feeds on fruit, seeds, and insects. Two carnivores had arrived, the tiger shrike, *Lanius tigrinus,* a migrant insectivore-carnivore, and the oriental hobby, *Falco severus* (Zann, Male, and Darjono 1990). Only single individuals of the house crow, thrush, and shrike were seen, and these species have not been recorded since. The pied triller, glossy swiftlet *(Collocalia esculenta),* and crested serpent-eagle *(Spilornis cheela)* were recorded in 1989 (Haag and Bush 1990), but the serpent-eagle has not been recorded since. The hobby was replaced by the peregrine falcon in 1989. In 1991 a male black-naped fruit-dove, *Ptilinopus melanospila,* an obligatory fruit eater and efficient seed disperser, was feeding on fruits of *Premna serratifolia* on the east foreland, and the black eagle, *Ictinaetus malayensis,* was seen over the island. In July 1992 there was a pair of fruit-doves, and the koel, which feeds on fruit and insects, was first recorded, on the east foreland.

On the same expedition two grassland, seed-eating weaver finch species

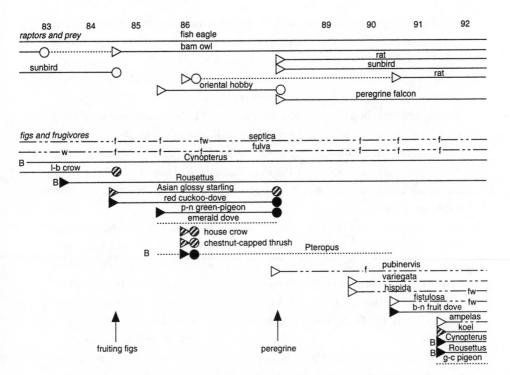

Figure 31. The incidence of fig, frugivore, avian raptor, and prey species on Anak Krakatau from 1983 to 1992. The black-naped oriole and yellow-vented bulbul, both facultative frugivores present throughout the period, are omitted for clarity. Fig species *(dot-dash lines)* are indicated by specific name only; bird species and rats *(solid lines)* by common name; bats *(B)* by generic name. Dotted lines indicate possible presence; triangles, first records; circles, last records. Solid symbols indicate obligate frugivores; hatched symbols, facultative frugivores. *Note:* b-n, black-naped; f, fruiting; g-c, grey-cheeked; 1-b, large-billed; p-n, pink-necked; w, pollinating wasps present. Common names, as in the text, follow MacKinnon and Phillipps (1994).

(munias) were encountered, the first to be detected on the archipelago. Van Balen saw and heard a small flock of birds whose calls identified them as scaly-breasted munias, *Lonchura punctulata*. Two days later the Javan munia, *Lonchura leucogastroides*, was mist-netted in the same area. We do not yet know whether these, the first grass-seed specialists to be discovered on the island, have successfully colonized, but their arrival has coincided with a rather sudden increase in the number of species of (largely introduced) grasses (Thornton, Zann, and van Balen 1993).

The only reptiles on the island in 1992 were the monitor, the small house gecko, or *chechak*, and the paradise tree snake. The *Mabuya* skink, noted in 1949 by Hoogerwerf, has never been found on the island since. The snake and gecko both probably became established on Anak Krakatau as casuarina woodland developed. Like the monitor, the gecko is capable of withstanding considerable volcanic disturbance. Dammerman (1948) found one on Sertung under the bark of a casuarina killed by Anak Krakatau's August 1930 ash fall, and he heard its call on Panjang after a heavy fall of volcanic ash in January 1933. In August 1985 we found it living among rocks on the rim of Anak Krakatau's active inner cone. The paradise tree snake feeds mainly on geckos and small birds.

Of the 19 species of land molluscs found on the Krakataus in 1984, only one was present on Anak Krakatau. This is an inhabitant of ground litter, and Smith and Djajasasmita (1988) suggested that two other litter dwellers will be the next land molluscs to colonize the island. Similarly, of the 18 types of plant gall present on the archipelago in 1982, only one occurred on Anak Krakatau (Yukawa et al. 1984). In contrast, Tim New and I showed that in 1984 about a third of the archipelago's insect fauna was represented on Anak Krakatau's limited vegetated area, and the island carried a good proportion of the archipelago's complement of soil nematodes, woodlice, land crabs, pseudoscorpions, collembolans, dragonflies, cockroaches, termites, earwigs, orthopterans, barklice, bugs, thrips, lacewings, flies, butterflies, ants, other aculeate Hymenoptera, and braconid and chalcid hymenopterans (Thornton and New 1988b).

No koinobiont braconid parasitoids of Diptera (koinobionts are internal predators that do not kill their host insect until they emerge from it) were collected on Anak Krakatau in the 1984–1986 surveys, although 22 species were collected on the other islands. Their absence may be due to a paucity of their principal hosts, leafminers and flies with larvae that are associated with fungi or fruit, on the young island (Maeto and Thornton 1993). This explanation is supported by Yukawa's finding there, in 1982, of only one of the archipelago's six species of tephritid fruit flies (Yukawa 1984b). The situation had already changed somewhat by 1990, however, when we collected nine fruit-fly species, four on Anak Krakatau (Schmidt, Thornton, and Hancock 1994), and when more recent braconid collections have been worked up it will be interesting to see if this island's braconid spectrum has also changed in favor of parasitoids of Diptera as the vegetational change has accelerated. Lepidoptera, however, particularly butterflies, are relatively well represented on Anak Krakatau, the archipelago's butterfly fauna being heavily biased toward species feeding on plants of coastal and near-coastal habitats, which of course predominate on Anak Krakatau. Not surprisingly, koinobiont braconid parasitoids of Lepidoptera are well represented on the island.

Butterfly diversity on Anak Krakatau, which has been well monitored, has risen rapidly over the past decade. The archipelago's butterfly fauna consists largely of early-successional species which, once established, may rapidly colonize other islands, and Anak Krakatau offered prime early-successional habitat. For example, the eight species of lycaenids that have colonized Anak Krakatau are all associated with strand-line vegetation, although two butterflies typical of secondary forest, *Troides helena* and *Euploea modesta,* invaded as suitable habitat began to form and they quickly became established (New and Thornton 1992b).

Seventy-seven genera of soil nematodes were present on the Krakataus in 1985. Anak Krakatau appeared to lack four orders found on the other islands, and the 19 genera that had colonized Anak Krakatau were, in general, widely distributed on the rest of the archipelago. The group best represented on the island was one comprising bacterial feeders and saprophages that are regarded as having pioneer characteristics. The island is also surprisingly rich in aculeate Hymenoptera, and the colonization of braconid Hymenoptera, Neuroptera, and land crabs also appears to have been faster than was the colonization of Rakata by these groups after 1883, when, of course, potential source faunas were much more distant. Anak Krakatau's fauna, like its flora, was about the same size as Panjang's in 1984. Although Anak Krakatau has received more than twice the survey coverage, as measured in zoologist-days, better surveillance cannot be the only explanation for the difference. Although the vegetated area covered only about 25 hectares in 1992, the island has 235 hectares of surface on which airborne arthropods may make a landfall and some 10 km of shoreline along which sea-dispersed animals may move after arrival until they reach habitable areas. It is thus possible that the whole island acts as a "catchment" for land arthropods and that those that survive the initial landfall move to and congregate in the vegetated area. In support of this theory, the density of insectivorous birds in the vegetated area of Anak Krakatau is surprisingly high (117 pairs per 10 hectares), more than twice that predicted by comparison with Australian woodlands (Zann and Darjono 1992).

Researchers from the Institute of Technology in Bandung (Ibkar-Kramadibrata et al. 1986) had heard the screech of the barn owl, *Tyto alba* (Figure 32), on the island in 1982 but it was not until 1986 that we first observed a pair of owls, hunting on the east foreland. The owl hunts in fairly open country, mainly for rats, although we found that it also takes bats. No rats had been recorded on the island, and intensive trapping for them during our 1985 visit had met with no success. Shortly before we left the island in that year, however, I found an immature male of the house rat, *Rattus rattus,* in the bucket of hand-washing sea water of our field toilet. We believed that this individual probably did not represent an established population. In spite of the lack of rats on the island, however, in both 1985 and 1986 we had found a large number of owl pellets, almost all containing rat remains, in an area adjacent to the southern lava flow that we called "Bone Alley." An owl pellet, about the size of

Figure 32. The barn owl was first seen in 1982 on Rakata.

a bantam egg, is a mass of bones, hair, feathers, and other indigestible parts of an owl's prey, which the owl regurgitates some eight hours or so after feeding, when at its roost. Evidently the owl was foraging and feeding on rats on the other islands and using Anak Krakatau as a conveniently central base.

In 1990, as soon as we stepped ashore on to the black sand of Anak Krakatau, it was obvious that rats had now colonized; their footprints were everywhere. Our trapping program established that populations of both the light-bellied country rat, *Rattus tiomanicus,* and the house rat had become established since our 1986 visit. The result of this influx of the owl's preferred prey was soon evident in the predator population. Shortly before his tragic death on the western lava in 1991, Peter Rawlinson discovered three owl nests, two with eggs and young (Figure 33), in crevices and overhangs of the southern and northwestern lava fields, clearly confirming the establishment of a thriving resident population of owls. By this time the screech calls were heard regularly every night, and hunting owls were seen quite frequently. Foraging radius depends on prey availability. On the Malay peninsula, two hectares may support a pair of barn owls (Anak Krakatau's area is about 235 hectares, only about 25 hectares of this vegetated), and recent estimates of foraging radius in Europe have been from 1.5 to 3 km (from 1 to 2 miles), which on the

Figure 33. In 1991 at least two pairs of owls were breeding on Anak Krakatau. This nest, formed mainly of rat bones, has seven eggs and two hatchlings.

Krakataus would mean that all islands would be within range of a home base on Anak Krakatau.

Seven of the archipelago's 20 species of bats have been recorded on Anak Krakatau. Lesser dog-faced fruit bats, *Cynopterus sphinx,* were first seen on the island in 1982 and in 1984 were quite common, over 40 individuals being captured in mist-nets in one night (Tidemann et al. 1990). In 1986 a large Malay flying fox, *Pteropus vampyrus,* which in some years congregates on Sertung, paid a short visit, and in the same year a third fruit bat species, the rousette *Rousettus amplexicaudatus,* was found. The first insectivorous bat was recorded when a male of the tomb bat, *Taphozous longimanus,* was captured in 1990 by chance, in a furled-up bird mist-net. Craig Wilson, a student examining owl pellets collected in that year, found the bones of two individuals of the nectarivorous bat *Macroglossus minimus,* although of course the animals may not have been taken on Anak Krakatau. On our visit two years later we found two more fruit bat species, *Rousettus leschenaultii* (one individual) and the large short-nosed fruit bat, *Cynopterus brachyotis.* There is no evidence, however, that any but the two *Cynopterus* species have established populations on the island, although clearly other bat species now do occasionally arrive and stay for at least a short time.

Anak Krakatau as a Successional Mosaic

As noted above, by analyzing photographs taken over more than two decades, Walsh and I showed that the "ecological landscape" of Anak Krakatau has been a dynamic patchwork mosaic (Thornton and Walsh 1992). The three forelands have carried different stages of the vegetational succession simultaneously, the north foreland "lagging" behind the east foreland by about 12 years. These forelands were formed at about the same time; if anything, the north foreland is slightly the older. Ash fall, and perhaps noxious fumes, in the severe eruptions of 1972–1973 may have affected particularly the north foreland's vegetation and imposed a successional asynchrony that persisted during subsequent decades. In other words, different areas on the island have exhibited different stages of succession, from grassland through casuarina woodland to the beginning of mixed forest.

Zann and Darjono have shown that the avifaunas of the forelands are also developing with a time delay between them of from 10 to 15 years and that, like the floras, they are nested subsets. Although the number of bird species on the island increased between 1984 and 1990, by 1991 the avifauna of neither the north foreland (7 species) nor the northeast foreshore (14) had reached the total of 17 species found on the east foreland in 1984. The east foreland, with the most advanced vegetational succession, was believed to have acted as a "bridgehead" for colonization of the island by birds, species gradually moving north to the two other vegetated areas as these habitats became optimal for particular species. It was not until the July 1992 survey that the rapidly changing avifauna of the north foreland "caught up" to that of the neighboring northeast foreshore (about 12 species each), but each of these areas supported only rather more than half the east foreland's fauna in that year, which seemed to have stabilized at about 22 species.

Somewhat surprisingly, in view of the fact that by 1990 the three forelands were connected by a strip of almost continuous vegetation, movement of birds between them was rather restricted. Of a total of 57 mangrove whistlers banded on either the east foreland or northeast foreshore in 1990, only one was found to have moved after 8 days, from the east foreland to the northeast foreshore about one kilometer away, although casuarina woodland provided continuous cover between the two areas. One individual of each of the collared kingfisher, yellow-vented bulbul, and mangrove whistler was recaptured in 1990 (kingfisher and bulbul) and 1991 (whistler) at the same site at which it had been banded in 1986. Also, two mangrove whistlers and a mangrove blue flycatcher that had been banded and released on the east foreland in August 1990 were recaptured there in July 1992. A magpie robin, however, banded and released on the northeast foreshore in August 1990, was mist-netted in July 1992 on the east foreland, indicating that occasional movement between forelands, even of quite territorial species like the magpie robin, does take place (Zann and Darjono 1992).

In 1990 and 1991 Bryan Turner studied the invertebrate communities feeding on microepiphytes (mainly algae and fungi) growing on casuarina twigs and branches. He found a pattern rather similar to that of the birds. Both species richness and number of individuals tended to increase from north to south over the three forelands. This distribution related fairly well to the age (time since establishment) of the trees but, more importantly, appeared to be linked to their surrounding floral diversity (Dr. B. D. Turner, personal communication). This pattern of nested subsets on the three forelands was also evident in spiders, tropical fruit flies (tephritids), and butterflies (Dr. W. Nentwig, personal communication; Schmidt, Thornton, and Hancock 1994; New and Thornton 1992b).

We were lucky to be studying the island during a decade when its woodlands were beginning to diversify, and we were able to follow the island's colonization by figs in some detail.

Raptors and the Diversification of Woodland

As discussed in an earlier chapter, fig trees are of great importance in the transition to mixed forest, and their incorporation into the east foreland's casuarina woodland in the last decade has been monitored by Partomihardjo. At the same time, zoologists have followed the incidence of pollinating fig wasps and the development of an association of insects within the fig syconia, as well as the colonization of the island by frugivorous birds, fruit bats, and avian raptors.

Two dioecious fig species, *Ficus septica* and *F. fulva*, both small trees, were first recorded on Anak Krakatau in 1971. Both are known to be dispersed by birds and bats. Bats *(Cynopterus sphinx)* were common by 1984, when they were seen regurgitating and defecating fig seeds and carrying whole figs in their claws, although the island's figs were immature in 1979 and not fruiting in 1984. Clearly *F. septica* and *F. fulva* could have colonized the island through bats foraging for figs on the other islands and using Anak Krakatau as a feeding roost. Two generalist frugivores, the black-naped oriole and yellow-vented bulbul, were already well established, and a pair of omnivorous large-billed crows, as well as stray temporary visitors from the other islands (see below), could also have been seed vectors. In 1984, although the island's fig trees were not fruiting, by sweeping vegetation we netted four species of agaonine pollinating fig wasps, including the pollinator of *F. septica* (*Ceratosolen bisulcatus*) and three pollinators (*Blastophaga, Liporrhopalum,* and *Platyscapa* species) of figs not yet present on the island. Fig wasps were arriving, presumably from the other Krakatau islands, which by that time supported fig wasp communities, comprising both pollinating and non-pollinating wasps, comparable in diversity to those of the adjacent mainlands (Compton et al. 1988). These airborne pollinators would have been available to pollinate any appropriate *Ficus* species that were fruiting (none were). In 1985 a sycoecine fig wasp (a species of *Diaziella*), a non-pollinating seed-galler, was caught in water traps set up on Anak Krakatau's western, windward ash-lava beds far from vegetation, indicating that other fig wasps, including pollinators, also may have been carried to the island in the air.

The first recorded fruiting of both fig species, between 1984 and 1985, almost certainly triggered an acceleration of the diversification of the island's woodland, by providing a niche for more specialized fruit eaters, including efficient seed dispersers. Both facultative and obligate frugivores soon arrived, and further fig species followed. In 1985 the red cuckoo-dove and the Asian glossy starling were first recorded, and more fig wasp pollinators of absent figs, mostly a species of *Waterstoniella*, and the non-pollinator *Diaziella macroptera* were collected.

The successful pollination of both pioneer fig species was confirmed in 1986. The pollinating wasps of *F. fulva (Blastophaga inopinata)*, *F. septica (C. bisulcatus)*, and various non-pollinating fig wasp species (parasites and seed predators) were found in figs. Well-developed fig wasp communities had become established. On the same expedition the cuckoo-doves were seen with an immature, and two further frugivores, the pink-necked green pigeon (a pair and an immature) and Geoffroy's rousette, *R. amplexicaudatus*, were recorded. Many fig seeds had been found in the guano of this rousette in its roosting caves on Panjang a year earlier. Strays of two omnivorous birds, the chestnut-capped thrush and the house crow, and of the Malay flying fox were also seen. We thought that the diversification of the east foreland's woodland by the addition of further fruiting trees would now accelerate through positive feedback, and between 1989 and 1992 other animal-dispersed forest trees did indeed colonize: *Arthrophyllum javanicum*, *Villebrunea rubescens*, a *Glochidion* species, and *Canarium hirsutum*. The fig flora also increased: in 1989 *Ficus pubinervis* was present; in 1990 two more, *Ficus hispida* and *Ficus variegata*; in 1991 *Ficus fistulosa;* and in 1992 a seventh species, *Ficus ampelas.*

This further diversification of the woodland was accompanied by the arrival of additional fruit-eating birds and bats. An obligatory frugivore and seed disperser, the black-naped fruit-dove, arrived between our 1990 and 1991 visits, two more fruit bats *(C. brachyotis* and *R. leschenaultii)* were present, and two more fig species *(F. hispida* and *F. fistulosa)* were fruiting. In the seven years since figs were first recorded as fruiting on the island, twelve frugivorous vertebrates had appeared, only three or four of which were believed to be stragglers, and five more fig species had colonized.

We knew that the presence of raptors may impede colonization by seed dispersers. The owl pellets that we had been finding in the period 1984–1986 in Bone Alley and elsewhere on the outer cone testified to this. At the end of each expedition we cleared Bone Alley of pellets and skeletal remains so that we could monitor the appearance of new prey remains in the ensuing interval. Barn owls are opportunists, exploiting whatever prey is at hand, and will take birds (and bats) when their preferred prey, rodents, is less readily available. In 1984 we found remains of two birds (in owl pellets) and three *Cynopterus* bats and in 1985 four birds (two columbids in owl pellets and two waterhens), a *Cynopterus,* and a rousette (Thornton 1994).

The new niche opened up for frugivores by the fruiting of figs in 1984–1985 was exploited indirectly by the avian raptors. In 1986 we found as many pigeons taken by predators (at that time resident or semi-permanent predators comprised only the fish-eagle and owl) as we recorded living pigeons. We found remains of thirteen individuals, which Craig Wilson identified as a rousette, two flying foxes, four waterhens, one emerald dove, three cuckoo-doves, and two pink-necked green pigeons, mostly with long bones broken and presumably taken by the fish-eagle. Following the arrival of the country rat and house rat by 1990 and 1991, respectively, preferred prey was now readily available and owl numbers increased from one to three pairs, two of them breeding, by 1991. Pressure on pigeons and fruit bats, however, was maintained by a colonizing event that occurred in 1989.

A pair of oriental hobbies *(Falco severus)* had become established between our 1985 and 1986 visits. This small falcon takes its prey in flight, even small birds and large insects as fast and maneuverable as swallows, swiftlets, and dragonflies. The population of one of the smallest birds on the Krakataus, the plain-throated sunbird, fell to one pair in 1984, and in 1985 not one was recorded. Its demise coincided with the hobby's colonization, but it was thought that the hobbies would become established since the remaining population of small birds and other flying prey was sufficient to support a pair. Between August and September 1989, however, the hobby was supplanted by the peregrine falcon *(Falco peregrinus)* (Figure 34), also an aerial raptor but one that specializes on larger prey, particularly pigeons (hence its occasional establishment in cities). In the absence of the hobby, the sunbird recolonized the island. In 1990 there were 11 pairs, and two years later 45 pairs.

Peregrines often occupied the highest point of the inner crater, a good vantage point over a relatively clear foraging area, although hot and usually enveloped in fumes of sulfur dioxide. The peregrine pefers pigeon-sized prey but, like the owl, choice of prey is almost certainly influenced by its availability. Smaller prey is exploited when need arises; a peregrine nesting on a high building in Melbourne was seen in 1992 to go after sparrows in heavy traffic. One morning in 1992 on Anak Krakatau Seamus Ward and Michael Jaeger witnessed a peregrine strike and kill a plain-throated sunbird, which fell dead almost at their feet. In 1992 the plain-throated sunbird was fairly common on Anak Krakatau, pigeons extremely rare, and in these circumstances the peregrine will take even very small birds.

Figure 34. The peregrine falcon supplanted the oriental hobby on Anak Krakatau in September 1989.

Some of the frugivorous pigeons and doves bred on the island but in 1990, three years after our previous visit, the remains of 50 frugivores were found. Craig Wilson found them to be 12 birds (8 of them columbids) and 37 bats (23 *Cynopterus*, 7 rousettes, 2 *Macroglossus*, and 5 flying foxes). In 1991 the tally was 24 frugivores (4 birds, 16 *Cynopterus*, 3 rousettes, and a flying fox) and in 1992, 5 (4 birds, of which 2 were columbids, and a *Cynopterus*). As Diamond (1974b) found on Ritter Island in the Bismarks, which was being recolonized following its 1888 explosive eruption, the peregrine appeared to have checked the establishment of immigrant birds. It probably deterred or aborted the exploratory visits of flying frugivores arriving to exploit the figs, its impact on them being greater than the hobby, fish-eagle, or owl. No pink-necked green pigeon was seen or heard in the 1990s, and only remains of emerald doves were found, in 1986 and 1990. One cuckoo-dove was seen in 1990, but by 1991 only the fruit-dove could be found. The fruit-dove was still present 16 months later. It is rather more of a deep forest bird than the other pigeons and doves, and its preference for deeper cover may have enabled it to escape the attentions of the peregrine. It is possible that, like the jaguar and puma in South America (Wilson 1992), the peregrine, also a top predator, was indirectly affecting the diversification of the island's developing young forest community through predation on the dispersers of important canopy trees, although it is too early to claim keystone status for the peregrine in this case.

A find in Bone Alley in July 1992 demonstrated that the falcon's impact on plant dispersal may not be entirely negative. Remains of a green imperial pigeon, a fairly large pigeon first found on the other islands in 1983 but not recorded from Anak Krakatau, were found with the skull smashed, indicating a raptor kill, probably the peregrine. About a score of marble-sized, spherical seeds of the palm *Oncosperma tigillarium* were found within the skeleton. This palm occurs on the other Krakatau islands but not on Anak Krakatau. Thus the seeds had been eaten on another island, and either the pigeon had been captured there and carried to Anak Krakatau to be consumed or the post-prandial pigeon had ventured to Anak Krakatau, where it fell foul of the peregrine. We believe the latter is the more likely. Although the peregrine's range was known to include the other islands, the more open island Anak Krakatau was clearly its headquarters. In either case, seeds of a plant not occurring on Anak Krakatau were deposited on its surface through the combined agencies of the pigeon and the peregrine. Imperial pigeons do not have grit in their thin-walled gizzards, and the seeds would probably have been viable when they reached Anak Krakatau. When we found them they were not, but had they been washed to a more sheltered situation or better soil by rain runoff some may have germinated. The falcon may of course also act as a secondary vehicle for seed dispersal through a "seed predator" pigeon, such as a pink-necked green pigeon, if the prey is deposited on the island before the seeds have passed to its gizzard. Moreover, the falcon may consume seeds that are within its prey, thus acting as a dispersal aid through a rather more extended pathway. In Zimbabwe, Hall (1987) germinated 15 plants, including two species of fig, from seeds obtained from regurgitated pellets of the Lanner falcon, *Falco biarmicus*. The falcon's prey included a pigeon, a dove, and a *Pycnonotus* bulbul.

The Frequency of Animal Arrivals

There is no direct way of estimating the frequency of visits to Anak Krakatau by individuals of those species that are already present. Such animals, of course, would not be identifiable as immigrants, since they would not differ in any way from residents. Individuals of species that do not occur on the island, however, are immediately recognizable as immigrants.

The discovery of a dead hooded pitta in 1955 was mentioned above. More recently, a Danish bird observer, Torben Lund, reported seeing the oriental white-eye, *Zosterops palpebrosa,* on the east foreland in August 1990, the only record of a white-eye on the archipelago. This inhabitant of open scrub occurs on Sebesi but did not persist on the Krakataus. In 1983 the pied triller, serpent-eagle, and a single leaf-warbler were first seen on the island. Sunda Strait is included in the range of two migrant species of leaf-warblers, neither of which has otherwise been recorded from the Krakataus.

We have seen other individuals that, like the pitta and leaf warbler, were likely to be lone "stragglers." In 1986 a house crow, *Corvus splendens,* appeared at our camp site and for two days frequented the camp kitchen, picking up scraps, apparently unaffected by human activity. The bird was also seen to eat a fig. This species is a human commensal in the east Asian region and frequently travels with or on ships and boats. It appears to be extending its range eastward through its association with marine commerce and competing, often successfully, with indigenous residents. It has not been recorded otherwise on the Krakataus and this individual was almost certainly a lone straggler; it may have come from a passing ship, as Sunda Strait is one of the world's busiest waterways. A tiger shrike was caught in a mist-net in the same year. Like the house crow, this migrant to the Indonesian area has not been recorded on the Krakataus either before or since. The chestnut-capped thrush, a bird of dense undergrowth, had been discovered on Rakata in 1984 and was found on Sertung in 1992. In 1986 a single individual was seen walking through the casuarina woodland of Anak Krakatau's east foreland; the species has not been recorded on the island since. In that year we also witnessed the arrival on the island, in daylight, of a Malay flying fox, which stayed on the island for a day or so. It was probably a straggler from the "camp" of several hundred that was on Sertung at that time. Very large fruit bats, probably of this species, have been disturbed near the summit of Rakata on two occasions in the 1980s, and clearly bats often move around between islands. Our lucky discovery in 1990 of the tomb bat, *Taphozous longimanus,* was mentioned earlier. In the following year, after some practice on the mainland of Java, I used an ultrasound detector on the island each evening, but I heard no calls of insectivorous bats. In 1992 we monitored ultrasound again, and again could detect no calls. The bat found in 1990 may have been a stray, but further investigation of the island by a bat specialist is now needed. In July 1992 a mature male of the rousette *Rousettus leschenaultii*, the first record of this species on the archipelago, flew to fruit bait on the east foreland.

Since 1984 we have noted a number of arrivals of "new" species to the island, some of which resulted in successful colonizations while others, although apparently not mere stragglers, arrived too recently for their status to be decided. For example, the distinctive calls of the Asian glossy starling were heard on the northeast foreshore and the east foreland on our 1985 and 1986 visits but the species has not been recorded on the island since. The fact that birds were present in two successive years indicates that they (a pair was present in 1986) were not strays but unsuccessful colonists that had, perhaps, arrived before sufficient fig trees were in fruit to sustain a breeding population. The barn owl, heard on the island in 1982, was evidently absent in 1984 but returned between 1985 and 1986. The oriental hobby had been seen flying and hovering around Rakata's summit by Alain Compost in 1982; a pair was present on Anak Krakatau in 1986 and still present in 1989. In that year the hobby was supplanted by the peregrine falcon, and the falcon was still present in 1992.

As noted earlier, the plain-throated sunbird probably became extinct on the island in 1984–1985 but recolonized some time between 1986 and 1990. A pair and an immature of the pink-necked green pigeon were seen together regularly on the island in 1986, and two skeletons were found in Bone Alley. A pair and an immature of the red cuckoo-dove were also seen several times in 1986. We have never observed the emerald dove or the green imperial pigeon on the island, only their remains after apparently having been killed by raptors. The arrival between 1990 and 1991 of the black-naped fruit-dove and its continued presence in 1992 probably indicates a colonization, but the status of the species, as indeed of many species on Anak Krakatau following the recent series of eruptions, is doubtful. The discovery of two species of *Rattus* on the island in 1990 was mentioned earlier, and evidence of the presence of an otter in 1990 and 1991 in an earlier chapter. Finally, the fruit bat *Cynopterus brachyotis*, two species of munias, and the koel were first recorded on the island in July 1992, and their present status is unknown.

No doubt there were many colonizations of the island by invertebrates during the past decade, most of which were unrecorded. In other cases there is good evidence of arrival. Two examples are the large black-and-yellow papilionid butterfly, *Troides helena,* between 1986 and 1989 and the ant lion, *Myrmeleon frontalis,* between 1986 and 1990. Both are conspicuous as adults, being quite large insects, the large swallowtail being so strikingly colored that even I, a color-blind entomologist, could recognize it, and the ant lion's presence is clearly signaled by the conical pits made by the larvae in ash or sand. Neither species was seen in 1984, 1985, or 1986. Comparison of the butterfly surveys of New and his colleagues with those of the Bush and Yukawa groups has also shown that four skippers (Hesperiidae) of grassland *(Polytremis lubricans, Telicota augias, Hasora taminatus,* and *Parnara guttatus)* arrived on the island in 1989 and another, *Pelopedas conjunctus,* in 1990, and there are many other examples. In fact, if we take the conservative view that butterfly species first recorded on the island by Yukawa in 1982 (Yukawa 1984c) and Bush in 1983 (Bush 1986) may all have arrived well before those dates, there are no fewer than 27 species discovered on the island since 1983 (New and Thornton 1992b). Eleven of these have not been found in surveys conducted since they were first seen and are presumed to have been stragglers.

Thus over a period of eight years we have records of some 54 species of animals, 25 of them vertebrates, arriving on Anak Krakatau. Fourteen of these (the white-eye, two munias, leaf-warbler, house crow, tiger shrike, serpent-eagle, otter, rousette, tomb bat, and four butterflies) were unknown on the other islands and thus were new arrivals on the archipelago. Over 20 of the 54 arrivals were stragglers that did not become established. Observers were present on Anak Krakatau over this period for a total of only about four or five months. Now take into account the following: arrivals of individuals of species already present would have been undetectable by observers; unsuccessful or very brief undetected colonizations could have occurred during the more than 90 months when no observers were present; and comparative data are not yet available for many groups of invertebrates. It is clear that hundreds of individuals must have arrived on the island over the eight-year period, many of them being of species new to the island and many probably representing failed colonization attempts.

The 1992–1995 Eruptive Episode: Prospect

Although we have evidence of the arrival of many animal species over the past decade, some of which have resulted in colonization, the development of a community on the island is clearly proceeding in fits and starts, with periodic checks as volcanic episodes exert their varying effects. Sometimes the whole island has been equally affected by these disturbances, and on other occasions particular areas and habitats have been damaged more than others. Volcanic activity may have been responsible for the interesting biotic heterogeneity— the presence of three early-successional stages in three different areas—evident on the island since the 1952–1953 eruptions. The very small proportion of the island's area supporting vegetation is unlikely to increase so long as periodic eruptions check any progress made in the intervals between them, and the island community is clearly developing in the face of extreme physical constraints. Perhaps the surprising thing is that community assembly has proceeded as far, and as quickly, as it has.

The decade following 1982 was the most progressive in the island's short history so far as its biota was concerned. During this time its figs first fruited and the small casuarina woodland became more diverse through increase in both animal dispersers of plants and animal-dispersed forest trees. By 1992 the island carried the most advanced community of its eventful young life. This short period of progress was then checked by a renewal of volcanic activity (Thornton, Partomihardjo, and Yukawa 1994).

The eruptive phase beginning on November 8, 1992, has been one of the most extensive, in terms of area affected, duration, and damage, since the beginning of the present biota in 1953. It is continuing at the time of writing (March 1995). By July 1993 at least three lava flows had been produced and a considerable depth of ash (more than 20 cm in places) was deposited on the island. The extent of damage has been only partially assessed as yet, as a result of necessarily brief visits to the island in July 1993 and November 1994, but components of the island's community appear to have been affected to different degrees. *Casuarina* trees free from climbing vines were relatively unaffected, apart from the scattering of isolated individuals on the eastern flank of the outer cone, which were dead. Vines climbing high on some of the casuarinas on the east foreland were dead, and although some host trees survived, others had fallen and were uprooted. Many specimens of *Hibiscus tiliaceus* appeared to have survived well and these and bushes of *Melastoma affine* and *Scaevola taccada* were flowering, although the *Melastoma* leaves were partially skeletonized by insect attack. Some of the figs were damaged and partially defoliated. The main damage seemed to have been to the sapling and shrub layer and the ground vegetation. In 1993 a small patch of the creeper *Mikania cordata* was found on the upper shore, with some surviving *Ischaemum*, together with another small area of *Canavalia*. The *Ipomoea pes-caprae* association, including the parasitic *Cassytha filiformis*, was virtually extinguished, at least above ground, both on the beach and at the ecotone on the landward edge of the forest, and most of the herbaceous vegetation was killed. In 1994, however, both *Ipomoea* and *Canavalia* creepers were thriving on the beach. Trees of *Macaranga tanarius* had fared badly and their saplings, which had grown thickly in a clearing near a conservation department shelter on the east foreland, were all killed. The shelter's corrugated iron roof had sagged and buckled under the weight of ash, and many *Pandanus tectorius* trunks were broken, and some trees felled, by the weight of ash accumulating in the large, tough rosettes of leaves. Many *Saccharum* clumps, and in 1993 almost all, and in 1994 all those on the slopes of the outer cone, were dead, at least above ground, and the lush grass meadow of *Ischaemum muticum*, the habitat of the white-breasted waterhen, was a straw-covered wilderness.

On these hour-long visits (with my wife in 1993 and Neville Rosengren in 1994) during eruptions every 5–15 minutes, tracks of the monitor and thick-knee on the beach advertised the survival of these species, and several species of land birds were seen. In 1993 these included a pair of white-bellied fish-eagles soaring around the erupting crater and patrolling the coast, where they were "mobbed" by white-breasted wood-swallows. Other birds present were the gerygone, collared kingfisher, black-naped oriole, magpie robin, yellow-vented bulbul, and one of the two species of sunbirds. We had a fleeting glimpse of a fast-flying falcon as it flew at about 4 m above the ground under the woodland canopy. At least three species of butterfly were also seen, the swallowtail *Pachlioptera aristolochiae*, a *Eurema* species, either *blanda* or *hecabe* (Pieridae), and the nymphalid *Ideopsis (Radena) juventa*. The pits of ant lion larvae were evident beneath the damaged shelter, although there was only about one-fifth of the number seen there a year previously. In 1994 there were still fewer pits, the fish-eagle was still patrolling, and the only other birds seen were the kingfisher and wood-swallows.

The effects of this eruptive episode on plant-dispersing bats and birds and their avian predators are not yet known. Although the peregrine (or perhaps the hobby) was still present in July 1993, it is possible that avian raptors were later completely expelled from the island. Other reports indicate that the type of Strombolian-Vulcanian activity observed during our visits, detonations with showers of hot ash and profuse lava bombs every few minutes, was typical of most of this eruptive phase. The raptors' roosting areas were high on the inner cone, which has now been subsumed by a new, higher one. If avian raptors have indeed been banished, and if appropriate plant species have survived, a period of unhampered colonization by pigeons may follow. Immigration

rates of bird-dispersed shrubs and trees may be restored for a time, until raptors recolonize. The differential effects of volcanic activity may thus be complex, with both negative and positive effects on the continued diversification
of the forest.

A lava flow in April 1993 cut a swath from 100 to 200 m wide through the
vegetation of the northeast foreshore, including one of our "permanent" study
plots. The northern flow in November 1992 moved through the western part
of the north foreland's casuarina woodland, and the southern flow (February
1993) covered the developing vegetation of the ash and lava to the north of
Bone Alley. The small patch of grassland on the north foreland, one of the very
few such areas remaining on the archipelago, appears to have narrowly escaped destruction, the lava passing within a few meters of it. The valley between outer and inner cones in the north has been filled with lava so that any
further flows from the eastern side of the cone are now likely to flow directly
to, and obliterate, parts of the island's developing community. Should the island become lava-based, protected from marine erosion by laval ramparts
along its entire coast, a new and different colonization process would begin,
primary succession on bare lava, which would almost certainly be much
slower than that on bare ash.

яэ

The future development of Anak Krakatau's community is difficult to predict.
At their present stage of development its major components are highly vulnerable, and continuing eruptions are likely to have at least transient effects on
them. Already, interesting questions arise. For example, since the herb and
ground layer, at least on the east foreland, has been almost extirpated in the
present eruptive phase, how will this community be renewed? Will the same
pioneer species return? The environment is now different, with trees that are
larger and more mature than when the previous herb and ground cover colonized. Have the fig trees suffered a significant depletion in fruit and therefore
also in fig-wasp production and, if so, what will be the effect of a reduced pollinator population on the future establishment of figs of these species? Clearly,
recent events have raised interesting questions and may have started a number
of instructive natural "experiments."

12 KRAKATAU AND ISLAND BIOGEOGRAPHY

They must often change who would be constant in
happiness or wisdom.
Confucius

The equilibrium theory of island biogeography proposed by MacArthur and
Wilson (1963, 1967) was an important conceptual advance in the field of biogeography. For several decades it has profoundly influenced the approach of
both ecologists and biogeographers to the assembly and dynamics of communities on geographical or ecological islands.

The theory states that as an island initially devoid of life is colonized by living things, the rate of immigration of additional species must gradually decline as fewer and fewer species on the mainland source area remain as
potential new colonists. Moreover, those species with good dispersal abilities
and attributes conducive to establishment will colonize the island early, so that
mainland species that have not colonized will be progressively less likely, and
will take progressively longer, to do so. The immigration rate curve, the number of immigrant species plotted against number of species on the island, thus
will not only be a declining one but it will also be concave. Extinction of
species from the island will gradually increase as more and more species ac-
crue and thus are available to become extinct. As the number of species increases, populations of at least some will contract, in some cases to sizes too
small for the species to remain viable on the island. Also, as ecological saturation is approached, increasing interactions between species will increase the

likelihood of extinctions. Thus the extinction rate curve will be an ascending one, and will also be concave.

Eventually, the theory concludes, a dynamic equilibrium will be reached when rates of immigration and extinction are equal. Although the complement of species will be continuously changing (this change in complement being known as "turnover"), the number on the island will remain more or less constant as immigrations on the average balance extinctions. Note that this does not necessarily imply a strict one-for-one replacement of species. The "equilibrium number" of species will be characteristic for a given group of organisms on a given island. It will depend on the island's size, which also often reflects environmental heterogeneity. Size largely affects the rate of extinction because larger islands are able to support larger populations and thus have a lower extinction rate and a higher equilibrium number than small, equally isolated ones. The equilibrium number will also depend on the island's distance from source populations, which affects predominantly the rate of immigration—less isolated islands have a higher immigration rate and a higher equilibrium number than more distant ones of the same size.

Equilibrium should thus be reached relatively quickly by groups of organisms in which both immigration and extinction curves are steep—good dispersers that are short-lived, allowing extinctions to occur rapidly—such as birds and flying insects. It will be reached more slowly by species that have limited powers of dispersal and that are longer-lived, persisting for longer once established, such as many trees of mature forest. Historical effects of climatic or even geological change, or the long-term effects of human disturbance, may dominate the course of colonization of the latter group. Not surprisingly, most proponents of equilibrium theory have studied birds and insects, and the theory has received little support from biogeographers concerned with groups such as forest trees.

There have been several additions to the original theory. The first concerns the effect of immigration on extinction and is known as the "rescue effect." Isolation, as well as island area, may affect extinction rate, through the higher likelihood of the "rescue" of declining populations on less isolated islands by further immigration of the same species (Brown and Kodric-Brown 1977). The second concerns "land-bridge" or continental islands, which carry a full complement of biota when they are cut off from surrounding areas, for example by rising sea levels. In such cases the newly isolated area immediately becomes "supersaturated," as an island, and its biota will "relax"—that is, it will approach an island-type equilibrium number of species from above rather than from below. The Krakataus, both those devastated in 1883 and the emergent Anak Krakatau, are "oceanic" islands biogeographically, since their biotas have been assembled from zero. Wissel and Maier (1992) have shown that the stochastic model of the relationship between number of species and island area, which is a corollary of the MacArthur and Wilson model, can only be derived theoretically if interaction between species is included. Competition thus appears to be a necessary component of the equilibrium model.

MacArthur and Wilson cited the Krakataus as an example of equilibrium having been achieved in the land bird fauna, on the three islands then existing, by the 1908–1919 intersurvey period, only three or four decades after the 1883 eruption. They drew attention to the apparent contrast with the colonization of plants, in which the rate of increase in species numbers showed no sign of declining. Surveys of the birds in 1951 (Hoogerwerf 1953) and more recently (Thornton, Zann, and van Balen 1993) have shown that although the colonization rate of birds (the rate of increase in number of species present) did not level off by the 1920s, as MacArthur and Wilson had believed, it did decrease markedly at about that time, as did the rates for plants (Figure 35d).

Dammerman's Early Contribution to Island Biogeography

Significant passages of Dammerman's classic monograph on the Krakatau fauna (Dammerman 1948) appear to foreshadow some aspects of the MacArthur-Wilson model. In chapter 5 of his book Dammerman compared the reconstituted Krakatau fauna of 1921, 38 years after the devastating eruption, with that of five other islands in the area. The islands he selected were Sebesi, the closest island to the archipelago, with area, height, geology, and location (in Sunda Strait) similar to Rakata, and also similar in that it, too, suffered the effects of the 1883 eruption, albeit not to the same extent as the Krakataus; Durian and Berhala, continental islands in the same region, one larger, one smaller than Krakatau, at distances from putative sources of colonists similar to that of the Krakataus; and Christmas and Cocos-Keeling, two oceanic islands, much more isolated than Krakatau but of greater area. Thus Dammerman was one of the first biologists to make comparative biogeographical analyses of island faunas, and in this respect he may be regarded as a pioneer of island biogeography (Thornton 1992b).

Dammerman had the idea of an equilibrium number of species on islands (he used the term *equilibrium*) and believed that Krakatau had not reached equilibrium but that the oceanic Christmas and continental Durian had. He also appreciated that different components of the fauna would be likely to equilibrate at different rates, determined in part by their relative ease of dispersal. In making this point he was implying that the isolation of an island must be considered with respect to the dispersal powers of the particular group of animals concerned. He noted the differences between the Krakatau and Christmas faunas in relative representation of Hymenoptera, spiders, land molluscs, earthworms, and wingless insect species, and explained them by the difference in isolation of the two islands: the much more isolated Christmas had a lower relative representation of such poorly dispersing groups in spite of the much greater time available for colonization on Christmas. The comparison with Durian was made in order to see what the Krakatau fauna might become, and here he showed an awareness that immigration rate would be higher early in the process and would decline toward the end.

Dammerman also foreshadowed another idea, one that was developed later by Heatwole and Levins (1972). He particularly remarked upon the general similarity, in the representation of carnivorous and parasitic forms, between the faunas of Krakatau and Christmas. On both islands mammalian predators were lacking, insectivorous bats were in the minority, and birds of prey on the land were few. The problem with the idea of a relatively fixed, deterministic pattern of faunal structure in this particular case is that, as Dammerman appreciated, Krakatau, unlike Christmas, was not yet at equilibrium and the comparison was not really valid. Indeed, recognizing the existence of major ecological links between components of the biota, he predicted that Krakatau's overall fauna would not reach equilibrium until the flora had done so. He also appreciated that the equilibrium number of species for particular islands will vary inversely with their distances from source populations.

Dammerman did not incorporate his ideas into a rigorous theoretical framework—he did not, for example, more explicitly characterize the role of island area—but in his study of the Krakatau fauna he appears to have appreciated most of the important features of island equilibrium theory. The important exception is the rise in extinction rate with species acquisition, a relationship, incidentally, that is not yet evident on the Krakataus.

Ecological Factors Affecting Colonization

Although MacArthur and Wilson acknowledged that, like all models of biological processes, theirs was a simplification, the model has been criticized on the grounds that it ignores the biological attributes of species and treats all species equally (for example, treating immigration as an essentially random process). Such criticism is unjustified. The model's purpose was to highlight the roles of island area and isolation in determining the dynamic equilibrium in the number of species inhabiting an island. As field biologists as well as theoreticians, MacArthur and Wilson knew that colonists are not always drawn at random from a mainland species pool, like marbles from a bag. In a general sense their model incorporated differences between species in dispersal and establishment capabilities, these partly accounting for the concavity of the curve in the relation of immigration to species richness. It also included competition, the increasing interaction within and between island species as their number increases being partly responsible for the concave shape of the extinction curve. MacArthur and Wilson also recognized that successional processes would affect the basic, theoretical colonization curves and even provided an example in which non-monotonic curves (with both rising and falling sections) may result. MacArthur (1972) later suggested that another factor, invasion sequence, might influence the colonization process and the equilibrium number of species. He suggested that a family of curves, reflecting different sequences in which members of the same mainland pool of species might invade, might apply to a given island. Not surprisingly, the equilibrium model did not explicitly cater for islands that increased or decreased in size during the colonization process (as did the Krakataus), nor for situations in which the developing biota on the island was subject to periodic bouts of drastic disturbance (also as on the Krakataus).

Many criticisms of the theory appear to stem from inappropriate uses of the model—for example, seeking a relationship between number of species and area of islands that are not at equilibrium, or attempting to apply it to highly heterogeneous groups of species, or to islands that have been highly modified by human activity. Other criticisms derive from a failure to appreciate that biological factors were incorporated into the model so far as was possible, and often from expectations of its application beyond those claimed by its authors. The model has nevertheless been the dominant paradigm in most island biogeographical studies, even those of some of its critics.

Colonization, immigration, and extinction rates may be calculated and departures from the model may then be considered in a quantitative way in relation to particular groups of organisms or particular islands. For example, it has been shown that different components (defined by either systematic or ecological criteria) of the developing Krakatau community may have different colonization characteristics. The curves for plants differ from that for birds, which differs from that for butterflies. Among vascular plants, the colonization characteristics of pteridophytes are not the same as those for spermatophytes (see Figure 35). Differences may be apparent at surprisingly fine levels

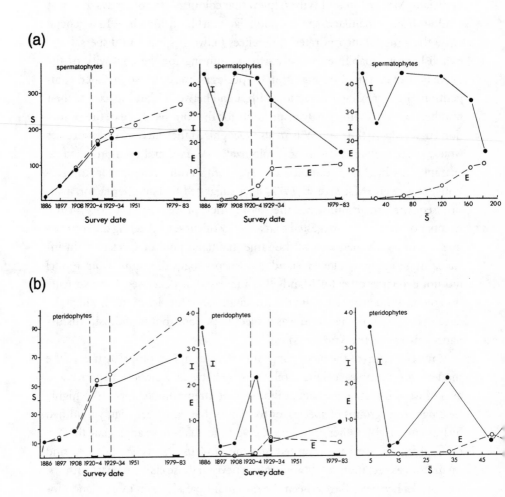

Figure 35. The immigration and extinction of species on the Krakataus: data for *(a)* spermatophytes, *(b)* pteridophytes, *(c)* butterflies, and *(d)* resident land birds. For each group of species, the graph on the left plots the accumulation of species with time (colonization rate); in these graphs the continuous lines and solid circles indicate number of species found at each survey, and the dashed lines and open circles indicate the cumulative number of species. In each middle graph immigration

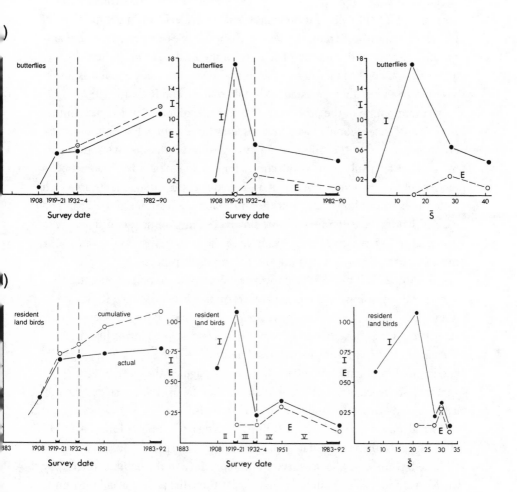

rate *(I)* is plotted against time as a solid line, and extinction rate *(E)* as a dashed line. *I* (or *E*) is the number of species per year immigrating (or becoming extinct) between surveys. In each graph on the right *I* and *E* are plotted against the average number of species *(S̄)* in intersurvey intervals. Vertical dashed lines demarcate the period of forest formation; grouped, integrated survey data are indicated by thick bars on the *x*-axis.

of discrimination. Japanese entomologists have shown differences even between groups of stinging wasps and bees (Yamane, Abe, and Yukawa 1992). Similarly, different colonization characteristics are evident in components of the flora having different modes of dispersal. Sea-dispersed seed plants reached saturation in number of species on the Krakataus only a few decades after colonization began, whereas the animal-dispersed component was slower to colonize and is still increasing (Whittaker, Bush, and Richards 1989).

An examination of the process at yet another ecological level of discrimination, that of successional phases or seres, may be revealing. For example, there are indications that the colonization curves of both pteridophytes and spermatophytes are really sums of the curves for separate phases of colonization: a pioneer phase involving largely sea- and air-dispersed coastal plants, in which a plateau is reached relatively quickly, and a later-successional phase of inland species that increase in number more slowly. The involvement of differences in dispersal ability in each of these examples indicates that dispersal is one of the most important features shaping the colonization process. In Chapter 7 we saw that dispersal is by no means always random. The sea barrier acts, in effect, like a sieve, imposing differing constraints on the dispersal to the Krakataus of the various organisms in the source areas, depending on their dispersal characteristics. It may be a broad highway for thalassochorous forms like the trees *Terminalia* and *Casuarina,* ineffective ("neutral") in the case of ferns, a partial barrier to birds, bats, the plants whose propagules they carry, and some anemochorous plants, and a complete barrier to organisms such as many large-seeded forest trees and terrestrial mammals.

Arrival, however, is only the first hurdle facing a potential colonist. Another vital challenge then immediately presents itself. The species must become established. It must be able to survive and reproduce in the habitats offered by the island, and if a community already exists there the potential colonist must be able to enter it.

Constraints on Establishment

We have seen in Chapter 11 that many species that reached Anak Krakatau did not become established. As was the case with dispersal, insights into the way that constraints on establishment may have affected colonization may be gained by a consideration not only of species that have successfully colonized but also of those that have not.

Habitat constraints

Most successful colonists of the Krakataus are widely distributed species that have good dispersal abilities and are able to survive in a fairly wide range of habitats (Thornton et al. 1990). For such species habitat constraints are minimal and this barrier to establishment is easily negotiated. However, many mainland species capable of reaching the islands have not become established, and lack of appropriate habitat on the islands is often a likely explanation.

A number of butterfly species now present on the islands are eurytopic (have wide distributions). Leps and Spitzer (1990) found that several of these Krakatau species were associated in Vietnam with transitional successional stages of vegetation and "ruderal" habitats, where there is recovery of natural vegetation after the abandonment of cultivation or following some other disturbance (*belukar* vegetation in Indonesia). About three-quarters of the Krakatau bird species are also eurytopic, including many which inhabit semi-urban areas such as gardens and other open, disturbed habitats in southeast Asia (Thornton, Zann, and Stephenson 1990). The bats present on the Krakataus are widespread in the Indo-Pacific. They are not restricted to primary forest and appear to tolerate disturbed habitats fairly well. Most do not have specialized roosting requirements (Tidemann et al. 1990); in the Philippines, for example, species found on the Krakataus are associated with disturbed habitats, and several occur on small, isolated islands (Heaney 1986, 1991; Heaney et al. 1991; Rickart et al. 1993). They are both good dispersers and good colonizers, the bat equivalents of Diamond's bird "tramps" (see below). There is a parallel also in psocopteran insects, which consume algae and fungi growing epiphytically on the leaves and bark of higher plants. The three families that are best represented on the Krakataus are those known elsewhere as transient exploiters of frequently encountered but ephemeral habitats, like dead or dying leaves (Thornton, New, and Vaughan 1988).

There is no mature forest on the islands and animals that are forest specialists are lacking. In 1986 15 bat species were collected in West Java that were not found on the archipelago, although two occur on Sebesi, only about a dozen kilometers away. At least four of the 15 species inhabit primary or tall forest. Two others are flat-headed bats of the genus *Tylonycteris,* which are specialist bamboo roosters, capable of entering the cavity of a bamboo through vertical slits just over half a centimeter (a quarter of an inch) wide. As yet the Krakataus have no bamboos, and thus no species of *Tylonycteris.* Many mainland snakes that are restricted to primary forest are also absent from the

Krakataus. *Dryophis trasinus*, the bamboo snake or green whip snake, is the analogue of *Tylonycteris*, being also a bamboo specialist. Nine mainland bird families comprise birds of mature forest. Trogons, parrots, nuthatches, and broadbills are all forest birds, hornbills require mature hollow trees for nesting, barbets are tree-hole nesters of the upper storey, pittas inhabit forest undergrowth, most of the leafbirds are birds of the forest canopy, and babblers are inhabitants of deep forest, feeding in undergrowth and on the forest floor (many babblers are also poor fliers). None of these families has colonized the Krakataus.

Many insects with broad food ranges but with other special needs associated with mature forest are also absent, although they are present in adjacent mainland areas. These include several groups of barklice, lacewings (Neuroptera), and aculeate Hymenoptera that are inhabitants of closed forest. Deep-forest butterflies are also poorly represented, only *Loxura atymnus* and one or two other forest "blues" (Lycaenidae) having become established. Many of the forest lycaenids cannot survive without the presence of particular ant species, and in these cases there would be an additional requirement from the Krakatau environment and another potential barrier to their successful establishment (New et al. 1988; Thornton and New 1988b; Yukawa 1984c).

The absence from the Krakataus of the mistletoe family Loranthaceae was discussed in Chapter 7 in the context of problems of dispersal. It is also possible that a habitat constraint has prevented its colonization. Mistletoe seeds may have been brought by the flowerpecker disperser but may have failed to become established because of a dearth of old, dying trees, which the plant parasite requires as hosts.

There are several "exceptions that prove the rule," bird families comprising predominantly forest species that are represented on the islands by an exceptional species that inhabits open scrub or mangroves. For example, of the 19 species of thrushes in the Sunda Strait area, 17 are forest species and, apart from the chestnut-capped thrush, the only other Krakatau thrush is the so-called magpie robin *(Copsychus saularis)*, an inhabitant of mangroves, open woodland, and gardens. Similarly, the Krakatau oriole is the only one of four in the area that is not a forest bird. The only Krakatau representative of the 21 flycatcher species in the Sunda Strait area is the mangrove flycatcher, one of only two that inhabit mangroves and coastal scrub.

Mangroves are in fact one of the most strikingly obvious absentees from the Krakataus among plant groups. Propagules of the Rhizophoracae, a mangrove family well represented on mainland coasts, do not appear to reach Krakatau.

Not one was found in the survey of drift flora of Anak Krakatau's beaches, although those of *Lumnitzera littorea,* a common component of mainland mangrove associations, were found stranded in enormous numbers (Partomi-hardjo et al. 1993). Thus colonization by *L. littorea* appears to be precluded not by dispersal constraints but by habitat deficiency, the lack of streams and rivers with brackish water outlets. *L. littorea* did become established when brackish lagoons occurred on Sertung's spit from shortly before 1919 to just after 1940, but the species persisted only for as long as the lagoons.

These lagoons were the basis of an unusual phase in Krakatau's colonization history. The islands have otherwise lacked large, permanent bodies of water, and a number of aquatic species became established only when this "aquatic habitat window" was open. Many, like *L. littorea,* were lost with the lagoons.

The dearth of substantial permanent freshwater habitats has precluded the establishment of grebes, cormorants, storks, ducks, and jacanas, all bird families known from the Ujung Kulon peninsula, for example. Many groups of insects having aquatic larvae have not colonized (for example, stoneflies, psychodid and simuliid Diptera, and gyrinid beetles) or have been recorded only recently in small artificial pools on Sertung, Panjang, or Anak Krakatau. Mayflies were not found until recently, and these were of a family (Baetidae) which includes species with unusually long-lived adults (up to 14 days) and some that are associated with brackish water. The Sertung lagoons, and a fisherman's well that survived on Rakata for a year in the 1930s, were the habitats of almost all the dytiscid and hydrophilid beetles and hydrometrid, gerrid, corixid, and notonectid bugs that have been recorded. Aquatic molluscs, too, were present only during the period of the lagoons, apart from one species found in the small Sertung pool in 1986.

Dragonflies and damselflies (Odonata) are the only aquatic insects that are at all well represented on the islands (Yukawa and Yamane 1985; van Tol 1990), and most of these are probably adventive and non-resident, although we have found a few nymphs in the small artificial pools. Almost all are of groups that breed in bodies of still water rather than streams or rivers. Small, man-made freshwater ponds have appeared (and disappeared) on the islands, and small, natural water bodies held for a time within suitable configurations of plant architecture, such as leaf axils, are temporary. It is likely that dragonflies and damselflies monitor the islands for potential breeding habitats on a regular basis, those with requirements for flowing water always being unsuccessful.

In 1908 both the collared kingfisher and the small blue kingfisher were present on the Krakataus. The former normally inhabits open coastal country, often mangroves, and is an aggressive pioneer colonist that has been highly successful on a number of small islands in the western Pacific. It is successful on the Krakataus in spite of the absence of mangroves: it occurs on all islands and there is a thriving population on Anak Krakatau. The small blue kingfisher inhabits forest streams, brackish ponds, and mangroves, none of which is available on the Krakataus. Evidently unlike the collared kingfisher, it could not easily adapt to the early Krakatau environment, for it has not been reported since 1908. There is little doubt that it arrived on an island devoid of its preferred habitat, but competition with the collared "supertramp" may also have been involved.

Although ants have been available on the archipelago for many decades, the ant lion, *Myrmeleon frontalis,* which feeds on ants, was able to colonize Rakata and, later, Anak Krakatau only after human activity had provided the dry, sheltered, sandy substrates necessary for construction of the pits by which its larva obtains food (Chapter 14). Food is of course an immediate problem for those species that are fortunate enough both to arrive and to find a suitable habitat. Unless they are able to sustain themselves, new arrivals have no chance of reproducing and becoming established.

Food requirements

The notion that hierarchical relationships, including food-web controls, are partly responsible for governing the formation and maintenance of island communities has recently been given prominence. Dammerman considered this aspect of island ecology over 50 years ago and discussed it at some length.

He recognized a pattern in the way the Krakatau fauna assembled. Early colonists comprised particularly those species that subsist on vegetable or animal debris, scavengers and detritivores, which have no specific requirements concerning the animal or plant on which they can feed. Such species included collembolans and tenebrionid beetles, which fed on natural flotsam washed ashore, and carrion feeders, such as muscid and sarcophagid flies and silphid, dermestid, and staphylinid beetles. Consumers of dead and rotting plant material included termites, earwigs, cockroaches, some crickets, millipedes, some psocopterans, several fulgorid bugs, and several families of flies and beetles. Thus the millipede *Spirostreptus* was found on Sertung before the carnivorous centipedes were found there, and earwigs, cockroaches, and crickets were well represented on Rakata by 1908, before any of the lacewings (the larvae of which are predaceous) were found. Many of the dipteran families found on the first animal survey in 1908 were feeders on decaying matter, and the first permanent insect populations encountered on Anak Krakatau were species living on stranded marine organic matter.

Dammerman's next colonizing category, which now would be called a trophic guild, comprised omnivorous animals or those with a wide range of food (euryphagous species). This category includes many ants, the monitor lizard, and birds such as the large-billed crow and those which, like the bulbuls and the oriole, feed on insects as well as plants. All these were present at the first survey, in 1908. Indeed, the first terrestrial vertebrate seen on the islands, having colonized within the first six years, was a carrion feeder that also eats shore crabs, turtle eggs, birds' eggs, scorpions, centipedes, birds, fish, and lizards—the monitor. Dammerman reasoned that euryphagous species would colonize before those with a narrow feeding range (stenophagous species), since they would be the more likely to subsist on whatever was available at the time of their arrival. Euryphagous owlet moths (Noctuidae) colonized the islands before oligophagous herbivorous insects (having a range of only a few food-plant species). Evidence from birds also supports Dammerman's contention. The earlier land bird immigrants in the first 50 years were predominantly generalized feeders and insectivores (Table 12.1), and the frugivorous birds that colonized early were largely species that also take insects or seeds. Ten of the islands' 12 facultative frugivores colonized within the first 50 years. In contrast, 6 of the 9 raptors colonized since 1933 (4 of them since 1953).

The highly vagile butterfly species *Anapheis java, Phalanta phalantha, Papilio demoleus,* and species of *Delias* have not become established on the Krakataus, although at least the first two reached the islands. Their food-plant

Table 12.1. Number of species of resident land birds, by feeding habit, that colonized the Krakatau Islands in the first and second 50-year periods since 1883. Percentages of pre- and post-1933 colonists in parentheses.

Feeding habit	Before 1933	After 1933
Omnivores	1 (3%)	0 (0%)
Fruit/insects/seeds	10 (29%)	2 (15%)
Nectar and insects	2 (6%)	1 (8%)
Seeds (granivores)	1 (3%)	0 (0%)
Insect (insectivores)	15 (44%)	2 (15%)
Fruit only (frugivores)	2 (6%)	2 (15%)
Predators	3 (9%)	6 (46%)
Total	34	13

Sources: Feeding data from Hoogerwerf (1949a,b), Frith (1982), and Zann, Male, and Darjono (1990); colonization data from Thornton, Zann, and van Balen (1993).

families (Capparidacae, Rutaceae, Santalaceae, and Loranthaceae) were lacking. The *Delias* butterfly that feeds on Loranthaceae, although migratory and thus a likely colonist, is absent. Dammerman (1948, p. 227) predicted that "as soon as this plant *[Loranthus]* makes its appearance we may be sure that the *Delias* will appear also." Perhaps all three, the plant, its bird disperser, and the butterfly, have reached the islands occasionally, but only the flowerpecker, which, evidently unlike the butterfly, does not rely solely on these plants for food, has become established.

Parasites, such as scoliid wasps, tachinid flies, and the bombyliid flies that parasitize the scoliid larvae, and the majority of braconid parasitoids, were relatively late colonizers. They presumably became established only after populations of their host insects were available. Koinobiont parasitoids of Diptera and Lepidoptera are both now well represented, about equally, on the older islands, but on Anak Krakatau the former are absent, the latter predominant. On the young island there is a paucity of dipterans but butterflies have been successful early colonists. The koel is a brood parasite of the large-billed crow and the oriole, which rear its young. It was not found until after its hosts were well established.

Just as parasites cannot colonize before their hosts, so predators cannot survive without prey. Becker (1975, 1992) showed that the proportion of carnivorous to herbivorous beetles was substantially greater on several island groups than on the presumed mainland source areas, and noted that the proportion found on the Krakataus by Dammerman was also higher than known mainland values. In beetles, carnivores are largely generalist feeders and thus likely to become established more easily than herbivores, which tend to have a more restricted range of acceptable food and may be less resilient physiologically. In general, however, predators are likely to be the later colonizers. Many groups of predaceous insects, such as vespid wasps, robber flies (Asilidae), tiger beetles (Cicindelinae), ladybird beetles (Coccinellidae), reduviid bugs, mantises, orthopterans (*Gryllacris* species), and lacewings (Neuroptera), were not found until the surveys of the 1920s. A species of coccinellid found in 1908 was unusual in being a plant feeder. The first avian raptors (three species) were seen in 1919. Two of them, the white-bellied fish-eagle and brahminy kite, take fish, sea snakes, and carrion and, like the monitor and beach skink, are not dependent on the land fauna for prey. Other raptors that depend on land animals for food—eagles, hawks, falcons, and the barn owl—were later colonizers. The serpent-eagle and oriental hobby were first recorded in 1951, and the barn owl, black eagle, and peregrine falcon not until our series of surveys in the

1980s. Whereas 3 of the 34 bird species to colonize within the first 50 years (9 percent) were raptors, 6 of the 13 to colonize since then (46 percent) are raptors (Table 12.1).

The predaceous reptiles of the islands are the reticulated python, the paradise tree snake, and the *tokay* and king geckos. Food may not have been an immediate imperative for a colonizing python, for these pythons can go without food for two and a half years. The species was seen in 1908 and until bats and rats colonized, it probably fed largely on monitors and birds. It is known to take nightjars in the Philippines (Neill 1958), and these groud nesters were early colonists of the Krakataus. The paradise tree snake may have been seen by Dammerman's party in 1933 but was not definitely recorded until 1982. It probably required the presence of trees for its successful establishment, although we have found it in the *Ipomoea pes-caprae* zone on the shore as well as quite high on the bare ash of Anak Krakatau's outer cone. Geckos are its principal prey but it also takes frogs (absent from the Krakataus), small birds, and bats. *Gekko gecko*—the *tokay*, one of the world's largest geckos (up to 350mm, about a foot, in total length; Figure 36)—is known to feed on insects, spiders, smaller geckos, mice (absent from the Krakataus), and small birds. It is conspicuous by its distinctive call (both its scientific and common names are onomatopoeic)

Figure 36. The carnivorous *Gekko gecko* was first heard on Rakata in 1982. It still has a limited distribution on that island and has not colonized the others.

and almost certainly was not present in 1951, but my wife and I immediately recognized its call when camping in the Zwarte Hoek area of Rakata in 1982. The species appears to be restricted to this area, indicating that it is a recent immigrant. *Gekko monarchus* is somewhat smaller than the *tokay* but also feeds on smaller geckos as well as insects. It was first found in 1984. The burrowing blind snake, *Ramphotyphlops braminus,* feeds largely on termites and was not recorded until 1984. There is only one predaceous species of bat present on the Krakataus, the Malay false vampire, *Megaderma spasma* (Figure 37), which feeds on large insects and sometimes small geckos and bats. It was first noted only in 1982, when six other species of bats were already present.

The colonization of two groups of animals did not follow the general sequence of arrival: plants, detritivores, omnivores, herbivores, parasites, and predators. These animals are important in the food web because although they

Figure 37. The carnivorous Malay false vampire bat was first recorded in 1982.

constitute consumer levels for exploitation by higher trophic levels they can colonize without the prior presence of producers (plants) on the island. Both groups exploit energy coming from outside the islands themselves. One is the guild of animals able to subsist on stranded marine detritus, and the other may be thought of as its aerial analogue, those animals feeding on the constant stream of wind-borne insects falling on to the islands. Spiders, most of which are good dispersers, are a good example of the latter group. The first living thing to be seen on the islands, only three months after the 1883 eruption, was a spider, spinning its web. It may well have survived, because invertebrate fall-out on to the islands is considerable and probably fairly constant. As detailed in the previous chapter, at least three levels of consumers subsist on the aerial fallout of invertebrates in barren areas, and together these organisms form a conduit through which energy derived from outside the islands flows into the archipelago's community without the involvement of plants.

The need for pollinators

The colonization of many animal-pollinated plants is constrained by the avail-ability of pollinators. We have noted in Chapter 9 that specialist nectarivorous bats arrived late and that plants with bat-pollinated flowers were correspond-ingly scarce on the islands. We have seen that fig species each require a contin-uous population of their specific pollinator if they are to become established. The rigor of this requirement for animal-pollinated plants is not always im-mediately apparent, however, and considerable field study may be needed to assess this need. One such case, besides the example of the fig species and their wasps, has been studied in some detail on the Krakataus (Gross 1993).

Ipomoea pes-caprae and *Canavalia rosea* (= *maritima)* are successful pio-neer beach creepers that spread vegetatively. In July 1992 they formed a zone of ground cover on the landward edge of the casuarina mixed woodland of the east foreland of Anak Krakatau. Gross found that only about one in seven flowers of *C. rosea* set fruit, an unexpectedly low ratio in a successful pioneer plant. She showed that in this environment cross-pollination is necessary for fruit-set in this species. The low fruit-set might thus result either from insuffi-cient pollinators or from the pollinator being "unfaithful," pollinating also other plant species competing for its services. *I. pes-caprae* and *C. rosea* have flowers of similar size, shape, and color. There are two species of solitary car-penter bees on Anak Krakatau. The large, black *Xylocopa latipes,* although an occasional visitor to *M. affine* flowers, was mainly seen to visit *C. rosea,* of which it was the chief pollinator (Figure 38). This bee is about twice the size of

Figure 38. The strongly flying large carpenter bee, *Xylocopa latipes,* is a faithful pollinator of *Canavalia rosea,* a creeping legume, on Anak Krakatau.

Xylocopa confusa, which is a catholic forager that visits flowers of *Premna serratifolia, Scaevola taccada, Melastoma affine, Eupatorium odoratum,* and, occasionally, *C. rosea,* in its collecting trips. By field observation and manipulation experiments involving hand pollination and the exclusion of pollinators by bagging flowers, Gross demonstrated that fruit production in *C. rosea* was limited neither by access to its pollinator nor by levels of pollination. In the mixed stand of creepers, *X. latipes* remained constant to *C. rosea,* and *I. pes-caprae* had a different pollinating bee. Poor fruit-set of *C. rosea,* therefore, did not result from either lack of pollinators or competition for them. By manipulating the fruiting load that the plants sustained, Gross showed that it was endogenous resources, such as water and nutrients, that were limiting fruit set. She concluded that on a substrate with adequate resources *C. rosea* would achieve successful establishment, with sexual reproduction, only in the presence of an established population of *X. latipes*. These bees are strong fliers and have been seen at sea flying between islands of the group.

Succession

Succession is sequential change in community characteristics, such as species composition, dominance pattern, and structure of vegetation, within which recognizable stages in the community's development, known as seral stages, are evident. In a developing ecosystem, succession is a prime determinant of both the habitat and food sources of animals.

As noted in earlier chapters, colonization of the Krakatau archipelago by plants was not random. A general pattern, already discernible to Docters van Leeuwen (1936), was confirmed by later studies (Whittaker, Bush, and Richards 1989). Sea-dispersed and wind-dispersed species arrived before those that are animal-dispersed, the beach vegetation and grasses before forest trees. Successional changes opened up a sequence of different "habitat windows" through which appropriate potential colonists could enter the system. For example, as woodland developed, open-country grasses, heliophilous orchids, and xeric, pioneer ferns declined, but shade-tolerant plants, such as forest ferns and orchids, adapted to more mesic conditions, were able to become established. Each stage of the succession has provided a physical and biological environment optimal for only a subset of the large number of species on the mainlands. This "tracking" of the gradually changing island environment by the pool of potential colonizers was also evident among animals.

There is little doubt that most of the turnover in both plants and animals in the past century has been successionally induced, and successional change has clearly affected groups of the biota differently. For example, species numbers of coastal plants appear to have reached a plateau quite early and have since remained stable, although without the balancing turnover expected according to theory. In contrast, on a broad scale the inland vegetation has undergone substantial successional change, which has influenced many other biotic components.

The vegetational changes of most significance to animals were the formation of open savanna grassland beginning in the mid-1890s, the change to woodland and forest formation by the early 1920s, and the closure of the forest canopy during the following decade or so. Floristic change (in the complement of plant species) affects many other components of the community, notably phytophagous animals, particularly those with a narrow range of food plants or specific food requirements, and in many cases their parasites and predators. Floristic change often involves vegetational change (changes in the structure of the plant component of the community), which alters the physical habitat and microclimate for animals (and plants) in a number of ways. For example, lowered wind velocity, light intensity, and temperatures, and increased humidity accompany forest formation and canopy closure.

For most groups of animals, the period of forest formation was the time of the highest rate of immigration of species, and the period of canopy closure the time of highest extinction rate. The data for butterflies, ants and other aculeate Hymenoptera, thrips, and resident land birds exemplify these changes, and Dammerman stressed the differences in representation of invertebrate groups between the 1908 "grassland survey" of Jacobson and his own surveys from 1919 to 1934, when the forest was forming. During the latter period the numbers of species of fulgoroid homopterans feeding on palms and grasses declined along with their food plants, and the number of jassid species (a group of sap-sucking homopteran insects) was halved. The density of ground-living tenebrionid beetles, which prefer dry to moist surface habitats, was greatly reduced, and several species of ants normally associated with grassland had greatly declined in numbers by the 1920s. Forest closure and loss of open habitat almost certainly contributed to the extinction of open-country insects like the ant *Tetraponera rufonigra,* species of sapromyzid flies, several genera of sphecid wasps, and some of the hesperiid Lepidoptera (skippers). Three species of skippers, one present since 1908, were lost in the decade following 1920, when their food plants, grasses and palms, were declining in importance. Three others seen in the 1930s were not found in the next invertebrate surveys, in the 1980s.

The zebra dove, *Geopelia striata,* an open-grassland ground feeder on the seeds of grasses and sedges, and the long-tailed shrike *(Lanius schach)* and large-billed crow, which also frequent open spaces, grassland, and scrub, are further examples of extinctions probably resulting from loss of habitat during successional change to forest. The dove was present in 1919 but was last recorded in 1951 on Sertung's spit, one of the few areas of grassland then persisting. The shrike became established before 1908 and was breeding on both Rakata and Sertung in 1919 but was not recorded in the surveys of 1928–1934, when the forest canopy was closing, nor in any thereafter. The crow was a very early colonist (by the first survey) but by the 1980s was confined to Anak Krakatau, where it is now thought to be extinct. Another loss was the coastal gecko *Lepidodactylus lugubris,* which inhabits open habitat such as rocks and tree trunks in open situations and the crowns of palms and coastal vegetation. It may have been affected by the overall reduction in extent of beach associations as well as by canopy closure.

The opportunity for establishment of some species may be limited to the short period of the succession when the island habitat offers their optimal ecological requirements. This successional "window of opportunity" closes for these species as the habitat moves away from their optima toward those of other species. Conceivably, the environment on the target island may be changing so quickly that there is insufficient time, even for appropriate members of the effective available pool, to exploit the transient opportunity to colonize. There is some evidence that this has been the case on the Krakataus. Dammerman believed that the grassland fauna had not reached maximal development by the time that forests began to form. Grassland was first evident in 1897 and had reached its greatest extent by 1908, when "mixed forest was beginning to supplant the grass." By 1919 the grassland had been "to a large extent replaced by mixed forest, at any rate on the lower areas." In 1933 Rakata was "almost entirely covered by trees even up to a great altitude and the grass jungle, at any rate as covering large areas, had almost disappeared" (Dammerman 1948, p. 149). The grassland window had closed, having been open for just over two decades. There are indications that this brief open-habitat phase was too short for the establishment of several animals that might have been expected to colonize at that time.

Sphecid hunting wasps of three genera characteristic of open sites on Java failed to colonize the Krakataus during the period when their optimal habitat was extensive (Yamane, Abe, and Yukawa 1992). The grassland phase was also missed by granivorous birds, such as sparrows and munias (Ploceidae), 13 species of which occur in coastal areas of West Java. Until 1992 only one granivorous bird species, the zebra dove, had ever been recorded from the Krakataus, and this was encountered after the time when grassland was extensive. In all, 22 families of non-aquatic birds present on the adjacent mainlands have never been represented on the Krakataus, and 12 of them comprise 31 species typical of grassland and open country. In addition, 14 species of Old World warblers (Sylviidae) inhabit reeds, scrub, and open grassy country in the Sunda Strait area, but only one (the flyeater or golden-bellied gerygone, *Gerygone sulphurea*)—which is, exceptionally, a bird of coastal scrub and mangroves—has colonized the Krakataus (Thornton et al. 1990).

It was during forest formation, of course, that many ecologically important animal-dispersed trees and shrubs first appeared. A large proportion of these are dispersed by fruit-eating bats or birds and so, once established, their presence was conducive to the immigration of further frugivores, their arrival setting in train a feedback that seems to be continuing. Before the forests were formed there was a preponderance (93 percent) of coastal birds of mangroves, scrub, and open country. Since forest formation began in the early 1920s, 62 percent of all bird colonists have been forest birds (Table 12.2). Since canopy closure, extinctions have been few and have involved species that may not have become fully established. For example, two bats, the insectivorous *Hipposideros diadema* and the nectar-feeding *Macroglossus sobrinus,* have not been encountered since they were first recorded, in 1928 and 1974, respectively.

It has been argued that the effect of successional change on the immigration and extinction of plant species and the resulting non-monotonicity (with both rising and falling sections) of the colonization curves, the existence of a core group of sea-borne plant species with minimal turnover, and the fact that most of the turnover up to now has been successionally induced—renders the equilibrium theory model "fundamentally inappropriate" to the Krakatau flora (Whittaker, Bush, and Richards 1989, p. 103) and, by extension, to the whole Krakatau case and the lowland tropics generally (Bush and Whittaker 1991, 1993). A comparative study of colonization curves for groups for which there are good data, however, does not support these claims.

The plant survey data up to 1983 (the primary data of the 1989 survey, in which a further 49 species were recorded, have not yet been published) do not yield the smoothly falling, monotonic curve of immigration rate that is predicted by theory. Instead, the rate first falls, then rises, and *then* falls monotonically toward the extinction rate curve (Figure 35a,b). The initial fall and rise were synchronous with successional changes, grassland and forest formation. The fall was greater for pteridophytes than for spermatophytes, and the subse-

Table 12.2. Number of bird species, by habitat preference, that colonized the Krakataus before and after 1921. Where more than one of the habitat categories applies, the score is shared. Percentages of pre- and post-1921 colonists in parentheses.

Habitat preference	1883–1921	1922–1992
Coasts, mangroves, open country	22 (73%)	4.0 (24%)
Scrub, secondary growth	6 (20%)	2.5 (15%)
Forests	2 (7%)	10.5 (62%)
Total	30	17

Sources: Habitat data from King, Dickinson, and Woodcock (1975) and Hoogerwerf (1970); colonization data from Thornton, Zann, and van Balen (1993).

quent rise in the rate for pteridophytes lagged behind that of spermatophytes, occurring not before, but during, forest formation.

As already noted, in both groups of plants the shapes of the curves suggest that we may be seeing the summation of two overlapping curves, that of an early pioneer phase, in which a pool of potential colonizers quickly became exhausted, and that of a subsequent phase of secondary colonizers during forest formation, the pool of which was depleted more slowly. The difference in the time of the peaks of the immigration rate curves suggests that the second phase was rather later in the case of pteridophytes than in spermatophytes. Possibly the beginning of forest formation, with its attendant changes in microclimates, was a prerequisite for the second wave of pteridophytes, which, unlike the first wave, was composed largely of shade-demanding species, many of them epiphytes. In both groups of plants there was a marked rise in extinction rate at about the time of canopy closure.

The data sets for birds and butterflies also involve a substantial number of species over time. In both groups immigration rates rose to a maximum, in contradiction of the basic theory; as for pteridophytes, the increases in bird and butterfly species coincided with forest formation. The expected decline followed. Peak immigration rates for birds, butterflies, and pteridophytes all lag behind the rate for seed plants by one survey. This suggests that changes in seed plant complement or vegetation structure may have influenced colonization by the other groups or, perhaps less likely, that seed plants responded to some changing factor in the physical environment earlier. Both of these explanations may apply to some degree.

Extinction rates of birds and butterflies have not risen steadily to approach the falling immigration rates, as they have in spermatophytes. The peak extinction rate for butterflies at about the time of canopy closure is not reflected in the curve for birds (Figure 35c,d). Ecological ties between plant species and birds are probably less binding, in general, than those between plant species and butterflies. Many butterflies have a limited range of larval food plants and would have greater difficulty than birds in coping with a rapidly changing floristic environment.

As noted earlier, the largely sea-dispersed coastal flora of the Krakataus has remained virtually constant for nearly 90 years. Yet although a large proportion of the archipelago's butterfly fauna is dependent on coastal food plants, butterfly species numbers have continued to increase. Bush and Whittaker (1991) suggested that this apparent paradox indicates that butterflies must arrive only rarely. Of 27 butterfly species newly recorded between 1979 and

1989, 16 feed on plants that have been available on the islands since 1920, and it was suggested that the "late" colonization of many butterfly species was due to failure to arrive rather than failure to become established. Butterflies were not surveyed between 1933 and the 1980s, however, so species newly recorded in the 1980s could have been on the islands for half a century or more before being discovered as new records. As demonstrated in the previous chapter, arrivals are much more frequent than is generally appreciated.

There are few butterfly species in the interior forests of the Krakataus, and for forest butterflies arrival may indeed be a rare occurrence, given that many of them are not particularly vagile. In this case, establishment is probably an additional problem because, in contrast to the coastal plant associations, the islands' forest associations are still slowly accumulating species and are highly simplified, quite unrepresentative samples of the mainland habitats of these forest insects.

MacArthur and Wilson appreciated that succession was likely to complicate the purely mathematical concept of biotic assembly. They suggested that later-successional species might colonize without the extirpation of those already present, so that, instead of rising steadily as expected from the theory, extinction rates may actually fall. In fact this does not seem to have occurred on the Krakataus. Rather, at about the time of forest closure extinction rates of butterflies and plants rose, whereas that of birds has remained fairly steady, at a low level. After an initial rise during the pioneer phase of colonization, immigration rate fell, and if it declines to the level of the extinction rate and is maintained there, there will be equilibrium without a rising extinction rate.

In contrast to the data for birds and plants, there is no evidence that the rate of increase in butterfly species is declining, either on the entire archipelago or on Rakata, which has been least affected by Anak Krakatau's eruptions. The curves for lycaenid butterflies taken as a group, and for ants, bees (except for *Apis*), sphecid wasps, braconid Hymenoptera, Diptera, Thysanoptera, Neuroptera, land molluscs, and bats also show that species numbers are rising at a rate similar to that in the first half century since 1883. In the case of bats this may be an artefact of recent developments in trapping and detection methods, however. Nymphalid and hesperiid Lepidoptera, cockroaches, dragonflies, reptiles, and vespid, pompilid, mutillid, scoliid, and eumenid wasps are similar to birds in that increase in species number has slowed markedly in the past 50 years.

As the environment of an island not yet at equilibrium changes through succession, so the set of appropriate colonizers in the mainland pool of potential colonizers will also change. Succession provides a changing ecological filter through which mainland colonizers must pass. Moreover, the putative equilibrium number of species will not be the same for all successional phases. It will be continuously changing too, and the mainland pool will be effectively "tracking" this changing equilibrium number. Certain segments of the mainland biota will be able to "keep up" better than others with the successional changes affecting their colonization chances. These will be groups with good colonizing attributes of dispersal and establishment, those with a broad range of optimal habitat, or a niche that is relatively unaffected by succession. It is in these groups, such as coastal plants, many birds, reptiles, cockroaches, and (were habitat available) dragonflies, that equilibrium numbers are likely to be achieved first. Relatively slow colonizers may be expected to approach equilibrium more slowly. These include groups with greater constraints of dispersal, such as land molluscs and insectivorous bats, perhaps, and those that are slow to become established, such as parasites, predaceous forms like Neuroptera and avian raptors, or that have special associations with other organisms, such as lycaenid butterflies.

Some of the stenophagous invertebrate groups associated with vegetation may now be near to a "watershed" between the colonization patterns of different ecological groups. For example, most members of the set of butterfly species with high vagility and early-successional food plants that is available on the mainlands are probably now established on the islands, but a more specialized group, with low vagility and late-successional food plants, is only beginning to colonize the interior forests (New and Thornton 1992a). The latter group will probably take longer to colonize. Their dispersal will be less frequent, and their food plants will arrive later and build up numbers more slowly.

Colonization and the development of Krakatau communities thus appear to have been punctuated by waves of immigration and extinction coinciding with major changes in habitat, including flora and vegetation type, resulting from succession. For example, when grassland was lost through succession, characteristic grassland animals were succeeded by those of woodland. This particular change appears to have been so rapid that the grassland fauna was still accumulating when the invasion of trees and the development of forest imposed a new set of conditions on colonizers, which tapped a different set of species in the mainland pool. There is some indication of a time lag between this habitat change and the immigration of the newly appropriate animal colonists.

Competition

One way in which the presence of a given species may affect the chances that another will become established is through competition. Competition is likely to be most intense between species that have the most similar demands on environmental resources that are limiting. Usually, but not always, these are species that are closely related phylogenetically (through evolutionary descent). The demands of two species may overlap to such an extent that the environment is incapable of permanently supporting both, and one of them will fail to colonize. Such an event is termed "competitive exclusion." Usually (but again with some very important exceptions), it is the incumbent species, which has had some time to adapt to the island environment and expand its population (and the amount of time it has enjoyed in this respect may be crucial), that prevails and the putative invader that is excluded. Competition may also be "diffuse," in the sense that a species may be affected to varying degrees by the combined activities of a number of species, rather than by any single competitor.

Competition is notoriously difficult to demonstrate directly. Its prior existence and its effects are usually implied by indirect evidence, such as striking instances of the exclusion of one of two competing species, thus obviating competition. Extreme caution should be exercised, however, in ascribing a given distributional pattern to anything other than chance; it must first be shown that chance is unlikely to have been responsible for the pattern. Often we do not know whether the species concerned can coexist in the same habitat on the mainlands, nor whether their populations are at carrying capacity (the maximum size the habitat will bear) on the Krakataus, and thus that resources are indeed limiting for them. There are a few examples on the Krakataus of the presence of only one of a number of potential colonists that have closely similar ecological requirements, and one or two cases that can be interpreted as instances of a species being replaced by a competitor.

In 1908 the wasps *Ropalidia artifex* and *Ropalidia variegata* occurred on Panjang and Rakata, respectively, but by 1982 both had evidently been replaced by a close relative, *Ropalidia fasciata* (Yamane, Abe, and Yukawa 1992). There is no obvious explanation for such a replacement. Perhaps competition was involved.

Two species of dog-faced fruit bats, *Cynopterus sphinx* and *Cynopterus horsfieldi*, were first recorded on the archipelago in 1919 and 1920, respectively, but whereas the former is now present on all islands, the latter is on Rakata and Panjang. Was the absence of *C. horsfieldi* from Sertung related to competition with *C. sphinx?* A third species, *C. titthaecheilus,* was established by 1974 and has been able to coexist with *C. sphinx* on Rakata, Sertung, and Panjang. Is this because its ecological requirements have little overlap with those of *C. sphinx,* or is it a question of the timing of its arrival? We do not know. A fourth species, *C. brachyotis,* was found in the last decade, coexisting with *C. sphinx* on Anak Krakatau and with both the other species on Sertung. It is important to re-member the dynamic nature of the flora and that the diversity of resources for these small fruit bats has been increasing. On balance, the case for explaining the absence of *C. horsfieldi* by competitive exclusion is weak.

The lesser coucal, *Centropus bengalensis,* a bird of grassland and open coun-try, had colonized the islands by 1908, when its optimal habitat was extensive. The greater coucal, *Centropus sinensis,* which inhabits open woodland and scrub as well as grassland, was heard once on Sertung in 1919, when the lesser coucal was already present on Sertung and Rakata, but has not been recorded since. The preferred habitats of the two species are closely similar, but on Java and Bali the greater coucal is the less common. Is this an example of competi-tive exclusion on the Krakataus? Perhaps.

Both the yellow-vented bulbul, *Pycnonotus goiavier,* and the sooty-headed bulbul, *Pycnonotus aurigaster,* had arrived on Rakata by 1908. *P. aurigaster* fre-quents open woodland and forest edges—wooded, bushy habitats—and in-cludes more fruit in its diet than does *P. goiavier,* which is a bird of scrub and more open country. The diet of *P. goiavier* includes a substantial proportion of worms, beetles, and crickets as well as fruit, and the species spends more time on the ground than other bulbuls. The savannah grassland with small patches of woodland that prevailed on the Krakataus in 1908 was closer to the pre-ferred habitat of *P. goiavier,* which has survived successfully, than to that of *P. aurigaster,* which was encountered only in 1908. A third species, *P. plumosus,* the olive-winged bulbul, was first seen in 1951 and may still persist on Rakata and Sertung. An inhabitant of the middle to lower canopy of forest edges and well-wooded country, it feeds on figs and berries as well as flies, caterpillars, and beetles. Its habitat thus overlaps with *P. goiavier* less than does that of *P. aurigaster,* and *P. plumosus* and *P. goiavier* now coexist. Did *P. aurigaster* arrive too early for its own good, before forest development, and thus, unlike *P. plumosus,* had to compete with its close relative *P. goiavier* on unfavorable terms in an environment which was not then heterogeneous enough to permit the survival of both? This appears to be a likely explanation.

The recent replacement of the oriental hobby by the peregrine falcon may be an example of exclusion as a result of competition loaded in favor of one of the species. We know that the peregrine was able to widen its ecological niche on Anak Krakatau by taking smaller prey than it usually prefers, but perhaps the hobby, a small-prey specialist, could not reciprocate by taking larger birds. If this were so, the hobby—in competition with the peregrine in an isolated habitat with a limited amount of potential prey—would have been doomed. Information on the size range of prey of these two falcons on the other islands or on the mainlands would be valuable in testing this suggestion.

Termites present an interesting possibility that competition may have an effect which is the opposite of competitive exclusion on islands. Species of the genus *Prorhinotermes* have an unusual distribution. They tend to be confined to tropical and subtropical islands and are not widely distributed on mainlands. On the Krakataus they are common in both coastal and inland forests, whereas on Java they have not invaded inland forests. They nest and feed in wet wood, and their colonies take about four years to mature. This is too long for pieces of wood of small diameter (less than 30 mm, or about 1 inch) to serve them as both food and nest. For successful reproduction they need pieces of wood thick enough to last four years, until their colony matures. In contrast, termite species that feed on wood but do not nest in it can exploit wood resources of any dimension, and thus they have a competitive advantage that usually enables them to exclude *Prorhinotermes* species from mainland forests. Abe (1984, 1987), in discussing this case, suggested that *Prorhinotermes* species are largely restricted to islands like the Krakataus because their competitors are scarce or absent there (owing to difficulties of dispersal). Islands would thus be relatively competitor-free havens for species of *Prorhinotermes*. This appears to be a case in which differences in both dispersal powers and competitive relationships may have determined the composition of the Krakatau fauna.

There is a rather similar case among flycatchers. The mangrove blue flycatcher, *Cyornis rufigastra*, is the only resident flycatcher on the Krakataus and one of the most abundant birds, in spite of the absence of mangroves. As noted above, its common name is apt. Of the 21 flycatcher species in the region it is one of only two (the other is the pied fantail, *Rhipidura javanica*) in the region that normally inhabit mangroves. It is listed as rather rare on Java, being confined to coastal and tidal forests, and its status is regarded as a matter of concern. The pied fantail is widespread and common in mangroves but also in open woodlands and gardens in the lowlands, and it is less of a mangrove specialist (MacKinnon and Phillipps 1994). Were the mangrove blue flycatcher not established on Krakatau, the fantail would be a likely colonist. Intensive and extensive investigations of the ecology of these species would be needed to test the possibility that the Krakataus offer an example of "ecological release" and "density compensation," terms coined for the expansion of habitat and increase in numbers, respectively, of one species (in this case the mangrove blue flycatcher on the Krakataus) in the absence of its competitor (the pied fantail).

"Outbreaks" of three insects have been noted on the Krakataus: the green cicada, *Dundubia rufivena*, a margarodid scale insect, *Cryptocerya jacobsoni*, and a species of fruit fly, *Batrocera albistrigata* (Yukawa 1984a,b; Yukawa and Yamane 1985). Colonization of a new environment in which reproduction is at first relatively unrestrained by predators or competitors may be a possible reason for such unusually high abundances. The scale insect was restricted to Rakata, where it appeared to be increasing its plant host range, suggesting, perhaps, ecological release. Although widespread in the Oriental region, it had never before been recorded from figs, but on Rakata it was found, sometimes abundantly, on four fig species. *B. albistrigata* made up at least 97 percent of individuals of all fruit fly species caught in lure traps on all four islands but was not taken in a comparable trapping program on the coast of Java. Evan Schmidt's group found that on the Krakataus the species was still preponderant in 1990, indicating that the 1982 outbreak was probably not an isolated "blip" (Schmidt, Thornton, and Hancock 1994).

The Effect of Landform Changes on Colonization

Since the great eruption a century ago "cleaned the slate" for a new community of species to colonize, the Krakataus have not remained physically constant. They have sometimes decreased in size, but sometimes they have increased, making part of the archipelago a later-starting component of the system. These changes have complicated the colonization of the Krakataus. The processes of cliff recession, the production of mobile cuspate forelands and spits, and the emergence of Anak Krakatau raise new biological questions. These include: How did the creation of new land affect the the communities that were already in place on the older terrain? To what extent are the biological processes occurring on the old islands and the new one similar? The first of these questions will be considered in the rest of this chapter; the second will be treated in the next chapter.

Both the birth and growth of the new island Anak Krakatau and the development and mobility of Sertung's spit probably affected colonization of the archipelago.

The emergence of Anak Krakatau in 1930 probably had a twofold effect. First, according to equilibrium theory, the increase in land area should result in a lower extinction rate and a higher equilibrium number of species. Second, it has also increased the diversity of habitats available on the archipelago. As early-successional plant communities developed on Anak Krakatau at a time when the forest canopy had already closed on the other islands, the new island, along with Sertung's spit, provided open-country habitats that elsewhere had long been subsumed by mixed secondary forest.

The existing low-lying sand spit to the north of Sertung is the dynamic remnant of a much larger feature that had developed soon after the 1883 eruption. A well-developed bluff marks its proximal end, evidence of a former sea cliff eroded into the pumice and ash that blanketed the island after 1883. By 1906 the spit carried *Ipomoea pes-caprae* and *Barringtonia* associations as well as *Casuarina* woods, and 23 years later it was covered in casuarina woodland. The spit suffered considerable damage from Anak Krakatau's 1930 eruptions, but aerial photographs taken in 1946 show that it was again clothed in casuarinas throughout its length, at the time, of 4 km (2.5 miles). The present vegetation is still casuarina woodland, with some *Hibiscus tiliaceus* and other plants of the *Barringtonia* and *Ipomoea pes-caprae* associations along the eastern coast. In 1984 there were small patches of grassland, but by 1992 these had gone, largely as a result of increasing overwash.

Because of the erosional regime operating over the past four or five decades (Chapter 5), the life span of any microsite on the spit cannot have been greater than about 10–20 years. Land newly formed by accretion on the east would have become part of the eroding west coast within a decade or so, preventing its plant succession from proceeding beyond the casuarina woodland stage to mixed forest. The spit has thus been a highly mobile, ever-young feature, with its vegetational succession held at an early stage by its short period of physical turnover. For the past several decades Sertung has been a successional composite, the main island undergoing forest development along with the other older islands but the contiguous spit being held, Peter Pan like, at the earlier grassland and casuarina woodland stages of succession. In this the island is unique on the archipelago. The contrast between the casuarina woodland and the adjacent mixed secondary forest of the more stable, older, main part of the island is abrupt (Figure 14), and of course involves the whole community. For example, in 1985 18 genera of soil nematodes (roundworms) were collected at the base of the spit, compared with 41 in the more complex soils of the immediately adjacent mixed forest (Winoto et al. 1988).

Habitat refuges

The presence of more open, early-successional habitats on both Anak Krakatau and the Sertung spit has provided refuges in these areas for open-country animal species that might otherwise have become extinct on the archipelago as their optimal habitat declined elsewhere by succession to mixed forest (Figure 39). Compton and associates (1988) reported a similarity in the chalcidoid wasp fauna of these two areas, which they ascribed to the similar vegetation type. The ant *Tetraponera rufonigra* was abundant in the extensive grasslands of 1908 and was common in 1933. In 1982 it could be found only in open glades on Anak Krakatau and Panjang. Several other aculeate Hymenoptera of open country, such as *Bembix borrei*, *Ropalidia (Icariola)* species, and *Pithitis smaragdula*, in 1982 were confined to Anak Krakatau or the Sertung spit. A vespid and a sphecid wasp, *Ropalidia variegata* and *Bembix borrei*, and two scoliid wasps, *Campsomeris phalarata* and *Triscolea azurea*, were found on all the older islands in the 50 years following 1883 but are now confined to Anak Krakatau (Yamane 1988). The lycaenid butterfly *Cata-*

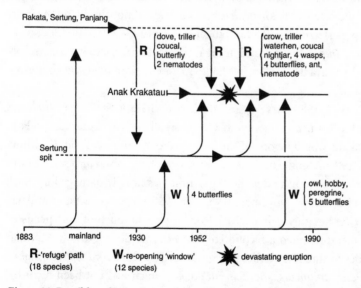

Figure 39. Possible colonization pathways for open-country animal species from the mainland to various components of the Krakataus. Anak Krakatau and the Sertung spit have served as habitat refuges and late-opening habitat windows for early-successional species. Open-country habitat *(solid line at top)* declined on the archipelago because of forest development *(dashed line),* but became available again on Anak Krakatau and the Sertung spit because of volcanic activity and erosional processes.

chrysops panormus, found on Rakata in 1933, is also now restricted to localized early-seral vegetation on the spit.

Mention was made earlier in this chapter of the zebra dove, which was last heard, on the Sertung spit, in 1951. Its confinement to this small refuge proved fatal during damaging ash falls from Anak Krakatau's 1952–1953 eruptions. Although the proximate reason for its extinction may have been volcanic disturbance, its survival was precluded by successively imposed habitat restriction, the refuge being so small that no part of it was free from ash fall. Another open-country bird, the large-billed crow, was present on every island surveyed up to 1934 but at the next survey, in 1951, it was seen only on the Sertung spit. By 1984 only a single pair remained, on Anak Krakatau. In this instance it seems that the refuges were more effective, but the crow has since become extinct. The white-breasted waterhen, *Amaurornis phoenicurus,* a frequenter of tree-shaded meadows, was confined to Anak Krakatau in 1992, as was the savannah nightjar, *Caprimulgus affinis,* an inhabitant of grassland and open country. These two species may not have survived the extensive 1992–1994 eruptions; in July 1993 their habitats had been severely damaged by ash fall. The lesser coucal, *Centropus bengalensis,* is now rare, if not absent, on the other islands but persisted on Anak Krakatau, at least until 1992. It may still survive; its habitat, the *Imperata* grassland of the north foreland, narrowly escaped being covered by a lava flow in April–May 1993 and in November 1994 appeared to be persisting.

After the 1984–1986 surveys, Zann, Male, and Darjono (1990) predicted that the pied triller, *Lalage nigra,* restricted to small areas of casuarinas on Panjang and Rakata, would be one of the next species to colonize Anak Krakatau. It did so by 1989. A single individual of the oriental hobby, *Falco severus,* was seen on Sertung a month after the vegetation was heavily damaged by Anak Krakatau's eruption in November 1952. In 1986 there was a pair on Anak Krakatau. Whether this pair derived from a population that had persisted since 1952 or represented a second colonization, we do not know. A similar doubt applies to three butterflies, the grass-feeding skippers *Polytremis lubricans, Pelopidas conjunctus,* and *Telicota augias,* which had not been recorded for over 60 years before they turned up on Anak Krakatau three years ago. These also could have persisted since the 1920s, perhaps using the Sertung spit's open habitat as a temporal bridge until Anak Krakatau's habitat became appropriate. The danaine butterfly *Danaus genutia* inhabits casuarina clearings and may also now be limited to Anak Krakatau and perhaps the Sertung spit. Thus some 18–20 species can be identified as being able to persist on the archipelago only or largely because of the provision of these two habitat refuges.

Reopening habitat windows

For mainland open-country species that missed the opportunity to colonize the Krakataus during the couple of decades of grassland phase at the beginning of the century, the new areas of early-successional vegetation on Anak Krakatau and the Sertung spit reopened windows of opportunity that decades earlier had been closed elsewhere by the development of forest.

The barn owl, *Tyto alba,* normally frequents open country and mangroves and was present on Anak Krakatau in the 1980s and at least up to 1992. It may not even have been available to take advantage of the transient opportunity to colonize at the turn of the century, as its spread to southern Sumatra and the Ujung Kulon peninsula of Java from the northwest was quite recent. A pair of peregrine falcons *(Falco peregrinus)* colonized Anak Krakatau in 1989. The peregrine's preferred habitat is high, rocky outcrops and open ground, and Anak Krakatau offered the only habitat approaching this on the archipelago, albeit somewhat pungent and hot underfoot. Five open-habitat butterflies— two nymphalids *(Precis orithya* and *Precis almana),* two skippers *(Parnara guttatus* and *Hasora taminatus),* and a swallowtail *(Graphium sarpedon)*—also probably utilized Anak Krakatau's late-opening habitat window to enter the archipelago from the mainland. Another skipper *(Pelopedas agna)* and three lycaenids *(Allotinus unicolor, Nacaduba beroe,* and *Prosotas lutea)* evidently colonized via the Sertung spit. As mentioned above, two other skippers may be either survivors from earlier Rakata populations or recolonizations from the mainland. The granivorous munias recorded recently on Anak Krakatau may not have become established, but if they did they would be a further example. Thus, not counting the munias, from 10 to 13 species may be examples of entry (or re-entry) to the archipelago through these reopening habitat windows, and in many cases entry was theoretically possible, but not achieved, 80 years or so ago.

The reopening of these windows may also have permitted the "rescue effect," mentioned early in this chapter, to operate. Populations of open-habitat species would have declined on the Krakataus to dangerously low levels as their preferred habitat was gradually reduced through succession. Population transfusions from the mainland, which might have saved them or delayed their extinction, would only have been possible when these windows reopened.

Thus the presence of early-successional stages on the emergent new island and the Sertung spit on balance may have reduced extinction on the archipelago, lowered turnover, and delayed any possible equilibrium (Thornton 1991, Thornton et al. 1989, 1990). Anak Krakatau has affected the colonization process in one other important way. The island has grown by a fairly regular series of island-building eruptions, one of which is occurring as these lines are being written.

The effect of Anak Krakatau's continuing volcanic activity

As described in the previous chapter, Anak Krakatau's volcanic activity, having aborted several early communities and checked the present community's development, has had a major effect on the development of its own biota. It has also had important effects on the other islands (Figure 40). We saw in Chapter 9 that ash falling on northern Panjang is thought to have been a major contributor to the extinction from the archipelago of the beach skink and zebra dove, and that two groups of plant geographers have recently invoked differential damage by fallout of Anak Krakatau's ash as one reason for the divergence of the archipelago's forests.

Although it is likely to be some time before magma evolution leads to it, disturbance from Anak Krakatau of very high intensity (a major Plinian or

Figure 40. An eruption of Anak Krakatau in December 1981. Ash fallout from the eruption column is moving east toward Panjang; Rakata in background.

ignimbrite-forming Peléan eruption) could, of course, set back the succession on the islands to zero.

Other Models of Biotic Assembly

Case and Cody (1987) have pointed out that the life spans of forest tree species are so long that any turnover that might occur within this group of species would inevitably be very slow. Because trees occupy the same microsite, the same part of a crucial environmental resource, space, for the whole of their lifetime, change in species composition of the forest-tree component of a community can occur only when trees of one species fill gaps in the canopy vacated by another species. Gaps may be small, resulting from the fall of a single tree, perhaps through age or lightning strike, or more extensive, perhaps as a result of landslip or ash fall, and the nature of the gap-filling succession depends, among other things, on the size of the gap. Certainly natural change within rain forests is slow. Wilson (1992) has remarked that the forest at Angkor in Cambodia dates from the abandonment of the ancient Khmer capital 560 years ago but still differs structurally from surrounding older forest. Case and Cody suggested that equilibrium of the forest-tree component of communities, if it occurs at all, may take so long that change in climate or even the size or isolation of the island may obscure any equilibration by constantly (and perhaps drastically) resetting the equilibrium number that the biota is tracking.

Bush and Whittaker (1991) have postulated that because ecological connections link the patterns of colonization and turnover of animals to the slow population and successional dynamics of forest trees, equilibrial systems in tropical lowland forests will be rare or absent (see above). The link with forest trees, however, is not of uniform strength throughout the community. For some animals the habitat or even microhabitat offered by a particular tree or a narrow range of tree species may be critical. Only bamboos satisfy the highly specialized roosting requirements of bats of the genus *Tylonycteris*, which may therefore respond to changes in the forest extremely closely. Fig wasps and other stenophagous insects also have a close, direct, and restrictive link, their survival depending on the presence of a particular tree species.

An example of a less restrictive link is that between the guild of small invertebrate grazers of microepiphytic algae and fungi on the surface of the leaves or bark of trees and their tree "hosts." The species of tree provides a particular microclimate for these insects, and the physical characteristics of its bark and

leaf substrates affect the nature and extent of their food supply of microepiphytes and, in many cases, their oviposition sites. But such indirect links are
less tightly specific (and in barklice have been shown to be so) than those involving animals feeding directly on the tree, for example by sucking its sap.
Near the extreme, the omnivorous monitor is not wholly dependent on a link
with forest trees, and even this link is a general one, mediated through a number of intervening links. The monitor thus has a considerable degree of inde-
pendence from the slow changes in floral composition and dominance
patterns of the forests. Indeed, its survival throughout the succession is testi-
mony to its freedom from such constraints.

Within the Krakatau community there is a continuum of species between
the extremes (for example, the fig wasp and the monitor), the median probably lying toward the monitor's end. Thus, although any equilibration of the
forest-tree component, and therefore of the total community, may well take a
very long time, it should not be assumed that all components would equilibrate in synchrony. Indeed, there is evidence to the contrary. Components of
the Krakatau community are moving toward asymptotes at different rates, and
such a differential dynamic pattern is probably likely to continue as forest succession and maturation slowly proceed.

On the basis of their conclusion that equilibrium will be rare or absent in
the whole lowland tropical forest community, together with the complications
of Anak Krakatau's emergence, the periodic disturbances from its eruptions,
and the observed non-monotonicity of the colonization curves of some components of the Krakatau community, Bush and Whittaker suggested that a
more appropriate theoretical framework may be developed from Lack's ideas,
especially in respect of habitat determinants. Lack (1976) proposed that the
number of species on an island is determined not by a balance between immigration and extinction but by the number of available habitats or resources.
He believed that because of competitive exclusion of immigrants by incumbent species with similar ecological requirements, the number of species is determined by the number of available ecological niches. Since potential
competitors are effectively excluded, extinctions will be few and there will be
little or no species turnover once the full complement of species is in place.
Lack thus argued that competitive exclusion would result in species assemblages that, over the short term (in ecological rather than evolutionary time),
are balanced, stable (because the number of available niches on the island is
fixed), and resistant to invasion by further species. According to this model
turnover would be almost completely restricted to transients, which are not

functional components of the community over any ecologically significant period of time.

Competition was also stressed by Diamond as an explanation of the distributional patterns of land birds in island communities of the western Pacific, although, in contrast to Lack, not at the expense of the dynamic equilibrium concept. Diamond found many examples of "checkerboard" patterns of distribution on archipelagos, which he interpreted as the result of competitive exclusion from individual islands by related species or those of the same ecological guild. Such exclusion was seen as an important basis for a set of "assembly rules," conditions for the assembly of island communities, which included the idea of assigning to species "incidence indices" on the basis of the size and complement of the island communities in which they occurred. The Krakatau collared kingfisher is one of Diamond's supertramps; this status is borne out by its obvious success on Anak Krakatau. Other Krakatau species, the olive-backed sunbird, white-bellied fish-eagle, Pacific swallow, white-bellied swiftlet, and the red cuckoo-dove, were classified as Pacific tramps of various ranks (Diamond 1974a, 1975).

In a study of over 30 islands in the Sea of Cortez, between Baja California and the Mexican mainland, Case and Cody found that the distribution of land birds was a pattern of nested subsets, the avifauna of one island simply being a sample of the avifauna of the next larger one, and so on. This type of distribution would be expected where there is little or no overlap in the ecological requirements of species (the presence of each species would be entirely dependent on the availability of its specific resources) but, unlike the model proposed by Lack, with no interaction between species. Many of the lizard species were also found to be non-interactive and their distribution was thought to be closely coupled to their individual resource requirements. Other lizard species, however, also with little overlap in ecological requirements, occurred together on islands more often than would be expected by chance. Still others were highly negatively correlated, the members of one species pair, for example, never occurring together on an island, a one in a million likelihood if chance were the only factor involved. Moreover, body size and population density of the most frequently occurring species on the islands declined with increasing numbers of other coexisting, related species. There was also indirect evidence of species interaction in the form of classical "character displacement" in two species that differ both in size and in size of prey: on "oceanic" islands where the larger lizard species occurred alone, it was smaller than it was on the mainland where the two species occurred together, and on those islands where the smaller species occurred alone, it was larger than it was when the two species coexisted.

Case and Cody thus showed that even within the same group of organisms (lizards) on islands within the same general area, the role of competition in shaping community structure may differ markedly. Both interactive and non-interactive models appear to have operated. They therefore cautioned against overgeneralizing, concluding that it was unreasonable to expect a single model to account for patterns of colonization on islands generally. As noted above, possible instances of competition, and even of competitive exclusion, can be identified on the Krakataus, although they are few and unproven. There seems to be no reason, however, why competition should not be a concomitant of the dynamic concept of MacArthur and Wilson rather than an alternative to it.

℮ɐ

Many critics of the equilibrium model have pointed to the absence of a working example. This would have to be a naturally occurring island, without environmental change or human influence, in which a dynamic equilibrium in species number over a period of time (i.e., with turnover of species), could be clearly shown by a series of comparable surveys. Such an island is difficult to find and it is likely to become more difficult in the future. Almost all islands investigated have either turned out to have been disturbed substantially by humans, or else there has been no evidence of significant turnover between surveys. The archeological work of biologists such as David Steadman has shown that on many Pacific islands the existing fauna is often an "unnatural," diminished one. Many unsuspected human-induced extinctions have been discovered when sub-fossil faunas of islands have been investigated, underlining the danger of taking extant faunas at face value, without knowledge of the island's human history (see, for example, Steadman 1989, 1993).

One of the very few instances in which turnover at equilibrium was demonstrated was the famous set of field experiments carried out by Simberloff and Wilson in the Florida Keys (Simberloff and Wilson 1969, 1970; Simberloff 1974). They "defaunated" small mangrove islets by enclosing them in plastic sheets and fumigating them. The islands varied in size, distance from a "mainland" source area of possible arthropod colonists, and number of contained arthropod species. After a year, although the islets' faunas had returned to, and leveled off at, the prefumigation numbers of species, they included only about a quarter of the particular species present originally. Species numbers stayed much the same until the experiment ended two years after fumigation, and turnover was quite high, about 1.5 species per year. Number of species bore the expected relationship to islet area in spite of the general homogeneity of all islets, so that area *per se* was an important factor. Of course these "miniature Krakataus" were not precisely analogous to the Krakatau situation, for although the islets were defaunated, they were not also "deflorated." The mangroves remained. Indeed, the plants actually comprised the islets and provided the substrate on which recolonization occurred. Simberloff (1978b) and Connor and Simberloff (1979) later reassessed the results and concluded that much of the turnover recorded was the inter-island movement of transients rather than the result of immigration and extinction in a resident population.

Krakatau biologists agree that turnover up to now has been predominantly successionally induced, with a substantial core of species, best identified in coastal plants, that has changed little in numbers or identity over the last half-century. Whitehead and Jones (1969) pointed out that it is possible, and perhaps useful, to treat this early-arriving, sea-borne plant group (Figure 27) as a separate colonization system with its own equilibrium. In a study of the avifauna of the Solomon Islands, Gilpin and Diamond (1976) also concluded that there was a core of relatively permanent island residents in which little turnover occurred, most of the turnover involving another set of species with high immigration and extinction rates. Williamson (1982), analyzing bird census data for the island of Skokholm off the Welsh coast, found species turnover to have no ecological bearing on the avifauna, which was virtually unchanged over a period of 50 years.

A situation rather similar to these natural examples was found in experimental communities. When Dickerson and Robinson (1985) assembled freshwater community microcosms in the laboratory by the sequential addition of species, most of their "colonists" persisted for only a short time, but there was a core of early-establishing species, the complement of which changed little. In some experimental microcosms the community became closed, resistant to invasion by further species, as Lack had suggested from his Caribbean island studies. Gilpin and Diamond (1982) suggested that such community cores may result from the dispersal attributes of the species concerned. Or it may be that the community is completely determined by habitat, as Lack proposed. The core would then represent that component of the biota that has filled all the available niches, and the rather transient turnover species may be those that fail to penetrate the now practically non-invasible communities because of competitive exclusion by incumbents. Caution must be exercised, however, in relating experimental studies to natural situations, and even in comparing natural situations themselves. For example, although most cases of turnover on the Krakataus have been related to succession, possible cases of replacement of species by others having similar habitat preferences have been noted above.

Use of the equilibrium model on the Krakataus as a theoretical framework within which to examine the colonization and equilibration of various biotic components can be illuminating, whether the components selected are systematic groups, groups defined also by dispersal mode, or even separate phases of the colonizing process. Such factors as succession, food requirements, competition, and disturbance are clearly relevant to the build-up of a community on the Krakataus, and the challenge now is either to build these into the dynamic model in some way or to construct a new, fairly rigorous, quantitative or semi-quantitative model. Intense disturbance is easily incorporated into the dynamic model—it obviously involves a sharp rise in extinction and a resetting of the process to zero or near zero—but the other factors may be more difficult to include.

Several explanations for the generally steady, low extinction rate of animals on the Krakataus can be suggested. The first is the non-random selection of a subset of the pool of putative colonists. One feature for which they appear to have been selected (see above) is broad ecological tolerance. Over the last half-century, the provision of late-opening habitat windows and habitat rescue through the development of early vegetational stages on Anak Krakatau and the Sertung spit has undoubtedly delayed the extinction of some species that otherwise would have been lost from the archipelago. Continuing disturbance by volcanic activity during this period may also have favored the maintenance of diversity by providing, through localized ash-fall damage, more extensive habitat patchworks than would have existed otherwise. These last two suggestions, though, would apply only at the archipelago scale, not to the island of Rakata, and if our observation of low extinction rate is borne out on Rakata by later surveys and analyses of other animal groups (several important invertebrate groups have yet to be fully analyzed), then some hard thinking will be in order.

Patience is a virtue in field studies of community dynamics and, given time, the truth will out on the Krakataus. Regular monitoring of selected components of the Krakatau community over the next couple of decades could settle the matter of equilibration for several groups. It seems likely, for example, that the number of land bird species will soon stabilize at perhaps rather less than 40 resident species, and it should be neither difficult nor expensive to determine whether the second condition (the existence of turnover at equilibrium) is occurring over several decades. There seems little to be gained by discarding a theory that serves a useful heuristic purpose without allowing the relatively short time needed for its testing to pass and before any more inclusive replacement, predictive model has been fully baked.

In the next chapter I turn to an important issue that must be considered in the context of any model of community assembly: the relative importance of chance and determinism in the assembly process.

13   COMMUNITY ASSEMBLY: LOTTERY OR JIGSAW PUZZLE?

Whether a particular species occurs in a given suitable habitat is largely due to chance, but for most organisms the chance is strongly affected—the dice are loaded—by the identity of the species already present . . . A community does not arrive on the shores of such an island [Krakatau] as a finished product. Instead, it is stacked like a house of cards, one species on another, loosely obedient to assembly rules.

Edward O. Wilson, *The Diversity of Life* (1992)

A community reflects both its applicant pool and its admission policies.

Jonathan Roughgarden (1989)

The relative importance of chance and determinism in the assembly of communities, long a central question of community ecology, has been the subject of increasing interest in the past decade.

If the presence of a species in a community were totally independent of the presence of any other particular species, its establishment on an island after arrival would depend only on the presence of a suitable physical environment. The composition of the community already present would not matter. If, at the other extreme, establishment were biotically determined, a colonist would succeed only if a particular combination of species were already present or, perhaps, if a particular combination were absent. An island community of this type could only be entered by certain species, entry or exclusion being determined by the composition of the community itself. The jigsaw picture of this community could only be extended if the piece to be fitted (the colonist) exactly coincided in shape with the empty space (ecological niche) to be filled, and the shape of this space would be precisely controlled by the shapes of the surrounding pieces (the species present) already fitted into the picture. Built up in this strictly deterministic way, the community would be a rigid, inflexible network of interconnected species with strict, tight constraints on which species could and could not enter. Knowing the composition of such a community, therefore, one should be able to predict successfully that a particular species would be able to enter, and that for other species the community would be closed.

Are island communities in fact relatively closed, with membership being strictly limited and entry being permitted only to species that fulfill specific entry requirements? Or do they accept any species that manages to reach them? Or is the answer somewhere between these extremes? Diamond's (1975) assembly rules for Pacific island avifaunas are in fact specific "entry requirements" that depend almost wholly on the complement of species already present. Are such rules of community assembly of general application and, if so, what are they and how flexible are they? These questions may be of some practical, as well as academic, importance. If there is a significant deterministic component in the community assembly process and we can recognize and characterize it, we might be able to make generalizations that would enable us to intervene in the process, where necessary, in a rational, considered way. The ability to control community assembly would be of enormous benefit in progams of ecological rehabilitation and reinstatement after human disturbance, such as clear-felling or mining, and in future decades we may even have to attempt the large-scale reassembly of tropical forest ecosystems.

Constraints on Stochastic Colonization

Dispersal is not always entirely random, as is often assumed. Of course there is a large element of chance in all dispersal modes in the sense that successful dispersal is not 100 percent assured either for any individual propagule or for any given species. The species making up the reservoir for colonization of an island, however, do not all have equal chances of colonizing. Some candidates are much more likely to arrive, to gain an interview for membership of the community "club," as it were, than others. Charles Darwin knew this, and so did Dammerman and Docters van Leeuwen. In an excellent review of this subject, Simberloff (1978a) cited studies of Galápagos plants, of arboreal arthropods recolonizing defaunated mangrove islets, and of birds in the West Indies, indicating that in these cases colonization was analogous to different darts (species) hitting a dart board (the target island). The darts are independent of one another (that is, there is no species interaction) but they are of varying quality (dispersal and population dynamics characteristics) and thus differing likelihoods of hitting the board and penetrating it (becoming established). A further example comes from a study of the rocky intertidal zone of Australian coasts. Species diversity was often found to be the result of what became known as "supply-side ecology," the apparently random settlement of disseminules on to the rocky substrate. In many instances neither disturbance nor any kind of interaction between species explained community diversity to any significant degree; successful colonization appeared to depend only on an organism's own dispersal and establishment characteristics (Underwood, Denley, and Moran 1983). Later in this chapter we shall see that the sequence of colonization and the intercolonization interval (the time between the colonization of one species and that of the next), both of which have stochastic and deterministic elements, may set the process on one of a rather limited number of courses toward a stable combination of species.

After arrival, food and habitat constraints further restrict the theoretical possibilities of a purely stochastic model. In Simberloff's analogy, some darts hit the board but fall off. In the club-joining analogy, searching questions asked at interview (on arrival at the island) floor some candidates, who are thus denied entry, but the interview is handled well by others, and they are permitted to join. The first important interview question for all candidates is, can your species survive in the physical conditions prevailing on this island or (if it is a plant propagule) at or near the site of your landfall? For animals an early question is, how will your species find food here? Not all candidates can answer these questions satisfactorily.

The period immediately after arrival is probably crucial for the colonizing species. The odds are almost always against the newcomer at this early stage, when its population is usually small and therefore highly vulnerable, as illustrated by the large number of unsuccessful colonizations of Anak Krakatau by animal species that we know reached the island. The putative colonist may make a landfall in an unsuitable habitat (a serious impediment to establishment in the case of plants) or at an inappropriate stage in the island's succession. This is one reason, in addition to the "rescue effect" mentioned in the previous chapter, why the arrival rate of a species is important for its establishment. When the environment of the target island is changing, the more trials at colonization a species makes, the more chances there are that one will be made when conditions are favorable.

It has long been known that interaction between species, particularly in successions, often provides a deterministic element in the colonization process and shapes the structure of the end community. Many studies have shown that competition between plant species is important in determining the course of succession, and the roles of habitat alteration and chance, as they affect colonization, have long been appreciated. In a review from which I have drawn freely here, Lawton (1987) summarized the models, each incorporating what may be termed sets of assembly rules, that have been proposed to explain observed patterns of plant succession. They include facilitation, in which early arrivers modify the habitat in some way in favor of later colonists but render it less suitable for themselves; inhibition, whereby priority rules, the first colonizer pre-empting a site and preventing subsequent colonization until it dies; and tolerance, in which earlier-arriving, faster-growing species are invaded by slower-growing ones that are more tolerant of the changing environment, and thus competitively superior, and these eventually supplant the specialist pioneers. Added to these is the extreme model, in which colonization of all species, both early and later in the succession, is determined purely by chance, interactions between species being of no importance. There is some disagreement on the relative contribution of these models in different circumstances, but not on the existence of some general rules of plant succession.

Unlike plants, most animals can select their habitat to some extent. Habitat preferences obviously form a set of assembly rules that will apply to different animal species as the habitat changes with plant succession. Some animals may both affect plant succession and respond to it. Herbivores in particular may alter the rate and course of succession. Various mutualistic relationships—such as those between mycorrhizal fungi and higher plants, plants and nitrogen-fixing microorganisms, plants and their pollinators and seed dispersers, plants and ants, and ants and lycaenid butterflies—may also do so. Lawton cited a number of studies showing that another general rule operates for animals in successional situations: the increased habitat stratification and species diversity resulting from plant succession promote diversity in animal species also, because the habitat and resources for animals become more diverse. Animal species richness thus generally (but not always) increases as plant succession proceeds.

The constraints imposed on community assembly by the life history characteristics and habitat requirements of individual species were stressed by Buckley (1982). He advocated breaking down island biotas into habitat components and treating these separately, rather than treating a complete island community as if habitat characteristics were the same for all components. In a successional situation the habitat is changing, of course, and we have seen that some species colonized the Krakataus only when a certain habitat, such as the mesic conditions resulting from forest formation and canopy closure, became available. Other species became extinct when this change occurred, and several were able to persist only because of the (chance) provision of habitat refuges by physical processes operating on the archipelago (such as the emergence of the new island Anak Krakatau). Such refuges may also have acted as late-opening windows of opportunity through which early-successional mainland species that for some reason could not colonize earlier, when there was extensive habitat suitable for them, could now enter the system. General habitat assembly rules appear to have operated on the Krakataus, in spite of the fact that habitat change has not been entirely uniform over the islands.

That hierarchical imperatives involving food have also applied on the Krakataus is indicated by the relatively early entry into the developing community of eurytrophic species and the later colonization of species with more specialized food needs. In some cases strict one-to-one relationships are clear. For example, the strict ties between some butterfly species and their larval food plants, like the *Delias* butterfly and the Loranthaceae, appear to have limited their chances of establishment. Others have become established in the absence of what were believed to be their plant foods; apparently they were able to survive on plants that were not thought to be their major food resources on the mainlands (Bush, Bush, and Evans 1990).

A colonist joining an existing trophic guild merely increases the number of species living at that trophic level, and its arrival may not perturb the existing community structure significantly. In contrast, it seems that a newcomer that fills a new role, or occupies a trophic level that was previously unoccupied, causes major community perturbation, and the trophic level concerned is usually high on the food chain. For example, changes resulting from the introduction of goats to islands, such as the Galápagos, that lacked browsing herbivores, are well documented. The depredations of goats may change both the structure and composition of the vegetation, affecting the erosional regime and thus the composition of soils. We saw earlier that the entry into the Anak Krakatau community of the peregrine falcon, an animal with a "new" speciality, predation on pigeons, may have repercussions on the development and diversification of that island's forest, although there has not yet been sufficient time to test this prediction, and the volcano's activity may prevent such testing.

Even when a colonizing species has survived the new physical conditions, found sustenance, and obtained an initial foothold on the island, it is still, as it were, very much on probation. It must now face other biological tests. Its arrival may be badly timed as far as the presence of other species is concerned, be they competitors, predators, hosts, parasites, pollinators, facilitators, or mutualists. Reproduction must be possible under the new conditions so that the initial success of the founding population can be consolidated. Plants that depend on pollinators must have them. For example, on the Krakataus it appears that the establishment of *Canavalia rosea* requires the presence of the carpenter bee *Xylocopa latipes*, in spite of the plant's obviously vigorous vegetative reproduction. Mutualists must have their partner species. The mutualisms between fig species and specific pollinating wasps are tight, specific, and obligatory (Ramirez 1969, Wiebes 1986), and if all relationships within the community were of such a nature there would be a highly deterministic system at the establishment stage of the colonization process. Other interrelationships, however, such as that between mistletoes and flowerpeckers, are clearly less tight.

The colonist population may also face negative relationships with other species: competition from ecologically similar species, and predation or herbivory by species already established. Anyone attempting to demonstrate these, however, must bear in mind the caveat stressed by Underwood and associates (1983): always first test the null hypothesis that random processes are responsible for observed patterns. In their experiments on the Florida mangrove islets, Simberloff and Wilson found that after a first, temporary, noninteractive equilibrium, as species populations increased further the number of species fell slightly, to a second, interactive equilibrium, presumably as a result of competitive exclusion. Diamond's "assembly rules" were largely based on the deterministic operation of competition, and such rules have also been proposed for desert rodents (M'Closkey 1978, 1985) and for assemblages of small mammals in the heathland and forest of southeastern Australia (Fox and Kirkland 1992).

Competition between species in the field has now been demonstrated by both experiment and observation. In an excellent review of studies on the structure and assembly of communities, Roughgarden (1989, p. 209) has pointed out that "one can, in the field, directly observe sharp species borders between ecologically equivalent species that are too similar to coexist." Forty years ago Edward Broadhead and I were able to recognize such boundaries between closely related species of psocopteran insects (barklice) living on the same larch trees in Yorkshire, England (Broadhead and Thornton 1955). The presumed avoidance of competition by coexisting, ecologically similar species through the fine partitioning of environmental resources has since been indicated in other insects as well as in plants and a variety of vertebrates. It is only by field experiment, however, that interspecific competition can be demonstrated, by comparing outcomes in otherwise similar situations, with and without the presumed competitor.

Possible cases of competition on the Krakataus were reviewed in the previous chapter. The evidence for the operation of rules of competition there is less convincing than that for rules involving habitat, and much weaker than the evidence—challenged by Connor and Simberloff (1979), but the challenge refuted by Diamond and Gilpin (1982) and Gilpin and Diamond (1982)—for competition rules in Pacific island avifaunas. Predation, like harsh physical conditions, can prevent competition in some communities. It is almost thirty years since Paine (1966) demonstrated that the activities of an intertidal predator (a starfish) are crucial in maintaining the diversity of its prey species (various sessile bivalves), the predator thereby being accorded keystone status. Since then other field studies (although, curiously, not those on freshwater benthos, the ecological analogue of the intertidal habitat) have confirmed the role of predation (and, incidentally, disturbance) in reducing competition and thereby maintaining species diversity. On the Krakataus there is no evidence that predators have exerted an influence on community assembly by maintaining prey diversity (which is not to say that it has not occurred, of course).

An increasing number of field studies indicate that for a given type of island a certain stable pattern that functions well as a community eventuates. A reanalysis of the data from Simberloff and Wilson's mangrove islet studies by Heatwole and Levins (1972) showed that as a new community was assembled on the islets by recolonization, the "trophic spectrum" (the proportions of broad trophic groups) characteristic of the original community was restored, although the species complement may have changed markedly. Although their conclusions have been criticized on the grounds that the result cannot be distinguished from the effect of chance, Heatwole (1971) had arrived at a similar conclusion from studying bare sand cays in the Coral Sea. Studies of the recolonization of Long Island in the Bismarks by birds (Diamond 1972) and of species turnover on a small sand cay in Puerto Rico (Heatwole and Levins 1973) supported the general conclusion.

Approximately constant predator-prey ratios have also been found both in model systems and in freshwater communities in the field, and Simberloff (1978b) noted that almost throughout the Americas small mangrove islets support arthropod communities with similar trophic compositions, albeit with basically similar source faunas. Similar remarkably stable faunal structural patterns, although not all relating to trophic groups explicitly, were found in American and Australian bird communities (Recher 1969) and in invertebrate communities of rotting logs (Fager 1968). Taken together, these studies appear to show a higher-level determinism: community composition in a particular habitat seems to be under the constraint that a particular pattern of energy flow be maintained, whatever the individual species involved.

Over the individual islands of an archipelago Becker (1992) showed that carnivores formed a surprisingly uniform proportion of the total number of carnivorous and herbivorous beetle species. He also found that the proportions of beetles and heteropteran bugs in two archipelagos at different sides of the earth, the Galápagos and the Canaries, matched. He observed that the proportion of predatory and plant-eating arthropods on British trees had been found to be constant, and that differences in colonizing ability between heteropteran families were also believed to be consistent. He concluded (p. 170) that "it begins to look as if arthropod communities on islands may be something more than a 'chance collection of organisms with similar physical requirements.'"

Simberloff (1974, 1978b) noted that in the experimental study of Florida mangrove islets, a sorting process took place as the interactive equilibrium was being reached, and perhaps afterwards. Adjustments occurred in the islets' species complements by the more frequent extinction of those that were ill-adapted to the presence of others, and the more frequent colonization of those that were well-adapted. This led to a third, "assortative" equilibrium, comprising species that formed a non-random, co-adapted community of "good mixers."

Roughgarden (1989, pp. 217–218) concluded that "a community is not simply a collection of all those who somehow arrived at the habitat and are competent to withstand the physical conditions in it . . . A community's membership *is* structured by the transport processes that bring species to it, and it *is* structured by the . . . species interactions of its members; and furthermore, it is rarely possible to focus on only one of these sides" (stress is Roughgarden's). Taken together, the studies mentioned above do indeed indicate that during the colonization process the multitude of possible chance combinations of species is limited in such a way as to produce an end-point community with a particular structure that is stable on the island concerned. Rather similar conclusions have emerged from work on experimental communities.

The Experimental Assembly of Communities

The approach to identifying factors operating in community assembly solely from comparative observational studies of natural communities suffers from the disadvantage, pointed out both by Simberloff and by Underwood's group, that it is almost impossible to distinguish random phenomena, such as which species of the source biota happen to reach a community, from deterministic mechanisms operating after arrival. The colonization process has been monitored and in many cases manipulated in field studies of small aquatic systems that involve the settlement and attachment of colonizing organisms on to natural or artificial substrate "islands." In these systems all resources save one vital one, space, are unlimited in the sense that they are available in the aquatic medium. Such studies have almost always shown that priority is all-important in determining the course of community development: early-arriving species occupy space on the substrate and thus preclude colonization by latecomers, in a manner redolent of the "inhibition" rule of plant succession, mentioned earlier. As Robinson and Dickerson (1987) pointed out, however, even in these studies there is still no guarantee that all species of the source pool have had the opportunity to colonize, and thus no way of distinguishing a failed colonization from a chance non-arrival, a "failed interview" from a "no-show."

Several groups of American experimental community ecologists (for example, Dickerson and Robinson 1985) have designed and performed colonization experiments in the laboratory, in which the immigration of species could be controlled. Robinson and Dickerson introduced microscopic freshwater organisms (Protozoa, bacteria, algae, rotifers), one species at a time, into artificial freshwater microcosms, laboratory "habitat islands," thereby assembling microcosmic island communities. Once perfected, the technique allowed them to change the order of "immigration" events (for example, to reverse it) using the same "mainland pool" of species, or to alter the intervals between immigrations (that is, to change immigration rate) while keeping the sequence of colonization the same. They found that, rather than interspecific interactions leading to a unique end-point community, there could be substantial differences between communities formed from the same species pool. The different end-points represented alternative states of the community, the outcome depending on differences in details of community assembly such as sequence of colonists and intercolonization interval (Robinson and Dickerson 1987; Robinson and Edgemon 1988).

MacArthur (1972) had already suggested that invasion sequence might influence the colonization process and the equilibrium number of species. He suggested that a family of colonization curves might apply to a given island, reflecting different sequences in which members of the same mainland pool of species might invade. Robinson and Dickerson (1987) found that changing either the immigration sequence or the inter-immigration interval (the equivalent of altering the distance between an island and its mainland source biota or changing the invasion rate) affected the species richness, species composition, and dominance patterns of the resulting communities. Although priority effects were important in some cases, the success or failure of a colonizing species could not be predicted solely by its position in the sequence. Order of immigration was more influential in affecting relative abundance patterns of species where immigration rates were relatively low (as on islands) than where they were high (as on continents). Robinson and Dickerson suggested that the arrival of species at a time when resource levels were suitable was more important than early arrival *per se*. As they put it (p. 593), "the early bird will only get the worm if there is a worm to be had, and if the bird eats fast." The number of possible end communities was surprisingly restricted. In 130 individual experimental communities assembled according to six different invasion schedules, only two basic community types, as defined by dominance patterns, resulted. This result has overtones of the field findings mentioned in the previous section. There appear to be deterministic constraints on the effects of random changes in immigration sequence and rate, constraints which are almost certainly related to interspecific interactions, including predation and competition.

Perhaps a note of caution should be introduced here. Most of the experimental systems, and many natural situations that have been studied in this context, have involved the assembly of freshwater communities or more or less sessile marine communities. Because of the often temporary nature of small natural freshwater bodies, the component species of this microbiota have evolved extremely good dispersal abilities. They are almost all specialist colonizers of a frequently ephemeral habitat. Similarly, many species of the intertidal marine communities that have been studied in this context are sedentary or fixed as adults, and most have excellent dispersal abilities, often restricted to a special dispersal phase of the life cycle and often involving the use of currents for passive dispersal. Paul Wellings, of the Division of Entomology of the Commonwealth Scientific and Industrial Research Organisation, Australia, has estimated dispersal rates for a range of terrestrial animal species, as kilometers per year in expansion of geographic range. He has told me that there is enormous variation, of several orders of magnitude, between species. Thus in many terrestrial natural communities the range of dispersal abilities is likely to be very much greater than in the aquatic systems on which most experimental work has been carried out; dispersal will not be as passive overall and dispersal rates not as uniformly high. Moreover, in many ecologically important terrestrial communities the full range of theoretically possible invasion sequences is biologically unrealistic. For example, neither a canopy tree nor a rhinoceros would be expected to be the first organism to arrive on the Krakataus. Thus perhaps we should be careful in attempting to extrapolate to communities in general findings from communities in which dispersal is largely passive and random.

Using nitrifying bacteria, algae, protozoans, and crustaceans as a species pool for colonization experiments, Drake (1991) found two types of "trajectories," or courses of development, in his experimental communities. What he termed indeterministic trajectories were not repeatable and led to alternative community end-points. Minor differences in growth rate and clutch sizes of founder populations in the replicates led to their divergence. Deterministic trajectories, in contrast, did not vary between replicates, and variation in characteristics of the founders had little effect on the result. These trajectories always led to communities with the same composition of producer species, the same dominance patterns, and the same susceptibility or resistance to the same set of consumer species. The difference between trajectory types lay in the constraints imposed on invading populations by properties of the community and environment at the time of their invasion. For example, large incumbent populations of grazers controlled the abundance of invading producers and did not permit divergence between replicates, whereas smaller grazer populations were incapable of controlling the variation in characteristics of invading producers and the subsequent divergence led to alternative community states. Alternative states, not necessarily produced in this way, have been recognized in a number of components of natural communities, including marine invertebrates, ants, Diptera, lizards, trees, and algae, as well as in the laboratory microcosms. As seen in Chapter 10, it is possible to regard the forests of the older Krakatau islands as alternative states resulting from the differential action of invasion sequence, interinvasion interval, and disturbance, on virtually the same complement of species.

Drake (1990a, 1991) concluded that the overriding rule of community assembly was that sequence of invasion determines the rules. In other words, the particular sequence determines which game is to be played, and hence which set of rules will apply. Although assembly rules are determined by invasion order, the sequence itself may be determined largely by chance. There seems to be little doubt from the work of the experimentalists that in many cases, and perhaps particularly on islands, patterns of invasion and colonization affect the structure of the resulting community by setting it on a particular trajectory. Moreover, there is some evidence that the trajectories thus determined may, in some way not yet fully understood, converge to one of a very few alternative end-point communities.

Drake made the point that both Simberloff and Underwood's group have strongly argued, that in natural situations on which we have limited information—in many cases nothing more than a "snapshot" survey of a community at a particular time—it is necessary to consider, as the null hypothesis, that the observed pattern is the result of chance. Only when random causes can be ruled out, and in such circumstances this is often difficult, should other possible explanations be preferred. Drake noted, however, that if one takes this approach it may often be concluded that community pattern and organization are the result of chance, when further information, particularly on the assembly history of the community, may show that this is not so. The status of species populations, for example, whether they are increasing or declining, and the persisting effects of environmental changes induced by species that are no longer present may have important influences on community development.

Current patterns may result from past processes that no longer operate, such as the invasion or extinction of ecologically powerful species, disturbance by human activity, or major perturbations of the physical environment (volcanic eruptions, for example); this kind of situation has been described as a community haunted by "the ghost of processes past." The mechanisms now controlling and maintaining community structure may not necessarily be those that controlled its assembly. The present mechanisms themselves may even result from earlier events that have left no other trace. Such events may have had a chance component—for example, the timing of the colonization of a particular species or of a volcanic eruption—but they may nevertheless define the set of assembly rules that will operate from that time. The assembly rules, therefore, appear to depend strongly on the historical context, and Drake argued that it is difficult to understand the way in which communities are assembled and the reasons for their pattern (or lack of pattern) solely on the basis of information on the current community. On the Krakataus, thanks to the work of earlier scientists, we are unusually fortunate in having a good deal of information about the current community's historical context.

Krakatau Colonization Patterns Compared

If the colonization process is predominantly deterministic, cases of repeat colonizations on the same or very similar islands from the same source pool of species should show basic similarities. The Krakataus have not provided us with the perfect experiment in this regard, for there have been differential changes in the island environments and, because of human activities, changes in the source pool. Nevertheless, volcanic activity has cleared the way for repetitions of the colonization process, some more precise repetitions than others, and it is of interest to examine these in this context.

Recolonizations of the Krakataus

There are several indications that chance has played a sigificant role in the assembly of Krakatau's biota. Recall that land molluscs are the only animals about which anything is known of the pre-1883 fauna. Five species, of a mollusc fauna of unknown size, were recorded before 1867, and all occur on the adjacent mainlands. Were there a strong deterministic component in faunal assembly, some or all of these may have been expected to recolonize the islands from what is still essentially the same reservoir of mainland species. The evidence of the surveys is to the contrary. Although land molluscs began to recolonize before 1908 and there are now 19 species, none of the five pre-1883 species has recolonized. It is highly unlikely, however, that these five constituted anything like a substantial sample of the pre-1883 fauna. They perhaps represented considerably less than a quarter of it. Some of the small, less conspicuous species now present may well have occurred, but may have been missed, in 1867. Nevertheless, what evidence there is certainly does not contradict a null hypothesis that the assembly of a land mollusc fauna on the Krakataus is random.

Very little is known of the pre-eruption flora either. Of perhaps 11 plant species that may have been present before the 1883 eruption, six, seven at the most, have been recorded since 1883. Unlike the case of land molluscs, this meager data set tends to oppose the null hypothesis, but it cannot be taken as disproving it, again because the pre-1883 sample is so small, probably less than 5 percent of the flora. Moreover, the identifications of four species from Webber's illustration (see Figure 15) are intelligent guesses only, and their reliability depends both on interpretative and artistic accuracy.

There is of course one important large-scale repetition of the colonizing process in which the baseline for comparison is much better defined. Anak Krakatau has had a number of early pioneer biotas in its turbulent young life, and the floral component of at least some of these was surveyed. Three floras were extirpated by eruptions in 1932–1933, 1939, and 1952–1953, the first being but a seedling flora. Of 32 identified species in these early floras, 30 (94 percent) recolonized the island after 1953 and 23 (72 percent) had already done so by 1971. A very damaging eruption in 1972 is presumed to have severely reduced this flora, and yet 95 percent, all but two of its 43 species, were present between 1979 and 1991. Anak Krakatau's 58 sea-dispersed plant species made up half of its 1989–1991 spermatophyte flora of 125 species. These species constitute the deterministic core in the colonization of plants that was recognized by Partomihardjo's group, and the fact that 42 of them have been found on all four islands confirms the core's high degree of predictability, which is presumably based on excellent colonizing (dispersal and establishment) characteristics (Partomihardjo, Mirmanto, and Whittaker 1992).

Colonization of the Archipelago and of Anak Krakatau

In an attempt to assess the degree of determinism in the assembly process, we have made another comparison, one that has less validity but is more extensive than that of Partomihardjo's group (Thornton et al. 1992). It has less validity because we have compared the colonization of Anak Krakatau since its extirpating eruptions of 1952–1953 not with that of earlier colonization sequences on that island (there is insufficient data on the earlier faunas) but with the colonization of the achipelago, and in particular of Rakata, since the eruption of 1883, over an equivalent period of time. We have treated the colonization of Anak Krakatau as a colonization model within a colonization model, and we have compared the two nested models. Our comparison is more extensive than a comparison of Anak Krakatau's plant recolonizations because we have considered a wider variety of community components.

The validity of this comparison may be questioned on a number of grounds. First, the pool of species available for the colonization of Anak

Krakatau included (perhaps largely comprised) the biota of the three older islands of the group, a subset of the mainland pool that had been selected for
colonizing ability. In contrast, immediately after 1883 the entire pool of potential immigrants to the Krakataus was some twenty times more distant, on
the mainlands of Java and Sumatra. As already noted, the roles of plant-
dispersing animals, and of dispersal mechanisms themselves, are not necessarily the same over seawater gaps of 3 km and 44 km, and the experimentalists
have shown that rate of immigration alone may affect the outcome of a colonization sequence. Also, frugivorous birds and bats were already established
on the three neighbouring Krakatau islands and were available at the outset as
dispersers of zoochorous plants to the new island. On the archipelago after
1883 such dispersers were not immediately available. In addition, Anak
Krakatau's biota was assembled in the presence of wide-ranging avian raptors
able to influence the colonization of the young island in ways that could not
have applied to the archipelago after 1883. Moreover, Anak Krakatau's colonization has occurred in the face of the considerable physical constraints imposed by the volcano's own activity, whereas after 1883 there were no further
volcanic episodes for almost five decades. Finally, Rakata is some 800 m high
and its uplands are unique in the archipelago—floristically, climatically, and in
vegetation type—and they certainly have no counterpart on Anak Krakatau.
Important floristic differences between Rakata's interior at and above 400 m
and its lowlands, however, were noted only in 1919. Its distinctive highland
biota has little relevance, therefore, in the time frame to which our analyses
were necessarily restricted (30–40 years after 1883, equivalent to the time since
1953 when the assembly of Anak Krakatau's present biota began).

Mindful of these differences between the two cases, we nevertheless believed
the comparison worth making. It is not the ideal one, but nature has provided
it, in the same geographical and climatic setting with only a small difference in
timing, and on a large scale. Moreover, we argued that if there was a substantial similarity in colonization pattern between the two models in spite of the
differences between them, this would indicate an important deterministic element in the colonization process. We therefore compared the sequence of colonization of various community components in the two models. In each case
we asked the question: Did the early colonizers of the Krakataus from the
mainland after 1883 also tend to be the early colonizers of Anak Krakatau during its colonization? Put another way, our question was: Has the colonization
of Anak Krakatau been biased in favor of the earlier-colonizing species of the
archipelago?

We examined this question over the following range of groups: spermato-phytes, pteridophytes, fig species, non-migrant land birds, butterflies, reptiles, spiders, braconid Hymenoptera, bats, and frugivorous birds. For seed plants, pteridophytes, figs, land birds, and butterflies, the data were sufficient to allow us to test statistically the null hypothesis that arrival of species on Anak Krakatau is unrelated to order of arrival on the archipelago since 1883. We also compared the dispersal-mode spectra of colonizing waves (immigrants be-tween successive surveys) of spermatophytes in the two models, and Maeto and I investigated the parasitoid mode of braconid immigrants in this context.

In the analyses of seed plants, pteridophytes, figs, non-migrant land birds and butterflies, we excluded species that were not present on the older islands throughout the period of Anak Krakatau's development up to 1983. In all analyses involving plants we omitted species introduced by humans. Our method was to compare the median arrival times on the archipelago of species that had colonized Anak Krakatau and those that had not. We found that in each case the biota of Anak Krakatau has been assembled with a bias toward earlier colonizers of the archipelago. The likelihood of this pattern occurring by chance was low: less than 1 in 20 (birds), about 1 in 100 (figs), and less than 1 in 1,000 (spermatophytes and pteridophytes).

In the case of butterflies there was not the same strong correspondence be-tween early arrival on the archipelago and occurrence on Anak Krakatau. Why should butterflies be exceptional in this? In general, butterflies are tied to the flora, through the feeding requirements of their larvae, more closely than many other well-studied animal groups, including birds. New and I had al-ready noted that the archipelago's butterfly fauna was ecologically "dishar-monic," heavily biased toward species of coastal or near-coastal habitats, with a very small proportion of deep-forest species (New et al. 1988, New and Thornton 1992a,b). The coastal flora of the archipelago, in contrast to the in-terior flora, has changed little since 1897. Anak Krakatau's flora comprises al-most entirely coastal species, and so representation, in its fauna, of butterfly species from different waves of immigration to the archipelago is fairly even. For example, 60 percent of the 1883–1908 archipelago wave is present on Anak Krakatau and 52 percent of the 1934–1989 wave.

The number of reptile species is too low for statistical treatment. Seven species arrived on the archipelago within 50 years of 1883. Four were present and available as potential colonizers of Anak Krakatau after 1953, and the two of these that were the first colonizers of the archipelago (the monitor and the *chechak* gecko) colonized the island. In contrast, of the five species to arrive in the last 50 years, only one (the paradise tree snake) has successfully colonized Anak Krakatau. This is not a striking difference, and may well be due to chance, but it is in the same direction as the bird and plant data.

Wolfgang Nentwig has told me that he found at least 90 species of spiders on Anak Krakatau in 1991, remarkably similar in number to the spider faunas of Sertung, Panjang, and Rakata 50 to 60 years ago. The same is true for family composition; in both cases the families Salticidae, Thomisidae, Araneidae, and Theridiidae were dominant and Theraphosidae and Mimetidae (which now occur on the older islands) were lacking.

Of the four bat species to arrive on the archipelago by 1933, two have been available to colonize Anak Krakatau, the lesser dog-faced fruit bat and Geoffroy's rousette. They were the first bats to colonize the island. In contrast, of the fifteen species that were first recorded on the archipelago in the last 50 years, only one has colonized Anak Krakatau (two others were recorded as single individuals). Thus the first bats to colonize the archipelago (frugivores) were also the first colonists of Anak Krakatau. The colonizing patterns of bats, therefore, are consistent with the trends seen in other components of the community.

Similar correspondences between colonization patterns of the archipelago and Anak Krakatau emerge when fruit-eating birds are considered (Table 13.1). In the first 25 years after 1883, one obligate frugivore colonized the archipelago along with five facultative frugivores, one of which was not recorded since. Of the five that persisted until Anak Krakatau's emergence, all have reached the new island (although the emerald dove failed to become established), and they constitute five of the nine fruit-eating birds recorded there.

Table 13.1.   Number of frugivorous bird species recorded on the Krakataus, by date of arrival. Facultative frugivores, which take insects and/or seeds as well as fruit, tended to be early colonizers of the archipelago and to predominate on Anak Krakatau. Number of species of each arrival wave known from Anak Krakatau in parentheses.

Species	Period of arrival				Total colonizing	Recorded on Anak Krakatau
	1883–1908	1909–1933	1934–1951	1952–1993		
Total	6 (5)	5 (3)	2 (1)	3 (0)	16	9
Facultative	5* (4*)	4 (3)	1 (0)	1 (0)	11	7
Obligate	1* (1*)	1 (0)	1 (1)	2* (0)	5	2

* Includes one "seed predator."

Of the five frugivores to arrive on the archipelago in the period between 25 and 50 years after 1883, three facultative frugivores have reached Anak Krakatau, and an obligate and a facultative frugivore have not. Of the five species to colonize the islands later than 50 years post-1883, only two have reached Anak Krakatau. Only one thrush has been seen on the island, but the black-naped fruit-dove, a specialist, may have been on the point of becoming established when the 1992 eruptive episode began. Nine of the eleven frugivores to arrive in the first 50 years, but only two of the five to arrive in the last 50 years, were generalists. They were also the most successful at colonizing Anak Krakatau (seven of eleven generalists, and one or two of the five specialists). Facultative frugivores, of course, may find sustenance other than fruit on arrival; the later-arriving fruit specialists may become more important once fruiting plants are established and their food is generally available.

Fruit bats and frugivorous birds were of course agents of the dispersal of figs both to the archipelago from the mainlands and to Anak Krakatau from the other islands, bats perhaps being more important in the latter case. Seventeen of the archipelago's fig species are bat-dispersed, and fourteen of these have been available on the three older islands as potential colonists of Anak Krakatau for the whole period since 1953. A comparison of the colonizing patterns of figs permits an additional test of the null hypothesis that colonization of Anak Krakatau is unrelated to order of colonization of the archipelago. Three of the fig species that arrived early on Rakata (in the first 14 years) were available to colonize Anak Krakatau. All have done so. Only one of the three species to arrive between 14 and 25 years after the eruption, and only two of the eight species of the 25–39-year wave, have colonized the island. None of the nine species that colonized the archipelago after four decades has colonized Anak Krakatau (but they have had less time to do so). Although seven of the last "wave" have not been available for the whole of the post-1952 survey period and have thus had less time to colonize, there are strong indications that Anak Krakatau's fig flora is also being assembled with a bias toward earlier arrivals on the archipelago.

As already noted, in the course of the colonization process there were changes in the representations of particular dispersal modes, the "dispersal-mode spectrum," of plants reaching the archipelago. When these changes are compared with the dispersal-mode spectra of succeeding waves of plant colonists of Anak Krakatau, basic similarities are apparent. In both systems the marine-dispersed proportion of plant immigrants declined over the first few decades and the wind-dispersed proportion has shown little consistent change. The animal-dispersed proportion gradually increased in the case of the archipelago's colonization, from zero in the first three years to about half of the immigrant plant species arriving in the fifth decade. In the case of Anak Krakatau, however, this fraction was over 10 percent of those arriving in the first decade after 1953, increased to about 38 percent of those arriving in the next decade, then declined markedly in the 8 years following 1971. The decline was correlated with a severe, although probably not extirpating, eruption of Anak Krakatau in 1972. The proportion then rose again to over 30 percent. The general pattern of change in dispersal-mode spectrum is thus similar in the two colonization sequences, and the differences between the two may be explained by the known history of the islands.

The much closer sources (the other three islands) to Anak Krakatau may be expected to affect the relative roles of fruit-eating bats in the two colonization sequences. The greater proximity of the colonizing pool surely favored a greater role for the smaller, less vagile, dog-faced fruit bats (*Cynopterus* species) in the colonization of Anak Krakatau than in the archipelago's earlier colonization. Moreover, one would expect generalist frugivorous birds to have a more important, earlier role where source and target are 2–3 km apart than when the sea barrier is 44 km in extent. After the 1883 eruption there was probably a latent period of a few years before any flying frugivores reached the archipelago. In marked contrast, Anak Krakatau emerged in 1930 into the middle of an archipelago with 11 species of frugivorous animals (nine of them frugivorous birds) already established on the closely surrounding islands and immediately available for the dispersal of plant disseminules. The guild of flying frugivores now included four more very important potential plant dispersers, the glossy starling, the pied imperial pigeon, and two *Cynopterus* bat species. By the time of the 1952 eruption there were 15 frugivores on the surrounding islands, and there were now two more fruit bats and an efficient seed-dispersing bird, the black-naped fruit-dove. This difference is reflected in Anak Krakatau's better early representation of animal-dispersed plants.

There is correspondence between the two colonization models in another group of organisms, insect parasitoids of the family Braconidae. Between 1984 and 1986 we collected 18 braconid species on Anak Krakatau, 14 of them (78 percent) koinobiont parasitoids of Lepidoptera. The caterpillar hosts of koinobionts are able to feed and move normally (and may thus act as a dispersal vehicle) until the adult parasitoid emerges; koinobionts are thus likely to be rather better colonizers than idiobionts, which kill or paralyze their host when they oviposit. Remarkably, the percentage of koinobionts present on the three older islands between 1908 and 1933 was also 78 percent (seven of the nine species present). The exact match with the Anak Krakatau percentage is of course coincidence, but the similarity confirms that koinobiont braconids are relatively good colonizers of early-successional habitats and demonstrates that they have been the predominant pioneer braconids in both models (Maeto and Thornton 1993).

Thus the colonization of Anak Krakatau shows similarities to the colonization of the archipelago over an equivalent period of time not only in the pioneer, largely sea-dispersed plant core but also in many of the community's other components. It is unlikely that this combination of similarities is due to chance. Rather, it strongly suggests the existence of a substantial deterministic component of the colonization process up to the stage of the beginning of forest diversification.

Does the Importance of Chance Decline?

Order of colonization is partly, but not always wholly, a matter of chance in nature. MacArthur (1972) realized the theoretical significance of colonizing sequence in the equilibration process, and it was found to be important in the development of experimental laboratory communities. There is some indication that it may have been important also on the Krakataus.

Passive dispersal mechanisms are important early in the colonization process, and groups having this type of nearly exponential dispersal characteristic reach asymptotes (flattening of the colonization curve) relatively quickly (such as the sea-dispersed coastal plants). From a simple model of colonization Seamus Ward and I conclude that stochasticity resulting from variation in either sequence of arrival or colonization interval is unlikely to be high in these early pioneer stages, when species' arrival rates will be high and have low variation. As colonization proceeds, new arrivals will be progressively poorer (later) dispersers. When arrival rates differ markedly, the same species will always colonize first and stochasticity will again be low. The high degree of determinism in the early stages of colonization, discussed in the preceding section, lends support to this view. Later in the process, most new arrivals will be active dispersers with dispersal characteristics nearer to the normal than the exponential. Arrival rates of newcomers will now be more similar, but distance from the source will now be of much greater importance, and some species will still be likely to arrive before others. Variation in colonizing interval (as between the trees *Neonauclea* and *Dysoxylum,* perhaps) will now provide the stochasticity. Species colonizing toward the end of the process will have such low immigration rates that arrival may be best characterized as single stochastic events rather than as a rate. Priority will now vary a great deal between cases and variation in colonizing sequence will now produce very high stochasticity. Thus over the whole colonization process the stochastic element of arrival should increase. It should also increase with distance from the source and, other things being equal, the increase should begin earlier in a very isolated community than in one close to the source.

For very early colonists, resources may be plentiful and competitive interactions relatively unimportant. Later, more or less stochastic factors like the incidence and size of forest gaps may also affect the chances of successful colonization. Tree falls may be caused by ecosystem events that are unrelated to the biotic community, such as lightning strike, landslide, or ash fall, or they may be due to the ageing of trees. When a tree falls in a forest, its fall immediately makes a number of basic resources available that were previously appropriated, the most important perhaps being space and light. Not only incumbent species but also the right type of colonizers—for example, heliophilous plants that may have a seed bank in the soil waiting for the opportunity—may now exploit the gap. Even minor disruption, whether natural or man-made, increases the invasibility of the community to the right kind of colonizers and provides a chink through which additional species may become established. Species with characteristics enabling them to exploit the transient opportunities provided by disturbance make good colonizers.

Once the effects of the disturbance have dissipated, competition for resources in the normal forest environment makes it a whole new ball game. Other criteria now determine the colonist's continued success. Some species are good colonizers but lack these secondary qualities. These are the shifting, nomadic exploiters of temporary opportunities, with good dispersal powers and high reproductive rates, the specialist colonizers, which must quickly seek another such opportunity or become locally extinct as succession proceeds. Examples are Diamond's (1974a) bird supertramps, early-successional trees such as *Macaranga* and *Casuarina* on the Krakataus, and the guild of barklice that specializes in ephemeral habitats by feeding on molds developing at a certain stage of the microsuccession in piles of fallen leaves. Preliminary data of Bush, Whittaker, and Partmihardjo (1992) indicate that small tree-fall gaps may be necessary for the continued presence of early-successional trees, larger gaps being occupied by species of more mature forest.

Chance events (disturbance, the timing of arrival of colonists) occurring early in the process may deflect the course of colonization into paths having widely different end-points. As the ecosystem grows in complexity and number of species, however, ecological constraints on establishment will increase in number and variety and become more restrictive and should progressively reduce the relative importance of the chance element in establishment. Moreover, as the ecosystem develops, chance events will have progressively less effect on the resultant condition simply because they occur at a later stage in the process and thus operate within progressively narrower bounds set by what has gone before. Although the stochasticity of arrival should increase, that of establishment should decrease, and establishment constraints are likely to override the chance element in arrival. The net effect should thus be an increased determinism as the number of species increases.

The very broad pattern of ecosystem development that I suggest here involves a gradual increase in the stochastic component of arrival as that of establishment declines. The deterministic component of the process of colonization derives not only from the different dispersal and establishment abilities of potential colonizers, but also from the effect of successful colonists on the establishment of later arrivals. The particular course of the process has a stochastic basis—colonization sequence, intercolonization interval, and perhaps disturbance—but is also affected secondarily and on an increasing scale by deterministic factors.

This general hypothesis of the colonization process suggests more specific questions in the Krakatau case. For example, will a close competitor happen to have preceded a given species, or will the island be free of competitors at the time of its arrival, thus greatly increasing its chances of colonization? Which potentially dominant animal-dispersed tree will arrive first, on which island, and for how long will it have the island to itself before a competitor arrives? How will its presence affect subsequent community development? What will be the severity and frequency of volcanic episodes, and will their timing be such as to affect certain key species?

To return to the much-used jigsaw analogy, the factors considered above would make the puzzle a strange (and difficult) one. To begin with, in the community jigsaw one cannot start to build up the picture anywhere: it is only possible to start with certain pieces. Moreover, not all the available pieces will be part of the picture—most will be unused after the puzzle is completed. It is difficult to accommodate turnover events into the community jigsaw analogy except by stipulating that a proportion of the pieces already fitted into place may disappear over time and others, with slightly different patterns, may be added as replacements, changing the picture slightly without destroying its overall cohesion. Further, there is not just one but a limited number of alternative final pictures, and which of these eventuates depends on which of certain key pieces happen to be fitted before others. I also think it likely that (like this analogy!) the puzzle becomes more difficult as the picture nears completion. Any one of a number of pieces of the community jigsaw will fit easily during the early stages of the puzzle, but as the picture is assembled the number of pieces that will fit into a given gap will decline, eventually to just one.

I have covered Drake's important contribution in some detail because I believe it is particularly relevant to the Krakatau situation, where we have, if not a perfect historical record of the community, at least one that is much more complete than is usual. If the present and future generations of island biologists are to grasp the opportunity offered by the Krakataus to pursue the necessarily long-term investigation of these central questions of community ecology, then, particularly in view of Drake's call for a historical perspective, we must ensure that the record for future researchers remains as complete as possible. Future studies must also concentrate on the particular interrelationships between individual species of animals and plants. It is only through such basic studies on the islands themselves that an understanding of the processes involved in the development of this young community will gradually emerge. Herein lies the importance of conserving the islands in a state as little disturbed by human interference as possible.

14

THE HUMAN PRESENCE,
PAST AND FUTURE

"You can see now why Krakatoa was always considered unfit
to live on," said Mr. F.
   "I couldn't be more completely convinced," I groaned.
   "That's the peculiar thing about nature," explained Mr. F.,
"it guards its rarest treasures with greatest care."
William Pène du Bois, *The Twenty-one Balloons* (1947)

If the Krakatau habitat were destroyed tomorrow, either by volcanic activity or by "development," probably not one species of animal or plant would be lost to the world. The loss would be of a different nature, and perhaps greater. It would be the loss of a very rare natural situation that provides insights into one of the most difficult and important questions in ecology, the way in which a natural ecosystem (in this case tropical forest) is formed.

Safeguarding the earth's biological diversity is vitally important. Krakatau, however, is a reminder that not all conservation needs are covered by the "biological diversity" umbrella, unless the meaning of this term is broadened to such an extent that it becomes all-inclusive and thus loses all meaning. The earth's biological wealth cannot be measured merely in terms of diversity, either ecological, species, or genetic diversity, important as these are, any more than the value of Shakespeare's contribution to literature can be measured by the extent of his vocabulary or the number and variety of his sonnets or plays. There is another important reason why certain special parts of the earth should be conserved with minimal human interference—they are essential for the elucidation of important biological processes. Some areas, such as the Galápagos, provide us with insights into the operation of evolutionary processes. Others, like Surtsey and Krakatau, may hold the key to crucial ecological questions if they can be effectively shielded from human disturbance.

Human Influences

The influence of much human activity on the process of community assembly could, and perhaps should, be categorized under the general heading of disturbance. A disturbance might take the form of habitat alteration, which may facilitate or impede the establishment of particular immigrants, or the accidental or purposeful transport of plant and animal propagules to the island. One of the many attractions of the Krakataus for biologists is that such disturbance has been small: the islands had never supported a long-term, substantial, resident human population, although visitors—fishermen, scientists, and others—have made their presence felt in various ways.

Habitat alteration

There has been minimal habitat alteration by humans on the Krakatau archipelago. A grass fire was started on Rakata in October 1919, when a young Dutch botanist threw away a match after lighting his pipe at Zwarte Hoek. The resulting spectacular conflagration spread over much of the western end of the island, then largely covered in grasses, but Docters van Leeuwen, who was present, did not regard it as having any serious effect on the vegetation. Handl asked to be evacuated and claimed financial compensation for damage to his concession. A naval vessel, sent out to check on the claim, reached the island four days later to find the fire out, the vegetation on the western side scorched, and all Handl's possessions intact—the fire had never reached the eastern side. When Docters van Leeuwen inspected the area in January 1922, there was no trace of fire damage.

Two small, concrete water-storage ponds on Panjang's northern ridge, one on each side of a small concrete bunker, were probably constructed when Petroeschevsky built a hut there with a concrete base in order to monitor Anak Krakatau's emergence, although they are not mentioned in Bristowe's description of the area in 1931, nor by Docters van Leeuwen or Dammerman. The tanks, each of about a cubic meter, now constitute one of the very few sizeable standing bodies of rainwater on the islands. During the 1980s they contained veliid water bugs and the larvae of baetid mayflies, chironomid midges, ephydrid Diptera, and mosquitoes.

At the base of a 20-m-high cliff on north Sertung is a small wet-season plunge pool, used for ablutions and as a water supply by visiting fishermen and others. In the early 1980s this had been edged with concrete and dammed to provide a small permanent reservoir, 2 by 1.25 m by about 0.3 m deep. The pool supported abundant small aquatic snails of the species *Melanoides tuberculata* (which occurred also in the surrounding area), dragonfly (libellulid), caddis (Hydropsychidae), and chironomid larvae, ostracod crustaceans, and dytiscid water beetles (Thornton and New 1988a). Since 1986 the water has been piped for about a kilometer to a post on the beach established by the PHPA (the Indonesian authority for nature conservation), and the pool has lost most of its freshwater fauna.

Another, even smaller concrete trough was found in 1990 on Anak Krakatau's east foreland. It had been roughly constructed since our 1986 visit and contained freshwater to a depth of about 10 cm, covering an area of 1 by 0.6 m. This pool supported dragonfly larvae, both anisopteran (several large individuals of one species) and zygopteran (one exuviae found), as well as notonectids (water boatmen), chironomids, and mosquito larvae. In April 1991 it contained only chironomid and culicid larvae, in July 1992 it was dry and overgrown, and a year later it was covered by ash fall and could not be found.

Some tree-felling and pumice removal has occurred in recent years, but this activity is believed to be under control. During our 1990 visit there was a flotilla of boats on the western beaches of both Sertung and Panjang, with well over a score of workers ashore gathering pumice for removal. The crews and collectors were apprehended by the PHPA, the skippers' licences and the pumice confiscated, and charges laid. The offenders were subsequently punished.

We have one detailed record of the rather unexpected results of human activity. On Anak Krakatau, Turner (1992) found that the ant lion *Myrmeleon frontalis* had become established around and beneath a shelter built by the PHPA between 1985 and 1986 in the trees just off the beach of the east foreland. It is the nymph of this large, weak-flying lacewing fly that is the "ant lion." In loose sand or ash it constructs a conical pit into which passing insects slide and from which they cannot extricate themselves. The ant lion sits at the bottom almost completely buried and grabs in its large jaws any prey sliding to within striking distance. The jaws pierce the prey, its fluids are sucked out, and the empty skin is flicked out of the pit. In 1991 about 750 nymphs and an unknown number of adults occurred on Anak Krakatau. Most of the pits were under the shelter's raised floor. Others were in the bases of *Saccharum* clumps on the lower slope of the cone, where they were sheltered from the east wind in the lee of high casuarinas but vulnerable to destruction by rain. Ant lions are typically found in permanently dry, dusty sites such as cave entrances or under huts. They were first noted on the archipelago on Rakata, near Handl's house, in 1920. The extensive *Saccharum* habitat, and presumably ants, have been present on Anak Krakatau for at least ten and possibly as many as thirty years. Yet the insect colonized Anak Krakatau only after the shelter was built in 1986. Its dry sub-floor area, which appears to have provided just the right conditions for establishment, was filled with ash in 1992–1995 and in 1995 ant lions were absent. Colonization of both the archipelago and Anak Krakatau by this insect appears to have been facilitated by human activity.

Human-assisted dispersal

Fortunately, dispersal of organisms by humans has not been of great significance on the Krakataus, at least until recently. There have been a few purposeful introductions (for example, of crops and poultry) and several accidental ones (such as insects, geckos, rats, weeds, and grasses).

Sunda Strait is a quite important fishing area. Small craft from Lampung province of southern Sumatra and from Labuan, Carita, and Sukanagara in West Java fish around the Krakataus mostly during the dry east monsoon season, from April to October, when seas are more easily negotiated. The rock pinnacle Bootsmansrots is a favorite shark-fishing area (I was apprised of this fact *after* swimming 100 m out to the rocks from a boat with Neville Rosengren and David McLaren). The archipelago's roughly circular configuration provides fairly sheltered anchorages whatever the direction of the weather, and fishermen disembark quite frequently for short periods to find wood or to obtain water from Sertung. In the coastal forest we have encountered temporary camps with fish-drying racks made from locally cut branches.

The likelihood of certain insects, rats, skinks, and geckos gaining access to local craft is quite high, and the inadvertent transport of snakes is a possibility. In fishing villages cargo is usually stacked in the supralittoral zone, and on the west coast of Java boats are routinely moored some way up coastal creeks and rivers, at the back of houses. Smaller boats are pulled high on the beach for safety, as are larger vessels needing repair. Thus supralittoral, coastal, and domestic animal species may easily enter boats directly, or gain access to cargo, and those occurring around fishing villages must have a quite high probability of dispersal to the Krakataus by boat traffic.

Scientific expeditions have become much more frequent since 1979, and scientists bring stores and equipment ashore and stay for periods of at least a few weeks, and much longer visits by individual scientists have now begun. In 1990 a wooden PHPA building was erected on Sertung, and this has been manned for periods of months at a time, with supply and relief vessels commuting fairly regularly. Up to 1993, when a second building was added, the staff there had no vessel.

There has been only one short period of permanent human presence on the islands. Handl's group was present on Rakata from 1916 (1914 according to Backer, 1929) to 1919 and made regular visits until 1922. The legacy was the black rat and 17 species of weeds and cultivated plants. All but the rat were quite quickly almost totally eradicated through competition with the natural plant community. Ten years after Handl's house had been abandoned, Docters van Leeuwen (1936) could find only four of the plant species, the rest having been shaded out, and by 1983 only three persisted.

Docters van Leeuwen also recorded nine introduced weed species on Panjang between 1928 and 1934, all from the area in which the Volcanological Service had camped from 1928 to 1931 when monitoring the birth of Anak Krakatau. These have all disappeared—they were probably shaded out soon after the post was abandoned. Evidently there were other, unknown medium-term visitors to the archipelago. In 1951 a house built of local materials was found on Sertung, with a garden containing 11 newly introduced crop and weed species. Deposits of ash from Anak Krakatau's 1952 eruptions destroyed all these (van Borssum Waalkes 1954, 1960). In the same year there was another cottage of natural materials at Handl's Bay, Rakata, with a well and a *ladang* (clearing) for the planting of coconuts.

Of the main plant-dispersal groups, the one dispersed accidentally by humans, often by hooked or barbed seeds adhering to clothing or taken along as food, has suffered by far the greatest losses. Whittaker and his colleagues listed 32 plant species that had been "introduced" to the islands up to 1979, only three of which were found between 1979 and 1983 (Whittaker, Bush, and Richards 1989). In recent years the introduction of plant species has accelerated, at least on Anak Krakatau. Its volcano has become a tourist attraction, particularly at weekends in the east monsoon season. Most visitors usually climb to the outer cone, volcanic activity permitting (and once, with tragic results, even when it did not), or, when the volcano is quiescent, to the crater rim, staying for only a few hours. Occasionally, however, overnight camps are set up on the beach. All but three of the 20 plant species, including 11 grasses, for which these excursions are probably responsible, were surviving in 1992. Thirteen species, including 9 grasses, were first discovered in Partomihardjo and Mirmanto's 1989–1991 surveys, mostly along trails or close to camp sites. Of the 26 species that now occur on Anak Krakatau but not on the other islands, 1 is assessed as having been brought by wind, 4 by birds or bats, 9 by sea, and 12 (including 6 grasses) by humans (Partomihardjo, Mirmanto, and Whittaker 1992).

The house rat, or black rat, *Rattus rattus,* perhaps the most important animal introduction to the islands, almost certainly arrived through the agency of boat traffic during the sojourn of Handl's group. Handl first noted it in 1918, and by 1920 it was present on the other side of the mountain, at Zwarte Hoek. Still the only rat species known on Rakata, it has recently been found also on Sertung and Anak Krakatau. The cream-bellied country rat, also known as the field rat or plantation rat, *Rattus tiomanicus,* was extremely numerous on Panjang when the Volcanological Survey team landed there in 1928, which means that a breeding population was established well before then. A Topographical Survey team had spent five months on the island in 1896, and the rat may have been introduced during that visit; other animals, but not rats, were mentioned in the Topographical Survey's report, implying that rats were absent before 1896. Dammerman (1948, p. 316) noted that the country rat "is easily introduced as it is found even on the smallest islands and especially those which are uninhabited," but the species evidently took several years to reach Sertung. Rats were absent from Sertung in 1921 and in 1933 the island was "still very poor in mammals . . . and rats have so far never been met. During each visit to the island numerous rat-traps were set but no rat was ever caught" (Dammerman 1948, p. 50). Between 1933 and 1982 Sertung was visited only in 1951 by a zoologist (Hoogerwerf) who did not stay overnight. The rat was discovered there in 1982, and since Iwamoto (1986) found morphological differences from the Panjang population, he considered that the Sertung population resulted from a second introduction, probably by fishermen from Sumatra, rather than by dispersal from Panjang. He agreed with Dammerman that *R. tiomanicus* was probably dispersed through boat traffic, citing its delayed colonization of Sertung as evidence against dispersal by natural means. A thriving population was found on Anak Krakatau in 1990.

As noted in Chapter 7, several of the geckos present on the Krakataus are associated with humans and some, particularly the house geckos, *Gekko gecko, Hemidactylus frenatus, Cosymbotus platyurus,* and *Lepidodactylus lugubris,* are likely to have arrived on boats or among baggage and provisions. Probably the commonest skink in the Sunda Strait region is *Mabuya multifasciata* (Figure 41), which occurs at very high densities along the Javan coast. Its colonization of Rakata was fairly precisely documented by Dammerman as being between January 1922, when it was not found during intensive collecting in the coastal region of southeastern Rakata, and July 1924, when large numbers were present in the same area. This is the area in which Handl's group stayed and was also a favorite camp site of Dammerman and Docters van Leeuwen. Although Dammerman believed that natural rafting was the most likely method of its dispersal, it seems likely that boat traffic was also involved.

There are a few instances of the possible dispersal of invertebrates as a result of human activities. These include a pseudoscorpion, the cockroaches *Periplaneta americana* and *Periplaneta australasiae,* a tenebrionid beetle, two centipedes, and several household spiders, all thought to have arrived with camp

Figure 41. The coastal skink, *Mabuya multifasciata,* arrived on Rakata between 1922 and 1924 and was confined to that island until 1991, when a population was found on Sertung.

stores, and three "semidomestic" culicine mosquito species. A dorylaimid soil nematode of the genus *Xiphinema* is suspected of having been brought to the islands with plants, and the present ant fauna includes a number of well-known domestic species.

Apart from the pigs on Panjang, which may or may not result from human introduction (Chapter 7), the few purposeful introductions of animals to the islands have not resulted in established populations. A small black dog was seen in the vicinity of what is now known as Owl Bay on Rakata in December 1933, and the lonely animal was still present in April 1934 (Dammerman 1948). In 1978 a mystic and his cat took up residence on Anak Krakatau. Both were removed by the Indonesian authorities after the mystic had (correctly) forecast the volcano's next eruption. Fishermen occasionally tether goats ashore to sacrifice and consume on feast days, and tourists as well as fishermen occasionally bring live domestic poultry to the islands so that fresh meat will be available. Fortunately, there have been no accidental escapes of potential population founders.

When one considers the frequency of visits to the islands and the very densely settled areas of Java and Sumatra only 44 km away, the situation could have been very much worse. The great value of the islands as natural laboratories could have been seriously eroded. That this has not eventuated is due in large measure to the awareness of the archipelago's scientific importance by Indonesian authorities, such as the Directorate of Forest Protection and Nature Conservation (PHPA) and the National Institute of Science (LIPI), and to their wise use of the limited resources at their disposal to control visits to the islands by both tourists and scientists.

The Krakataus as a National Park

Krakatau became part of Indonesia's national park system by riding on the glamorous coat-tails of Ujung Kulon, the justly famous national park on Java's western tip which is a sanctuary for the Javan rhinoceros, one of the world's really rare large mammals.

In 1914 plans were made to set aside Rakata as a nature reserve. In October 1916, however, in spite of strong community objections, Handl was granted a 30-year lease of 870 hectares, virtually the entire eastern half of the island, with a concession to extract pumice for use as building material. Evidently the economic rewards of development overrode conservation considerations in those days, as so often today. Handl left Rakata in February 1917 when the lease was canceled because he was unable to keep its terms, and in July 1919 the western half of the island, together with Sertung, was designated a National Monument. In 1925 the eastern half was added, and Rakata and Sertung were included in the Ujung Kulon reserve, some 45 km to the south, which had been set aside four years earlier as a "Strict Nature Reserve" administered by the Department of Internal Affairs. In 1937 the islands of Panaitan and Peucang and the Handeulum islands were added to the Ujung Kulon reserve, as was a large area of the adjacent Mount Hondje Forest Reserve, and administration was transferred to the Forest Service. This entire area was declared a forest sanctuary and prohibited area.

Ujung Kulon was declared a National Park *(Taman Nasional)* in 1982, with the Krakataus as a Nature Reserve *(Cagar Alam)* within the park. Until recently, this combination resulted in the paradoxical situation that the political control of the Krakataus, which are part of the Lampung province of Sumatra, was separate from the conservation control of the national park of which it was a part. Moreover, the Krakataus were very much the lesser half of the combination. They carried no endangered species, whereas Ujung Kulon—well known for its tiger (until the 1970s), leopard, banteng (a wild ox), deer, gibbons, crocodiles, some 200 species of birds, and virtually the last remaining natural population of the Javan rhinoceros—attracted international attention and concern. By the early 1990s, visitors' accommodations at Ujung Kulon had been upgraded to a state of luxury that some feel to be inappropriate for a wilderness area of such excitement and attraction. Thus while the peninsula was fully manned and patrolled by armed rangers to deter rhinoceros-horn poachers and the staff there had a power boat at their disposal at times, the Krakataus were unmanned until the late 1980s and the staff, with no vessel of their own, had to rely on radio contact with Sumatra when transportation was required. The control of access to the islands from Java by local or international tourists was theoretically through the PHPA office in Labuan, West Java, but the procedure was routinely circumvented without penalty. Control of scientific visits, however, was administered from Tanjung Karang in the Lampung district of Sumatra.

The paradoxical administrative situation was resolved in 1990, when the Krakatau Archipelago became a Nature Reserve in its own right, separate from Ujung Kulon. The declaration of 1982 making Ujung Kulon a National Park was passed into law in 1992. Now both PHPA and political administration of the Krakataus are from Tanjung Karang in Lampung. The West Java office of the PHPA in Labuan, however, continues to sell tourist permits. Park rangers from Tanjung Karang are stationed on Sertung for duty periods, but they are still isolated and without a vessel, so their control over visitors to the other islands of the group is difficult, to put it mildly. There is no lack of personnel with the knowledge and training to care for the islands effectively; the urgent need now is for funding to permit the purchase of two reliable seagoing power vessels, one of which could then be stationed on the islands at all times.

Indonesia has not followed the example of Iceland, which restricts access to its "new island," Surtsey, to scientists only. The Krakatau volcano has recently become the focus of a burgeoning tourist complex on the western shore of Java, and tourist dollars are welcome in Indonesia. With a staff of wardens in place and the practical means to carry out their duties effectively, the Krakataus could be advertised to the same extent as Ujung Kulon. Some of the profits from tourism in West Java, as well as park fees, may provide the funding necessary for its improved management. Properly organized visits, under the compulsory, close guidance of trained PHPA staff, as at Ujung Kulon, would have little or no impact on the ecosystem; indeed, much can be seen and appreciated without even going ashore. Although the archipelago lacks Ujung Kulon's large mammals, a visit to the Krakataus has an excitement of its own. After some fourteen visits I still get a shiver as the boat rounds the eastern point of Rakata and enters the great submarine caldera. Even if Anak Krakatau is quiescent, Rakata's great cliff scar impresses upon the visitor the enormous volcanic forces that were released in 1883. One cannot fail to be reminded who is in charge. At the same time, the mixed forest clothing the three older islands testifies to the relentlessness of the gradual process of natural recuperation and recovery. The visitor has a glimpse of two sides of the same coin, both the destructive and the healing powers of nature, each seeming to exceed by far anything that human power, with all its technological aids, can yet achieve.

Future Research on the Islands

The archipelago's young forests are much simpler than the mature lowland forests of the mainlands. They comprise far fewer species and, rather like temperate forests, are characterized by only one or two dominant tree species. Thus they are highly suitable systems for detailed studies of tropical forest dynamics, development, and succession. Eight permanent botanical plots were established by Whittaker's group on the three older Krakatau islands: two on each island in 1989, and two further sites on Rakata in 1992. A "control" plot was marked out on Panaitan Island off Ujung Kulon and another on the lower slopes of Gunung Galunggung on the adjacent Sumatran mainland in 1992. Three permanent plots on Anak Krakatau were set up in 1990 by Partomihardjo. All these were carefully selected by the botanists concerned as being representative of important plant associations, and their long-term monitoring (except two on Anak Krakatau that were covered by lava in 1993 and 1994) will provide basic qualitative and quantitative data on the successional dynamics of these developing associations.

Susanne Schmitt, of Whittaker's group, has begun full-time studies of forest dynamics on the islands. These will involve a comparative ecological study, in forests both on the Krakataus and the adjacent mainlands, of the filling of light gaps formed in the forest canopy by natural tree fall, the major mechanism of tree replacement. Data will be gathered on the composition, species diversity, rates of growth, and turnover of forest plants. Comparative floristic, vegetational, and ecological data should allow the identification of differences in composition, structure, and dynamics between the recovering forests of the Krakataus and the mature mainland forests.

The monitoring of colonization, both of the archipelago as a whole and of Anak Krakatau, a long-term extension of the work of earlier investigators, needs to continue in order to answer several controversial questions concerning the process of community assembly. Will the fauna ever reach a dynamic equilibrium of the type envisaged by MacArthur and Wilson? If so, how long will this take? Will some components of the biota equilibrate before others, on some islands before others, and, if so, which and when? What will these components have in common, and what will their similarities tell us about the process as a whole? We already know that the islands have different forest types, albeit assembled from virtually the same component species, and following the development of these over the next few decades will be an absorbing study. Will the forests diverge further and, if so, how and to what degree? Or will they converge, so that the forests of the various islands come to be more or less the same? How closely will animals respond to these changes and what part, if any, will they play in helping to bring them about?

In recent years one or two more butterflies of the interior forests have been noted, and a transition to a slower colonization by forest-dwelling species may now be beginning. Although species surveys should be continued in order to monitor such changes, it is information on species interactions that will provide evidence of the progressive ecological integration of the biota as its diversity increases. Each of the major habitat types on the archipelago, and each of the principal plant species, is progressively building up a suite of animal species comprising herbivores of various degrees of specialization and the predators and parasites that depend on them to various degrees. The complexity of these associations is increasing annually, and they are expected to equilibrate in broad terms. The relative simplicity of the Krakatau ecosystem (which becomes less simple every year) may permit this process to be followed. Studies of the dispersal and pollination of key plant species, particularly dominant trees, including their relationships with land crabs, rats, nectarivorous and frugivorous bats and birds, moths, ants, and bees should also continue to be rewarding. Louise Shilton, of Leeds University, U.K., with Compton and Whittaker, is beginning an extensive investigation of the role of animals as dispersers of plants to the Krakataus and between islands of the archipelago.

The nature of the first species assemblages is important both in the colonization process and in the further development of the island communities, and Anak Krakatau's periodic eruptions may provide opportunities to repeatedly observe and compare these initial associations. In this way insights may be gained into the way their composition may constrain both colonization and succession. Certainly the eruptive episodes will allow assessments of the comparative susceptibility of community components to fairly frequent disturbance, and their rates of recovery. One such question, mentioned in Chapter 11, concerns the effect on the developing vegetation of the relative recovery rates of vertebrate plant-dispersers and their avian predators, and there are others. For example, on Anak Krakatau the pioneer ground creeper *Canavalia rosea* appeared to have been almost completely obliterated in July 1993 by recent eruptions. The year before, Gross (1993) had found that its specialist faithful pollinator there was the large carpenter bee *Xylocopa latipes*. The bee nests in *Casuarina* trees, which had survived fairly well. Will the bee also survive the volcanic episode and, if so, on which plants will it forage? If it is a rigid specialist, unable to adapt to other floral resources in the face of a drastic decline of *Canavalia*, it may become locally extinct. Other, smaller, anthophorid carpenter bees (of the genus *Amegilla*) nest on the ground in driftwood and fallen wood and in the understorey (which also appears to have been seriously damaged). These more opportunistic, generalist pollinators serve many plant species, and their survival would be important in the recovery of the plant community.

Other questions involving pollination concern the diversification of the forest. In 1992 we found, unexpectedly, that on Anak Krakatau pollination rates of two very rare fig species, *Ficus hispida* and *F. fistulosa*, were higher (approaching 100 percent) than those of the two common, earliest-arriving, dioecious pioneers, *Ficus fulva* and *F. septica* (18 percent and 81 percent, respectively, compared with 93–97 percent on the older Krakatau islands). The very low pollination rate of *F. fulva* was not due to an excess of female (non-wasp-producing) trees, nor to the action of predators or competitors of the pollinators within figs. *F. hispida* and *F. fistulosa* are well below the critical population size at which they can sustain populations of their own pollinators (see Chapter 9). They must depend on wasps carried to the island in the air, but these are perhaps sufficient to effect complete pollination of their very few trees. The influx of pollinators may be inadequate for *F. fulva*'s much larger population, which, however, may not yet have reached critical population size, some trees not having neighbors producing wasps at the correct time. The airborne pollinators were perhaps not yet so limiting for *F. septica*, which had a smaller population and smaller average crop size than *F. fulva*. Stephen Compton's group hopes to use the "natural experiment" provided by the present eruptive episode to test this hypothesis. Any damage to *F. fulva* and *F. septica* may be expected to have reduced their pollinator populations, perhaps setting them back to the early stage of the colonizing process.

Several other questions concerning the fig-pollinator mutualism, highlighted by Bronstein (1992), could well be investigated on the Krakataus, where the fig flora is relatively small. As examples: How far do fig-pollinating wasps typically move on leaving the syconium? Are hermaphrodite and female syconia of dioecious fig species equally attractive to pollinators? One question particularly applicable to the Krakatau situation is: Is the pioneer flora of monoecious fig species biased in favor of species with long sexual phases and short intervals between successive reproductive episodes?

Other problems concerning colonization are likely to be solved by a combination of good old-fashioned fieldwork and the application of high-tech molecular biology. Simon Cook, of La Trobe University, Melbourne, has begun a research program that involves sampling island and mainland populations of selected species of land birds and bats. Mitochondrial DNA is extracted from samples of feathers, hair, or blood, where appropriate, and then amplified by a polymerase chain reaction (PCR) technique. A good deal of demanding laboratory work should permit the identification and characterization of segments of the DNA molecule of the individuals sampled from these populations for comparison with similar samples from the purported source areas on Java and Sumatra. It may then be possible to identify the sources of the island populations and perhaps even to determine how many successful colonizations there have been from each. Further than this, the PCR technique can now be applied to samples of hair and feathers from museum specimens, so that information can be obtained about past populations, both on the Krakataus and in source areas. The history of past colonization of the islands by the species concerned may therefore be even more precisely revealed.

The Krakatau studies have a fairly obvious relevance to problems of rehabilitation and reinstatement after major disturbances to the environment through human activity, such as clear-felling. The reinstatement of a natural ecosystem is an enormous, perhaps impossible challenge, and the difficulty is compounded when little or nothing is known about its natural functioning. If we are to turn rehabilitation and reinstatement into a science like the other applied ecological sciences such as forestry and agriculture, much hard, basic ecological research must be done first. Such research, alas, is often the Cinderella when it comes to obtaining funding, even when politically correct public statements would imply otherwise. In forestry or agriculture, the goal is usually a relatively simple one biologically, perhaps maximal production of timber or wheat. In reinstatement, the goal is the dynamic ecosystem before disturbance. This is usually unknown in detail, and if known it is likely to be much more complex and less understood than a monoculture of timber trees or a field crop. Thus basic studies of major ecosystems should be ongoing, whether there is a current problem requiring solution or not.

Ecological processes take time, and time must be given to their study, so that we can learn to distinguish the effects of what the medical people would call our "intervention" from the natural heterogeneity, the enormous natural variation, and the natural changes that are occurring all the time. The goal may in fact be unattainable in practice, but this does not mean that we should not strive to reach as close to it as possible. Reinstatement is an embryonic, not even a fledgling science, and we must learn by good basic and experimental research so that we can answer such questions as: What is the succession likely to be at this site? Which stage of the succession could our intervention best accelerate? What would be the natural recolonization and succession if we did not intervene at all? Krakatau shows that natural recovery can occur, at least to some degree, given time. After a century, and without any human intervention, a new ecosystem has been assembled across a sea barrier 44 km wide: from the shore to the peak of Rakata is a tropical forest community comprising over 400 vascular plant, 50 butterfly, 40 land bird, 14 bat, 9 reptile and hundreds, perhaps thousands, of invertebrate species.

Thus the scientific case for conservation of the Krakataus rests not on the need to save a rare or endangered species but on the need to learn from a unique and long-running natural experiment. It is the developing ecosystem, an example of the natural recovery of lowland humid tropical forest from the most extreme natural disturbance, rather than any "flagship" endangered species, which attracts the interest of the world's biologists. The case for a well-funded conservation effort to safeguard this archipelago—unparalleled elsewhere in the world as the cradle of a developing isolate of one of earth's most important, complex, least understood, and most endangered ecosystems, the tropical forest—is surely outstandingly strong. Added to its biological importance are other compelling reasons for conservation: the historical interest and aesthetic appeal of the archipelago, and its undoubted continuing interest to the earth sciences.

There can be few areas of the earth with a stronger case for World Heritage listing, yet the case is much more difficult to advocate and publicize than more obvious and appealing (and certainly important) ones, such as the need to conserve rhinoceroses, tigers, golden tamarinds, giant pandas, or birdwing butterflies. If this book has done anything to further the cause of an international effort to assist Indonesia in its valiant efforts as custodian of this remarkable group of islands, so signally touched by nature, it will have been well worth the effort.

GLOSSARY

aa   Lava that has cooled as an unstable jumble of sharp-edged, irregular blocks. Beds of *aa* lava are difficult and often painful to traverse.

acid rocks   Rocks having a high proportion of silica (more than 65 percent). Acid rocks have relatively low density and high viscosity and thus tend to form **pyroclastic rock**, rather than **lava**, during a volcanic eruption.

aeolian   Wind-borne.

andesites   Rocks of moderate acidity (silica content about 60 percent), density, melting point, and viscosity. Commonly produced in Andes-type volcanoes, they are formed at least partly by partial melting of the **oceanic crust** of a descending plate.

anemochory   Dispersal by wind.

anthropochory   Dispersal of organisms as a result of human activity.

apterous   Lacking wings.

arboreal   Living in trees.

assembly rules   Conditions determining the combinations and sequence of **colonization** of species that form a **community**.

association   A group of plant species repeatedly found together in a particular habitat and characterized by dominant species. An association is usually named after one of the dominants; for example, the *Ipomoea pes-caprae* association.

asthenosphere   Layer of the earth's **mantle** below the **lithosphere**, which consists of crystalline rock with a low percentage of liquid. Within this layer seismic waves travel more slowly than they do in the upper mantle; also called the *low-velocity layer*.

barklice   Insects of the order Psocoptera living on trees and shrubs.

basalts   Dark-colored, basic rocks of low (about 50 percent) silica content, high density and melting point, and low viscosity as molten rock. Basalts are characteristically formed at mid-oceanic ridges from partial melting of upper mantle material.

base surge Surface cloud of dilute, low-density material rolling away from the base of an eruption column down the flanks of the volcano at high velocity as powerful horizontal blasts; first recognized in test explosions of nuclear weapons. Also known as a *ground surge*.

basic rocks Rocks that are usually dark and relatively dense, with low silica content (about 50 percent) and low viscosity, thus tending to form lavas during a volcanic eruption.

belukar The local word for young secondary growth on disturbed land in Malaysia and Indonesia; a precursor of secondary forest. See **secondary succession**.

biogeography The study of the patterns of distribution of living organisms and the processes responsible for the patterns.

biota The fauna, flora, and microorganisms of an island (or region or area).

blast Violent effect that may accompany volcanic explosions. Sudden decompression of magma, for example as a result of a crack in the overlying rock, results in the rapid vaporization of water; if this occurs on a volcano's flank, the resulting blast is directed laterally over a narrow sector.

bryophytes Mosses and liverworts.

buttresses Plate-like extensions of a tree's trunk radiating from the base between the trunk and the horizontal roots.

caldera A large, usually circular depression of the surface of a volcano; caused either by explosion or, more usually, by sinking of the area due to the emptying of the **magma chamber**.

canopy The cover formed by the uppermost layer of branches and leaves in a forest; may be closed (continuous) or open (with gaps between the canopies of individual trees).

cauliflorous Having flowers arising directly from the trunk or major branches.

chalcids A group of small wasps, including parasitoids of other insects and the pollinators of figs.

character displacement Evolutionary change in the expression of a character of a species in the presence of a competitor (compared with its condition in the competitor's absence), such that the two species diverge in this respect.

character release Evolutionary change in the expression of a character of a species in the absence of a competitor (compared with its condition in the competitor's presence), such that the two species converge in this respect.

co-ignimbrite ash deposits Deposits resulting from fallout from high, voluminous ash clouds formed in association with an ignimbrite-producing **pyroclastic flow**.

collapse The sinking of land through the roof of a **magma chamber** after the chamber has been emptied by a large-volume eruption.

colonization The arrival and establishment of a species in an area from which it was absent.

colonization rate Strictly, the difference between immigration rate and extinction rate; rate of net gain in species of the group concerned.

community A well-defined assemblage of interacting species, clearly distinguishable from other such assemblages.

community ecology Branch of ecology concerned with the structure, development, characteristics, and functioning of communities.

competition Interaction between individuals of a species population (intraspecific) or of different species (interspecific) that have demands on the same limiting resources (e.g., food, water, living space), which is deleterious to the population or species concerned.

competitive exclusion Exclusion of a species from an island because of the presence of a close competitor. Competitive exclusion may be indicated by a "checkerboard pattern" of incidence of species on the islands of a group, with the competitors never co-occurring on an island (this needs to be distinguished from a chance distribution), or by the demise of a species on an island coinciding with the immigration or rise in population of its competitor. D. Wells (1982) cited a case on Redang Island, Malaysia: as the purple-throated sunbird, absent in 1950, became common, the olive-backed sunbird declined; the latter species had disappeared by 1977, except on a small islet not inhabited by the purple-throated sunbird.

consumer An individual that feeds on other organisms, whether these are dead or alive, for its source of energy and nutrients.

continental crust The earth's **crust** under continents, usually 30–45 km thick. It is composed of a variety of igneous, sedimentary, and metamorphic rocks that are usually light-colored, rather coarse-gained, and acidic.

crater The opening of a volcanic vent from which magma is erupted.

crepuscular   Active around dawn or dusk.

crust   The outer layer of the earth, above the **Moho boundary**.

cryptoturnover   **Turnover** that occurs but is not scored because of the immigration and extinction of the same species between surveys.

debris-avalanche   **Debris flow** resulting from the partial collapse of the edifice of a volcano.

debris flow   Very high density flow deposits.

declination   Angular deviation of a compass needle from true (geographic) north.

density compensation   The occurrence of higher densities of a species where fewer competitors are present. Cranbrook (1988) gave an example: there are fewer species of small mammals in the forest of Tioman Island, Malaysia, than in mainland forests, but their average densities on the island are higher.

detritivore   An animal that feeds on **detritus**.

detritus   Small pieces of dead and decomposing tissue of animals or plants.

dioecious   Having the sexes separate, as in humans. In plants, having male and female flowers on different individuals.

disseminule   Part of a plant or animal (such as a seed or fertilized egg) that is moved from one place to another and is capable of developing into another individual.

dominant   In ecology, describing a species that has major ecological influence in the community, perhaps through size (for example, canopy trees) or numbers (for example, a particular species of grass in a grassland).

drupe   A more or less fleshy fruit with one seed compartment and one or more seeds; also called a *stone fruit*.

ecological release   The broadening of a species' **niche**, which may or may not include habitat expansion, in cases where there are fewer competitors; also termed *niche expansion*. For example, Wells (1982) found that on the Langkawi island group, Malaysia, the collared scops-owl and the white-rumped shama have larger foraging ranges than their mainland counterparts, which have more competitors. Ecological release is often implied by **character release**.

ecosystem   The physical environment and biotic community of a discrete habitat or area.

endozoochory, endochory Dispersal of an organism by an animal that carries a propagule from one place to another inside its body, usually in its digestive tract.

epicenter That point on the earth's surface that is directly above the focus (point of origin) of an earthquake.

epiphyte A plant (such as an orchid or fern) growing on the surface of another plant (for example, a tree) without extracting nutrition or water from it.

equilibrium The dynamic state of balance in numbers of species on an island when immigration of additional species is roughly compensated for by extinction of species from the island; the number of species on the island would thus fluctuate within rather narrow limits.

eruption column Mass of gases, ash, and rocks ejected by a volcano. The eruption column is propelled by the explosive force of the eruption; later, buoyant gases and dust may rise and travel as an *eruption cloud.*

establishment Following arrival, the persistence of a species to the point of breeding and population increase; the second phase of colonization.

euryphagous Feeding on a wide range of food.

eurytopic Having a wide distributional range.

extinction In this book, the loss of a species from an island or archipelago.

extinction rate The number of species becoming extinct per unit time.

fall Material, such as volcanic ash, that has been ejected into the air and has fallen back on to the earth's surface as a fall deposit or fall layer.

fault Differential movement of rock on opposite sides of a fracture.

fauna The animal species of an island (archipelago, area, or region).

faunistic change Change in the composition of the fauna of a community.

flagelliflorous Carrying flowers at the tips of long, pendulous branches.

flora The plant species of an island (archipelago, area, or region).

floristic change Change in the composition of the flora of a community.

flow Ejected volcanic material that moves along the surface of the ground (or on or under the sea) from the eruption point; a flow may arise directly, from the vent, or indirectly, as a result of the collapse of the **eruption column**. As material settles, a flow deposit is formed.

focus   The point of origin of an earthquake.

food chain   A particular sequence of species, each being the food of the next, by which energy flows through a community.

food web   The combination of interconnected individual food chains in a community or ecosystem.

frugivore   A fruit eater.

fumarole   An opening in the earth issuing steam and gases.

graben   Depression in the earth's surface between two faults where a block of rock has sunk as a result of the faulting.

gravity anomaly   The difference between the measured value of gravity (corrected for elevation and the gravitational attraction of the rock between the measuring instrument and the surface of the earth's spheroid) and its theoretical value at that latitude; also *Bougher anomaly*. Gravity anomalies are generally negative over mountains, positive over oceans.

gravity survey   Plot of gravity anomalies over an area; gives an indication of the density contrasts and thus the structure below the surface of the earth.

guild   A group of species that have similar feeding niches (for example, leaf-eating or nectar-feeding) and thus similar roles in the community.

habitat   A particular environment, such as open grassland, closed forest, bare lava, in which an organism lives.

habitat refuge   In this book, an area at a stage of **succession** that is out of phase with nearby areas, creating a habitat no longer present in other locations because of successional processes.

habitat rescue   In this book, saving a species from successionally induced extinction by provision of a **habitat refuge**.

halophyte   A plant capable of thriving in substrates impregnated with salt; for example, a seashore plant.

heliophilous   Adapted to growing in high light intensities (e.g., full sunlight).

herb   A seed plant whose stem is green and fleshy, not woody.

idiobiont   **Parasitoid** which kills or paralyzes its host when the parasitoid lays its egg, preventing further movement or development of the host.

igneous rock   Rock formed by cooling and crystallization of liquid melts of **magma**; e.g., basalts (of volcanic origin) and granites (of subcrustal origin).

ignimbrite Literally, "fire-cloud" rock. Type of volcanic rock, formed from a pumiceous **pyroclastic flow** and composed of totally unsorted homogeneous, very fine fragments, mostly of pumice, but which may include fragments of lava crystals and may be partly welded.

ignimbritic Eruption style or phase similar to **Peléan** except that de-gassing occurs near the vent opening, the mass of fragments and hot gas frothing sideways as a **pyroclastic flow** over the vent, like the froth of beer after pouring. This type of volcanic eruption results in pyroclastic flow deposits of ignimbrite; often accompanied by **co-ignimbrite ash deposits.**

immigration Movement of a species into an area from which it was absent.

immigration rate The rate of addition of species to an area, in species per unit time.

insectivore An insect eater.

interspecific Between species.

intraspecific Within species.

keystone species A species that is of such importance to a community's integrity that its deletion would be assumed to cause drastic changes in functional interrelationships and alter the community in a fundamental way.

koinobiont **Parasitoid** that allows the host larva to develop, feed, and move until emergence of the adult parasitoid.

larva An immature stage in an animal's life cycle, differing from the adult in form, usually in food requirements and feeding habits, and sometimes also in habitat.

lava **Magma** that erupts and then cools to solid rock.

life cycle The various phases (e.g., dormant egg, larval stages, adult) through which an individual organism passes between the fertilized egg and death.

lineament An alignment of topographical or geological features in a straight line or a curve, which, because the arrangement of features is considered too precise to be fortuitous, is believed to represent crustal structure or geological history.

lithic Literally, "of rock"; used in volcanology to describe dense material ejected by eruptions (often pre-existing rocks).

lithosphere Solid outer layer of the earth, consisting of crust and often upper mantle, 50 to 250 km thick, above the **asthenosphere.**

magma Molten rock produced by the partial melting of upper mantle material.

magma chamber   A reservoir of magma in the **lithosphere** from which the material ejected by a volcano derives.

mantle   That layer of the earth, almost 3,000 km thick, between the **crust** and the core of the earth; of a pitch-like consistency, it can behave as a viscous liquid.

metamorphic rock   Rock of any origin that has been moved to some depth below the surface of the earth where it is subjected to high temperature and pressure and thus recrystallized, changing in texture and mineral content but not in chemical composition.

microclimate   The climate in a very restricted space or habitat, such as a tree hole or under loose bark; usually different from the climate of the surroundings.

microsite   The space occupied by a single plant.

Moho boundary   A discontinuity in the earth's outer layers at from 5–10 km beneath the ocean floor and 35 km or more below the continents, above which seismic waves travel more slowly than they do below it; also called the *Mohorovičić discontinuity*, after its discoverer.

monoecious   Bearing male and female sex organs (flowers) on the same individual (plant).

monsoon forest   The natural vegetation of undisturbed tropical areas where there is an annual dry season, during which many of the trees shed their leaves.

mycorrhiza   The close association between a fungus and plant roots which is essential for optimal growth and development of many plants. In some cases, if not in all, the fungus transfers nutrients from decomposing organic matter directly to the plant.

nectarivore   An animal feeding wholly or largely on nectar.

niche   The functional relationships, or role, of a species within an ecosystem, including its effect on other species and on the physical environment.

non-monotonic curve   A curve having both positive and negative slope.

nuée ardente   Literally, "glowing cloud"; an old term for an ignimbrite-producing **pyroclastic flow**.

oceanic crust   Crust under the oceans, 6–8 km thick, consisting of black, dense, fine-grained basaltic and igneous rock with silica content less than 50 percent.

oligophagous   Feeding on only a few, usually related species.

omnivore   A mixed feeder whose diet includes both plant and animal food.

ostiole The opening of a fig **syconium**.

pahoehoe Lava with a surface of smooth, plate-like, or ropy form.

paleomagnetism Magnetism produced in rocks at the time of their formation and retained; paleomagnetism thus provides information about the position of rocks in relation to the earth's past magnetic field.

pappus A ring or tuft of hairs or feathery processes.

parasite An organism closely associated for all or some part of its life with a living individual of another species (the host), from which it obtains part or all of its nourishment, usually without killing the host.

parasitoid A special type of predator—an insect parasite, the larva of which feeds on the internal tissues of another immature insect (its host), killing it.

Peléan Eruption style or phase characterized by highly viscous magma; accompanied by **pyroclastic flow**.

phenology Seasonal pattern of change in population biology or in other biological characteristics (e.g., flowering, breeding, migration) of a species.

phreatomagmatic Due to the interaction of water and magma.

phylogenetic Of or pertaining to evolutionary descent.

phytophagous Feeding on plant material.

Plinian Eruption style or phase characterized by self-accelerating de-gassing of rising viscous magma as the pressure on it decreases, similar to the opening of a champagne bottle. If de-gassing occurs at low levels of the vent, pumice may be blasted up the vent, shotgun-like. The cone's summit may be blown off or a collapse crater formed. The **eruption column** rises at about 500 m/s to about 40 km. Several cubic kilometers of fragmented pumiceous material are emitted, producing well-sorted fall deposits. Fine material stays airborne for weeks or months.

pollination The transfer of pollen from the male to the female organs of flowers, resulting in fertilization of the ovule.

pollinator An animal that in the course of its normal activities effects pollination.

primary succession Succession on an area previously devoid of life.

producer A species that synthesizes organic matter from inorganic materials; green plants, as producers, are the foundation of almost all food chains.

propagule A dispersing part, individual, or number of individuals of an organism, capable of colonizing a new area.

prothallus A small, thin structure produced in the life cycle of a fern, developing from a spore and from which the larger, spore-bearing plant grows.

pseudoturnover A component of recorded **turnover** derived from erroneous records of absence or presence of species on an island.

pteridophytes Vascular plants (such as ferns) producing spores rather than seeds.

pumice Glassy rock of low density that is highly vesiculated as a result of the expansion of large volumes of gas within magma; the gas bubbles are separated from one another by a thin film of glass. Many types of pumice float on water.

pyroclastic flow Very hot, incandescent mass of solid fragments cushioned on escaping gases and traveling like a hovercraft. The denser part of the ejected material hugs the ground, following topography silently and moving with great force and at high speed (up to 200 km/hr). A pyroclastic flow may arise from the blowing out of the side of a lava dome and be directed laterally, or it may move radially down a volcano's flanks as a "base surge." Pyroclastic flow deposits are unsorted, homogeneous. Those producing ignimbrites are turbulent, low-density clouds, largely of fine pumice fragments, and may extend for several kilometers.

pyroclastic rock Literally, "fire-broken" rock; rock ejected from a volcanic vent as solid fragmentary material rather than as liquid lava.

raptor A bird of prey, for example, a hawk, eagle, or owl.

reopening habitat window In this book, an area of early-successional habitat that re-appears late in the overall succession of a larger area and provides a second opportunity for colonization by early-successional species. It may also permit the **rescue effect** to operate for declining populations of such species or provide a **habitat refuge**.

rescue effect Reinforcement of a declining, isolated population of a species by immigrants, thus delaying or preventing the species' extinction from an area.

rhyolites Glassy rocks formed by extrusion of molten material by volcanic activity.

scavenger An animal that feeds on the dead remains or waste material of other animals or plants.

scoria Loose, rubbly material whose pieces, which contain many gas bubbles and vesicles, range widely in size but are mostly of diameter greater than 1 mm.

scoria cone A cone built up from successive layers of scoria from a Strombolian-type eruption. Scoria cones are usually symmetrical and rarely more than 200 m high; the angle of slope is 33 degrees, the "angle of rest" of loose scoria.

secondary succession Succession that follows interruption of the normal, primary succession, but not the complete extirpation of life, by human activity or natural events.

sere One of a series of recognizable stages in the **succession** of a community.

sorted deposits Deposits (usually pumice) formed from the sequential fall (due to the action of gravity on particles of different sizes) of airborne volcanic material, so that at any particular distance from the vent most of the fragments are of similar size.

spermatophytes Seed-bearing plants.

spore A small reproductive body, usually a single cell, from which a new organism arises.

stenophagous Having a narrow range of food.

stenotopic Having a restricted geographical range.

stratosphere That part of the atmosphere above the troposphere, extending from between 8 and 16 km to about 50 km above the earth. The stratosphere has little moisture and a fairly uniform temperature.

Strombolian Eruption style or phase in which gas escapes from the vent spasmodically in minor explosions every few minutes. Semisolid lava fragments are shot high into the air and fall back around the vent as light, vesiculated fragments of wide size range (but few smaller than 1 mm); see **scoria cone**.

sub-fossil fauna Animal species represented by remains that are of the Recent period (the last 10,000 years), usually incompletely mineralized or unmineralized.

succession The sequence of natural changes in structure and species representation of a developing ecosystem such that one community gradually replaces another.

Surtseyan Eruption type or phase characterized by steam and ash being blasted into the air; **phreatomagmatic** eruption type. Well-scattered, highly fragmented, vesiculated fragments form a low, broad cone.

syconium The inner cavity of a fig.

tephra Solid, fragmented matter ejected by a volcano, often extremely rapidly; includes ashes (less than 4 mm in diameter), lapilli ("little stones," 4–32 mm), bombs (rounded stones over 32 mm in diameter), and blocks (angular stones over 32 mm).

thalassochory Dispersal of organisms on the sea surface.

trophic level A particular stage of the food chain defined by the method of obtaining food; for example, "secondary consumers" feed on "primary consumers," which feed on **producers**.

tropical rain forest The natural vegetation of undisturbed tropical areas with over 200 cm of annual rainfall well distributed throughout the year.

troposphere The atmosphere up to a region at which temperature no longer falls with height, 8–16 km above the surface of the earth.

tuff Rock formed of small, compacted volcanic fragments (ash and small lapilli).

turnover Change in species composition of a community by the extinction of some species and the immigration of others.

unsorted deposits Flow deposits in which gravitational sorting has been prevented by turbulence within the flow; unsorted deposits may thus be composed of fragments of a wide range of sizes at any one time or place.

vagility The inherent power of dispersal of a species.

vascular plants Plants (pteridophytes and spermatophytes) with well-defined tracts of fluid-conducting supporting cells within the tissues.

vegetative change A change in structure of the plant component of a community, for example, from grassland to open woodland.

Vesuvian Eruption style or phase characterized by the ejection of magma in a fairly sustained (several hours) blast of escaping gas that sends a long-lasting "cauliflower cloud" several km into the air. The resulting fall deposits are typically well-sorted pumice. Also referred to as *sub-Plinian* type.

viscosity Resistance to flow of a fluid; for example, treacle, or molasses, is viscous. Molten magma becomes more viscous as it cools, but at a given temperature magmas of different compositions have different viscosities .

volcanotectonic depression Large surface depression formed over a long period by intermittent collapse of the roof of a magma chamber after it has been emptied.

Vulcanian Eruption style or phase characterized by intermittent or continuous violent eruptions sending a dark plume of steam, gas, and ash several km high and causing a fine ash rain downwind. The magma extruding from a Vulcanian eruption is more viscous than **Strombolian** magma. Fall deposits are highly fragmented, angular, non-vesiculated, and they contain a high proportion of fragments less than 1 mm in diameter.

welding   In a flow deposit, the consolidation of fragments and particles that occurs if they are still hot enough when deposited for their edges to fuse where they touch.

zoochory   Dispersal of organisms within (endozoochory) or on the outside of (ectozoochory) animals.

BIOGRAPHICAL NOTES

Cornelius Andreis Backer. Plant systematist whose 1929 monograph represented the chief critique of the theory of total extirpation of life on the Krakataus in 1883. Having emigrated to Indonesia from Holland in 1901, at the age of 27, to become a schoolmaster in Jakarta, he commanded such broad knowledge of the local flora by 1905 that Treub appointed him to the Bogor Herbarium staff. Backer's visits to the Krakataus were made soon afterward: in 1906 with Ernst and in 1908 with Jacobson. When Treub suggested he bring his vast plant collection to the Herbarium, Backer asked, only half in jest, if Treub was willing to pay for the railway truck. Appointed "Botanist with special responsibility for the Flora of Java" in 1912, he was said also to have been good at languages and generous with assistance to colleagues. H. C. D. de Wit wrote in 1949 that "Backer's phytography is composed with painstaking accuracy though he never loses himself in too much detail. Some of his manuscripts he kept unpublished for years, always reconsidering, improving, adding new facts ... He criticised bitterly when he believed authors to be hasty and careless, an attitude which has been judged by some as too severe." After receiving his pension in 1924 and studying the weeds of cane fields for the Javan sugar industry, Backer returned to Holland in 1931, where he continued to work on the ferns of Java. He must have been something of an eccentric, for he had a sun helmet made up in tweed to wear in Holland. He was awarded an honorary doctorate from the University of Utrecht in 1936, the year in which the monograph of Docters van Leeuwen, his chief protagonist, was published.

Jan van Borssum Waalkes.   Botanist. Born at Groningen, Holland, in 1922, and educated at Groningen University, he was appointed Assistant at the Bogor Herbarium in 1950. He visited the Krakataus both before and after the 1952 eruptions of Anak Krakatau and made important assessments of their effects. He returned to the Netherlands in 1957.

Karl Willem Dammerman.   Zoologist whose 1948 monograph is one of the three classic works on the Krakataus (along with those of Verbeek and Docters van

Leeuwen). Born at Arnhem in Holland in 1885, Dammerman was some five years younger than his botanical counterpart and colleague, Docters van Leeuwen. He obtained his doctorate at the University of Utrecht in 1910 and in the same year was appointed to the Netherlands East Indies Division of Plant Diseases, which was soon based at Bogor. He compiled a reference book on the agricultural zoology of the region, and in 1919, the year after Docters van Leeuwen became Director of the Botanical Gardens, Dammerman was appointed Curator (in charge) of its Zoology Museum. Now began Dammerman's studies of Krakatau zoology, which were to continue for two decades. As a zoologist with wide interests, including the soil and litter fauna of the tropics, cave-dwelling animals, and various aspects of animal ecology as well as entomology, he encouraged exploratory field work at the museum and devoted much time and effort to the establishment of nature reserves and the necessary protective legislation. In 1932 he succeeded Docters van Leeuwen as Director of the Gardens and guided the institution through a period of considerable financial and political difficulty. He inaugurated the Treub Foundation in an effort to support continued scientific research in Indonesia, before retiring to Leiden, Holland (and World War II), in 1939. He managed to continue to work during the

German occupation and became a Research Associate at the Leiden Museum in 1943, his Krakatau monograph eventually being published in 1948. Having been a widower for some sixteen years, he remarried in 1951, only to die some months later at the age of 66 in Voorburg, the Netherlands. As described in his obituary (1951), Dammerman was an unusual man of determined and upright character, retiring and shy in public and averse to publicity of any kind. His well-known reticence is immortalized in the name of an Indonesian mollusc species named in his honor, *Thiara carolitaciturni*. Off duty, however, he was said to be very witty, with a great sense of fun and an "extreme delight in experiencing awkward or perilous situations."

Willem Marius Docters van Leeuwen. Biologist whose botanical monograph is one of the great classic works on the Krakataus. Born in Jakarta in 1880, he re-

ceived his doctorate from the University of Amsterdam in 1907 and in the next year was appointed Entomologist at the Experimental Station at Salatiga, Java. From 1909 to 1918 he was a schoolteacher, first in Semarang, where he became headmaster, and later for three years in Bandung. His interests were in general biology, the Indonesian alpine flora, gall-forming insects, interactions between ants and plants, flower biology, pollination, and seed dispersal. He was also a first-class photographer. Today he would be described as a plant ecologist with a specialty in plant-animal interactions, the ideal type of biologist, in fact, to devote his energies to the study of Krakatau. In 1918, at the age of 38, he became Director of the Bogor Botanical Gardens (and Extraordinary Professor at the Medical College in Jakarta) and began the long-term studies of Krakatau plants and vegetation based on his many visits to the islands between 1919 and 1932. After retiring to Holland he became a very popular Lecturer at the University of Amsterdam (known to the students as "Uncle Doc"), and, from 1942 to 1950, Professor of Tropical Botany. Like his zoological counterpart and colleague, Dammerman, he was awarded the Dutch equivalent of a knighthood. He died in 1960, three weeks before his eightieth birthday.

Alfred Ernst. Professor of Botany at the University of Zurich, Switzerland. A visitor to the Bogor Gardens on a Swiss traveling research grant, he took over leadership of a short expedition to the Krakataus in 1906, at the age of 31, when Treub became ill. He also accompanied Docters van Leeuwen to Krakatau in 1931.

Ernst's research interests were in floral development, genetics, and the reproductive biology of plants. In 1934 he wrote a long review article on the biology of the Krakataus as a response to the criticisms of Backer.

Andrei G. G. F. Hoogerwerf.   Vertebrate zoologist and conservationist. Hoogerwerf arrived in Java in 1932, an athletic young man of imposing stature, as second taxidermist at the Zoology Museum of the Bogor Botanical Gardens, and made a number of visits to Ujung Kulon. In 1934 he joined the Board of the Netherlands Indies Society for Nature Protection, and in 1937 he was seconded to the Director of the Bogor Botanical Gardens with the special brief of nature protection and wildlife management. After the Japanese occupation, he returned to Holland in 1948 on sick leave, but in 1950 he again traveled to Java, this time to become the first Chief of the newly formed Nature Protection and Wildlife Management Bureau of the Forest Service. He made a vital visit to the Krakataus with van Borssum Waalkes in 1951 and retired to the Netherlands in 1954. His famous monograph on Ujung Kulon was published in 1970. He died a few years later, shortly after a visit to wildlife areas in East Java.

Edward Jacobson.   Entomologist, zoologist. Born in Germany of Dutch parents, Jacobson came to Java in 1892, at the age of 22, to work in his father's business. He made the first zoological survey of the Krakataus in 1908. In 1910, the year after his report on the Krakatau fauna was published, he became Director of the Bogor Zoology Museum. He received an honorary doctorate from the University of Amsterdam in 1932 and died in Semarang, Java, in 1944, at age 74.

Otto Penzig.   Mycologist and Professor of Botany at the University of Genoa, Italy. In 1896 and 1897 Penzig, who was born in Austria, was an invited visiting scientist at the Foreigners' Laboratory, Bogor Botanical Gardens. He studied slime molds (myxomycetes) at Bogor and at the Cibodas mountain garden and led a visit to the Krakataus in 1897. He died in Genoa in 1929, aged 73.

W. A. Petroeschevsky.   Geologist and volcanologist. This Russian scientist was on hand to study the emergence and growth of Anak Krakatau, which he named. He accompanied Bristowe to Panjang in 1931, reconnoitered Anak Krakatau from the air in 1947, and revisited the island in 1949.

Leendert van der Pijl.   Flower biologist with interests in pollination and dispersal. Born in Utrecht, he was a graduate of the University of Amsterdam, where he took his Ph.D. in 1934 while on study leave from his teaching post at the Christian Lycaeum in Bandung. He accompanied Toxopeus to the Krakataus, including the new island Anak Krakatau, in 1930. Like Toxopeus, he was interned by the Japanese during World War II. He was then evacuated to Holland but was able to return to Java in 1947 and revisit Krakatau in 1949.

Charles E. Stehn.   Geologist, Head of the Geological Survey, Bandung. Stehn made research trips to the islands in 1921 and 1928. On a visit with Docters van Leeuwen in 1929, he tied a rope round a *Neonauclea* tree on Rakata's summit, dropped the other end over the precipice, and proceeded to climb down so that, suspended almost 800 m (2,6000 ft) from the bottom, he could examine the then-pristine face of the cliff. He monitored the birth and emergence of Anak Krakatau, and his studies of the stratigraphy of volcanic deposits provided important information concerning the events of August 1883.

Johannes Elias Tejsmann.   Plant systematist. Born at Arnhem, Holland, in 1808, he arrived in Java as Gardener to the Governor-General and became Chief Curator of the Bogor Botanical Gardens. An indefatigable collector, he visited the Krakataus in 1857 by sailing *prahu*. He died at Bogor a year before the 1883 eruption.

Lambertus Johannes Toxopeus.   Entomologist. Born in Java in 1894, he was educated at the University of Amsterdam. Unusually tall, he was a teacher with considerable field experience, having participated as entomologist in the third Archbold Expedition to New Guinea in 1938/39. He was an acknowledged specialist on the lepidopteran families Lycaenidae and Hesperiidae and was described as a gifted and perceptive researcher who was somewhat withdrawn socially but enjoyed speaking at scientific meetings. During WWII he was interned in a Japanese concentration camp and on his release at the end of the war became Reader at the University in Jakarta. In the same year as his Krakatau visit (1949, with van der Pijl) he was appointed Professor at Bandung, where two years later he was tragically killed in a road accident.

Melchior Treub.   Plant morphologist and physiologist, organizer, leader. At Leiden, where he was a gold-medal student, Treub had demonstrated the true nature

of lichens by growing the algal and fungal components separately. In 1880, at the age of 29, he became Director of the Bogor Botanical Gardens, an office he was to hold for almost thirty years, presiding over the institution's golden period as its most effective and influential head. Traveling with Verbeek in June 1886, he was the first biologist to visit the Krakataus after the 1883 eruption. His report of this visit drew the attention of biologists world-wide to the natural "experiment" that had been set in train. Treub returned to the islands in November 1888 with Sluiter, and again in 1897 with Penzig. Under his leadership the Gardens became a world-famous institution. He greatly extended the research facilities by building new laboratories (including one at the mountain extension at Cibodas) and was instrumental in founding a number of experimental stations throughout Indonesia. He encouraged overseas visitors to carry out research at Bogor by forming a Foreigners' Laboratory (from 1914 the Treub Laboratory) and opening a "Buitenzorg Fund" for their financial support. He died in 1910, the year following his retirement, at St. Raphael in the south of France, aged 59.

Rogier Diederik Marius Verbeek.   Mining engineer, geologist. Born in 1847 in Doorn, Holland, the son of a clergyman, Verbeek graduated from Delft University

of Technology in 1866. Appointed as a mining engineer of the Mining Department of the Dutch East Indies in the following year, he worked in East Borneo (Kalimantan) and Sumatra and mapped the geology of Krakatau. He was awarded an honorary degree from the University of Breslau (Germany) for his Sumatran studies. Within a week of being appointed to study the cause, extent, and effects of the 1883 eruption, Verbeek was on his way to the volcano. His preliminary report was published less than six months after the eruption, and a translation appeared in *Nature* on May 1, 1884. His final 546-page monograph appeared in the following year and made him world-famous. Within two years of the eruption he had climbed from sea level on crumbling, unconsolidated ash deposits, along the edge of Rakata's great cliff to its summit, not once but four times. After the two years studying Krakatau, he made a ten-year geological survey of Java and Madura Island and then worked on the more eastern islands before retiring in 1901, but it is for his masterly study of Krakatau that he is remembered. He died of a stroke in Holland two days after his eighty-first birthday.

REFERENCES

Abbott, H. L., and E. E. Fowle. 1913. "Volcanoes and climate." *Smithsonian Miscellaneous Collections*, 60 (29): 221–229.

Abe, T. 1984. "Colonization of the Krakatau Islands by termites (Insecta: Isoptera)." *Physiological Ecology, Japan*, 21: 63–88.

———1987. "Evolution of life types in termites." In S. Kawano, J. H. Connell, and T. Hidaka (eds.), *Evolution and Coadaptation in Biotic Communities*, pp. 124–148. Tokyo: University of Tokyo Press.

Alexander, H. G. L. 1979. "A preliminary assessment of the role of the terrestrial decapod crustaceans in the Aldabran ecosystem." *Philosophical Transactions of the Royal Society of London* B, 286: 241–246.

Baas, P. 1982. "Appendix 4. Identity of the carbonised wood sample from Krakatau." In Flenley and Richards (1982), pp. 181–183.

Backer, C. A. 1909. "De flora van het eiland Krakatau." *Jaarverslag Topografische Dienst in Nederlands Indië over 1908*, 5: 189–191.

———1929. *The Problem of Krakatao as Seen by a Botanist*. Surabaya, Weltevreden: Published by the author.

Barker, N., and K. Richards. 1982. "The vegetation of Anak Krakatau." In Flenley and Richards (1982), pp. 142–164.

———1986. "Notes on the plant ecology and biogeography of Anak Krakatau." In M. Bush, P. Jones, and K. Richards (eds.), *The Krakatau Centenary Expedition 1983: Final Report*, pp. 167–181. Hull, U.K.: Department of Geography, University of Hull, Miscellaneous Series, 33.

Bartels, M. 1919. "Lijst van vogels en zoogdieren, waargenomen van 24–29 April 1919 op de eilanden Krakatau en Verlaten Eiland." *Handelingen van het Eerste Nederlandsche Indische Natuurwetenschappelijk Congress Welfeureden*, 1: 76–79.

Beaglehole, J. C., ed. 1963. *The "Endeavour" Journal of Joseph Banks*. Sydney: Angus & Robertson.

Becker, P. 1975. "Island colonization by carnivorous and herbivorous Coleoptera." *Journal of Animal Ecology*, 44: 893–906.

———1992. "Colonization of islands by carnivorous and herbivorous Heteroptera and Coleoptera: Effects of island area, plant species richness, and 'extinction' rates." *Journal of Biogeography*, 19: 163–171.

Bemmelen, R. W. van. 1942. "Krakatau." *Bulletin East Indies Volcanological Survey*, 1941, 18: 53–60.

Bemmelen, W. van. 1909. "Oplooding van het Krakatau-bekken." *Topografische Dienst Jaarverslag, 1908,* 4 (3): 176–188.

Berg, N. P. van den. 1884. "Vroegere Berichten Omtrent Krakatau. De uitbarsting van 1680." *Tijdschrift voor Indische Taal-, Land-, en Volkenkunde* (Batavia), 29: 208–227.

Berghaus, H. 1837. *Länder und Völkerkunde.* Stuttgart: Hoffmann'sche Berlags.

Bird, E. C., and N. J. Rosengren. 1984. "The changing coastline of the Krakatau Islands, Indonesia." *Zeitschrift für Geomorphologie,* new ser., 28 (3): 347–366.

Bishop, S. E. 1884a. "The equatorial smoke-stream from Krakatoa." *Hawaiian Monthly,* May 1884: 106–110.

——1884b. "The remarkable sunsets." *Nature* (London), 29: 259–260.

——1887. "Origin of the red glows." *History and Work of the Warner Observatory,* 1887, 1: 63–70.

Bödvarsson, H. 1982. "The Collembola of Surtsey, Iceland." *Surtsey Research Progress Report,* 9: 63–67.

Bois, W. P. du. 1947. *The Twenty-One Balloons.* New York: Viking Press.

Boon, P. B., and R. T. Corlett. 1989. "Seed dispersal by the Lesser Short-nosed Fruit Bat (*Cynopterus brachyotis,* Pteropodidae, Megachiroptera)." *Malayan Nature Journal,* 42: 251–256.

Bordage, E. 1916. "Le repeuplement végétal et animal des îles Krakatoa depuis l'éruption de 1883." *Annales de Geographie* (Paris), 25 (133): 1–22.

Borssum Waalkes, J. van. 1952. "Een bezoek aan de Krakatau-eilanden." *De Tropische Natuur,* 32: 35–44.

——1953. "On the state of the vegetation on the Krakatao Islands in 1951–1952." *Eighth Pacific Science Congress, Abstracts of Papers,* p. 210.

——1954. "The Krakatau Islands after the eruption of October 1952." *Penggemar Alam,* 34: 97–104.

——1960. "Botanical observations on the Krakatau Islands in 1951 and 1952." *Annales Bogoriensis,* 4 (1): 5–63.

Bréon, R., and W. C. Korthals. 1885. "Rapport sur une mission scientifique dans le Détroit de la Sonde." *Archives des Missions Scientifiques et Litteraires,* 12:433–437.

Bristowe, W. S. 1931. "A preliminary note on the spiders of Krakatau." *Proceedings of the Zoological Society of London,* 1931: 1387–1400.

——1934. "Introductory notes." In F. Reimoser, "The Spiders of Krakatau," pp. 11–12. *Proceedings of the Zoological Society of London,* 1934: 11–18.

——1969. *A Book of Islands.* London: G. Bell and Sons.

Broadhead, E., and I. W. B. Thornton. 1955. "An ecological study of three closely related psocid species." *Oikos,* 6 (1): 1–50.

Bronstein, J. L. 1989. "A mutualism at the edge of its range." *Experientia,* 45: 622–636.

——1992. "Seed predators as mutualists: Ecology and evolution of the fig/pollinator interaction." In E. Bernays (ed.), *Insect-Plant Interactions,* vol. 4, pp. 1–44. Boca Raton, FL: CRC Press.

Brown, J. H., and A. Kodrik-Brown. 1977. "Turnover rates in insular biogeography: Effects of immigration on extinction." *Ecology,* 58: 445–449.

Brown, W. C., and A. C. Alcala. 1957. "Viability of lizard eggs exposed to sea water." *Copeia,* 1: 39–41.

Brun, A. 1911. *Recherches sur l'exhalaison volcanique.* Paris: A. Hermann & fils.

Buckley, R. 1982. "The habitat-unit model of island biogeography." *Journal of Biogeography,* 9: 339–344.

Bush, M. B. 1986. "The butterflies of Krakatoa." *Entomologist's Monthly Magazine,* 122: 51–58.

Bush, M. B., D. J. B. Bush, and R. D. Evans. 1990. "Butterflies of Krakatau and Sebesi: New records and habitat relations." In Whittaker et al. (1990), pp. 35–41.

Bush, M. B., and R. J. Whittaker. 1991. "Krakatau: Colonization patterns and hierarchies". *Journal of Biogeography,* 18: 341–356.

——1993. "Non-equilibration in island theory of Krakatau." *Journal of Biogeography,* 20: 453–457.

Bush, M. B., R. J. Whittaker, and T. Partomihardjo. 1992. "Forest development on Rakata, Panjang and Sertung: Contemporary dynamics (1979–1989)." In Thornton (1992), pp. 185–199.

Camus, G., M. Diament, and M. Gloaguen. 1992. "Emplacement of a debris avalanche during the 1883 eruption of Krakatau (Sunda Straits, Indonesia)." In Thornton (1992), pp. 123–128.

Camus, G., A. Gourgaud, and P. M. Vincent. 1987. "Petrologic evolution of Krakatau (Indonesia): Implications for a future activity." *Journal of Volcanology and Geothermal Research,* 33: 299–316.

Camus, G., and P. M. Vincent. 1983a. "Un siècle pour comprendre l'éruption du Krakatoa." *La Recherche,* 14 (149): 1452–1457.

——1983b. "Discussion of a new hypothesis for the Krakatau volcanic eruption in 1883." *Journal of Volcanology and Geothermal Research,* 19: 167–173.

Case, T. J., and M. L. Cody. 1987. "Testing theories of island biogeography." *American Scientist,* 75: 402–411.

Compton, S. G. 1995. "Seed dispersal ecology of an African fig tree *(Ficus burtt-davyi)*." *Journal of Biogeography,* in press.

Compton, S. G., I. W. B. Thornton, T. R. New, and L. Underhill. 1988. "The colonization of the Krakatau Islands by fig wasps and other chalcids (Hymenoptera, Chalcidoidea)." *Philosophical Transactions of the Royal Society of London* B, 322: 459–470.

Connor, E. F., and D. Simberloff. 1979. "The assembly of species communities: Chance or competition?" *Ecology,* 60 (6): 1132–1140.

Cook, J., and J. King. 1784. *A Voyage to the Pacific Ocean to Determine the Position and Extent of the West Side of North America,* vol. 3, pp. 471–472. London: G. Nicol and T. Cadell.

Cooper, W. S. 1931. "Krakatao." *Ecology,* 12: 424–426.

Corner, E. J. H. 1952. *Wayside Trees of Malaya.* 2d ed. Singapore: Government Printing Office.

Cotteau, E. 1886. *En Océanie, Voyage autour du Monde en 365 Jours 1884–1885.* Paris: Hachette et Cie.

Cowles, G. S., and D. Goodwin. 1959. "Seed digestion by the fruit-eating pigeon *Treron.*" *Ibis,* 101: 253–254.

Cranbrook, Earl of (ed.). 1988. *Key Environments: Malaysia.* Oxford: Key Environments Series, Pergamon Press.

Crome, F. H. 1975. "The ecology of fruit pigeons in tropical northern Queensland." *Australian Wildlife Research,* 2: 155–185.

Dammerman, K. W. 1922. "The fauna of Krakatau, Verlaten Island and Sebesy." *Treubia,* 3: 61–112.

———1929. "Krakatau's new fauna." *Proceedings of the Fourth Pan-Pacific Science Congress, Java, 1929,* 37: 83–118.

———1948. "The Fauna of Krakatau, 1883–1933." *Verhandelingen Koninklijke Nederlandsche Akademie van Wetenschappen, Afdeling Natuurkunde* II, 44: 1–594.

Dana, J. D. 1890. *Characteristics of Volcanoes, with Contributions of Facts and Principles from the Hawaiian Islands.* New York: Dodd, Mead & Co.

Darwin, C. 1859. *On the Origin of Species.* 6th ed., 1872. London: Oxford University Press.

Darwin, F., and A. C. Seward, eds. 1903. *More Letters of Charles Darwin.* London: Murray.

Diamond, J. M. 1972. "Reconstitution of bird community structure on Long Island, New Guinea, after a volcanic explosion." *National Geographic Society Research Reports,* 13: 191–204.

———1974a. "Colonization of exploded volcanic islands by birds: The supertramp strategy." *Science,* 184: 803–806.

———1974b. "Recolonization of exploded volcanic islands by New Guinea birds." *Explorers' Journal,* March 1974: 2–10.

———1975. "Assembly of species communities." In M. L. Cody and J. M. Diamond (eds.), *Ecology and Evolution of Communities,* pp. 342–444. Cambridge, MA: Harvard University Press.

Diamond, J. M., and M. E. Gilpin. 1982. "Examination of the 'null' model of Connor and Simberloff for species co-occurrences on islands." *Oecologia* (Berlin), 52: 64–74.

Dickerson, J. E., Jr. and J. V. Robinson. 1985. "Microcosms as islands: A test of the MacArthur-Wilson equilibrium theory." *Ecology,* 66: 966–980.

Docters van Leeuwen, W. M. 1920. "The galls of Krakatau and Verlaten Island (Desert Island) in 1919." *Annales du Jardin Botanique de Buitenzorg,* 31: 57–82.

———1922a. "The vegetation of the island of Sebesy, situated in Sunda-strait, near the islands of the Krakatau-Group, in the year 1921." *Annales du Jardin Botanique de Buitenzorg,* 32: 135–192.

———1922b. "The galls of the islands of the Krakatau group and of the island of Sebesi." *Bulletin du Jardin Botanique de Buitenzorg,* ser. 3, 4: 287–314.

———1923. "On the present state of the vegetation of the islands of the Krakatau group and of the island of Sebesi." *Proceedings of the Second Pan-Pacific Science Congress, Australia, 1923,* 1: 313–318.

———1924. "Een vlinderzwerm in de Straat van Malakka." *De Tropische Natuur,* 13: 182.

————1933. "Germinating coconuts on a new volcanic island, Krakatao." *Nature*, 132 (3339): 674–675.

————1935. "The dispersal of plants by fruit-eating bats." *Gardens Bulletin, Straits Settlements*, 9: 58–63.

————1936. "Krakatau 1883–1933. A. Botany." *Annales du Jardin Botanique de Buitenzorg*, 46–47: 1–507.

————1954. "On the biology of some Javanese Loranthaceae and the role birds play in their life-histories." *Beaufortia*, 4 (41): 103–204.

Drake, J. A. 1990a. "Communities as assembled structures: Do rules govern pattern?" *Trends in Ecology and Evolution*, 5: 159–164.

————1990b. "The mechanics of community assembly and succession." *Journal of Theoretical Biology*, 147: 213–233.

————1991. "Community-assembly mechanics and the structure of an experimental species ensemble." *American Naturalist*, 137 (1): 1–26.

Edwards, J. S. 1987. "Arthropods of alpine aeolian ecosystems." *Annual Review of Entomology*, 32: 163–179.

Edwards, J. S., and P. Sugg. 1993. "Arthropod fallout as a resource in the colonization of Mount St. Helens." *Ecology*, 74 (3): 954–958.

Ernst, A. 1908. *The New Flora of the Volcanic Island of Krakatau.* London: Cambridge University Press.

————1934. "Das biologische Krakatauproblem." *Vierteljahrsschrift der Naturforschenden Gesellschaft in Zurich*, 79: 1–187.

Escher, B. D. 1919a. "De Krakatau-group als vulkaan." *Handelingen van het Eerste Nederlandsch Indisch Natuurwetens drappelijk Congress Welfeureden*, 1919: 28–35.

————1919b. "Veranderingen in de Krakatau-group na 1908." *Handelingen van het Eerste Nederlandsch Indisch Natuurwetens drappelijk Congress Welfeureden*, 1919: 198–219.

————1928. "Krakatau in 1883 en in 1928." *Koninklijk Nederlandsch Aardrijkskundig Genoot schap. Tijdschrift*, ser. 2, 45 (4): 715–743, 798.

Ewing, M., and F. Press. 1955. "Tide-gage disturbances from the great eruption of Krakatoa." *Transactions of the American Geophysical Union*, 36 (1): 53–60.

Fager, E. W. 1968. "The community of invertebrates in decaying oak wood." *Journal of Animal Ecology*, 37: 121–142.

Fisher, R. V., and H.-U. Schmincke. 1984. *Pyroclastic Rocks.* Berlin: Springer-Verlag.

Fitch, T. J. 1972. "Plate convergence, transcurrent faults, and internal deformation adjacent to Southeast Asia and the Western Pacific." *Geophysical Research*, 77: 4432–4460.

Flenley, J. R., and K. Richards, eds. 1982. *The Krakatoa Centenary Expedition: Final Report.* Hull, U.K.: Department of Geography, University of Hull, Miscellaneous Series, 25.

Fogden, M. P. L. 1972. "The seasonality and population dynamics of equatorial forest birds in Sarawak." *Ibis*, 114 (3): 307–343.

Forster, M. 1982. "A study of the spatial distribution of bryophytes on Rakata." In Flenley and Richards (1982), pp. 103–126.

Fosberg, F. R. 1985. "Botanical visits to Krakatau in 1958 and 1963." *Atoll Research Bulletin*, 292: 39–45.

Fox, B. J., and G. L. Kirkland, Jr. 1992. "An assembly rule for functional groups applied to North American soricid communities." *Journal of Mammalogy*, 73 (3): 491–503.

Francis, P. W. 1985. "The origin of the 1883 Krakatau tsunamis." *Journal of Volcanology and Geothermal Research*, 25: 349–364.

Franklin, B. 1784. "Meteorological imaginations and conjectures." *Memoirs and Proceedings of the Manchester Literary and Philosophical Society*, 2: 357–361.

Fridriksson, Sturla. 1975. *Surtsey: Evolution of Life on a Volcanic Island*. London: Butterworths.

———1994. *Surtsey: Lifriki i Motun* (Surtsey: The development of life on a young volcanic island). Reykjavik, Iceland: Society of Natural History and Surtsey Research Society.

Fridriksson, Sturla, and Borgthor Magnusson. 1992. "Development of the ecosystem on Surtsey with reference to Anak Krakatau." In Thornton (1992), pp. 287–291.

Frith, H. J. 1982. *Pigeons and Doves of Australia*. Adelaide: Rigby.

Furneaux, R. 1964. *Krakatoa*. London: Secker and Warburg.

Gadow, G. 1933. "Magen und Darm der Fruchttauben." *Journal of Ornithology*, 81: 236–252.

Gandawijaja, D., and J. Arditti. 1983. "The orchids of Krakatau: evidence for a mode of transport." *Annals of Botany*, 52 (2): 127–130.

Gates, F. C. 1914. "The pioneer vegetation of Taal Volcano." *Philippines Journal of Science*, sec. C, *Botany*, 9: 391–434.

Gilpin, M. E., and J. M. Diamond. 1976. "Calculation of immigration and extinction curves from the species-area-distance relation." *Proceedings of the National Academy of Sciences, U.S.A.*, 73 (11): 4130–4134.

———1982. "Factors contributing to non-randomness in species co-occurrences on islands." *Oecologia*, 52: 75–84.

Goodwin, R. E. 1979. "The bats of Timor: systematics and ecology." *Bulletin of the American Museum of Natural History*, 163: 73–122.

Gould, E. 1978. "Foraging behaviour of Malaysian nectar-feeding bats." *Biotropica*, 10: 184–193.

Green, P. T. 1993. *The Role of Red Land Crabs* [Gecarcoidea natalis *(Pocock, 1888); Brachyura, Gecarcinidae*] *in Structuring Rain Forest on Christmas Island, Indian Ocean*. Ph.D thesis, Monash University, Australia.

Griggs, R. F. 1930. "'The problem of Krakatao as seen by a botanist' by C. A. Backer." *Science*, 71: 132–133.

Gross, C. L. 1993. "The reproductive ecology of *Canavalia rosea* (Fabaceae) on Anak Krakatau, Indonesia." *Australian Journal of Botany*, 41: 591–599.

Guppy, H. B. 1906. *Observations of a Naturalist in the Pacific between 1891 and 1899*, vol. 2, *Plant Dispersal*. London: Macmillan.

Haag, W. R., and M. B. Bush. 1990. "Biogeographic implications of bird distributions in the Krakatau Islands, and Sebesi Island, Indonesia." In Whittaker et al. (1990), pp. 32–34.

Hall, G. 1987. "Seed dispersal by birds of prey." *Zimbabwe Science News*, 21: 9.

Hammer, C. U., H. B. Clausen, and W. Dansgaard. 1980. "Greenland ice sheet evidence of post-glacial volcanism and its climatic impact." *Nature* (London), 288: 230–235.

Harkrider, D., and F. Press. 1967. "The Krakatoa air-sea waves: An example of pulse propagation in coupled systems." *Geophysical Journal of the Royal Astronomical Society*, 13: 149–153.

Harvey, M. S. 1988. "Pseudoscorpions from the Krakatau Islands and adjacent regions, Indonesia (Chelicerata: Pseudoscorpionida)." *Memoirs of the Museum of Victoria*, 49: 309–353.

Heaney, L. R. 1986. "Biogeography of mammals in SE Asia: Estimates of rates of colonization, extinction and speciation." *Biological Journal of the Linnean Society*, 28: 127–165.

————1991. "An analysis of patterns of distribution and species richness among Philippine fruit bats (Pteropodidae)." *Bulletin of the American Museum of Natural History*, 206: 145–167.

Heaney, L. R., P. C. Gonzalez, R. C. B. Utzurrum, and E. A. Rickart. 1991. "The mammals of Catanduanes Island: Implications for the biogeography of small landbridge islands in the Philippines." *Proceedings of the Biological Society of Washington*, 104 (2): 399–415.

Heatwole, H. 1971. "Marine-dependent terrestrial biotic communities on some cays in the Coral Sea." *Ecology*, 52: 363–366.

Heatwole, H., and R. Levins. 1972. "Trophic structure stability and faunal change during recolonization." *Ecology*, 53: 531–534.

————1973. "Biogeography of the Puerto Rican Bank: Species-turnover on a small cay, Cayo Ahogado." *Ecology*, 54: 1042–1055.

Heithaus, E. R. 1982. "Coevolution between bats and plants." In T. H. Kunz (ed.), *Ecology of Bats*, pp. 327–367. New York: Plenum Press.

Herron, M. M. 1982. "Impurity sources of F⁻, Cl⁻, NO₃⁻ and SO₄²⁻ in Greenland and Antarctic precipitation." *Journal of Geophysical Research*, 87 (C4): 3052–3060.

Hill, A. W. 1930. "Botanical exploration of Krakatao." *Nature* (London), 125: 627–629.

————1937. "The flora of Krakatau." *Nature* (London), 139: 135–138.

Hill, J. E. 1983. "Bats (Mammalia: Chiroptera) from Indo-Australia." *Bulletin of the British Museum, Natural History*, 45: 103–208.

Hoogerwerf, A. 1952. "Verslag over een naar de Krakatau-groep gemaakte dienstreis van 5–15 October 1951." *Dienstrapport, Djawatan Penjelidikan Alam, Bagian Perlindungan Alam dan Pemburuan*, 33: 1–198.

————1953. "Notes on the vertebrate fauna of the Krakatau Islands, with special reference to the birds." *Treubia*, 22: 319–348.

————1970. *Udjung Kulon, the Land of the Last Javan Rhinoceros*. Leiden: E. J. Brill.

Hossaert-McKey, M., M. Gibernau, and J. E. Frey. 1994. "Chemosensory attraction of fig wasps to substances produced by receptive figs." *Entomologia experimentalis et applicata*, 70: 185–191.

Howarth, F. G. 1979. "Neogeoaeolian habitats on new lava flows on Hawaii island: An ecosystem supported by windborne debris." *Pacific Insects*, 20: 133–144.

Ibkar-Kramadibrata, H., R. E. Soeriaatmadja, H. Syarif, W. Paryatmo, E. Surasna, M. Sutisna, D. Galih, A. Syarmidi, S. H. Widodo, and I. Birsyam. 1986. *Explorasi Biologis dan Ecologis dari Daerah Daratan di Gugus Kepulauan Krakatau menjelang 100 tahun sesudah Peletusan.* Bandung: Institut Teknologi Bandung.

Itino, T., M. Kato, and M. Hotta. 1991. "Pollination ecology of the two wild bananas, *Musa acuminata* subsp. *halabanensis* and *M. salaccensis:* Chiropterophily and ornithophily." *Biotropica,* 23 (2): 151–158.

Iwamoto, T. 1986. "Mammals, reptiles and crabs on the Krakatau Islands: Their roles in the ecosystem." *Ecological Research,* 1: 249–258.

Jacobson, E. R. 1909. "Die nieuwe fauna van Krakatau." *Jaarverslag van den Topographischen Dienst in Nederlandsch-Indie,* 4: 192–211.

Janzen, D. H. 1979. "How to be a fig." *Annual Review of Ecology and Systematics,* 10: 13–51.

Joppien, R., and B. Smith. 1987. *The Art of Captain Cook's Voyages,* vol. 3. Oxford: Oxford University Press.

Judd, J. W. 1884. "Krakatoa." *Proceedings of the Royal Society of London,* May 1884: 85–88.

———1889. "The earlier eruptions of Krakatoa." *Nature* (London), 40: 365–366.

Junghuhn, F. 1853. *Java, Zijne Gedaante, Zijne Plantengroei en Inwendige Bouw,* vol. 1. 2d ed. The Hague.

Katili, J. 1970. "Large transcurrent faults in Southeast Asia with special reference to Indonesia." *Geologische Rundschau,* 59: 581–600.

Kaufmann, S., D. B. McKey, M. Hossaert-McKey, and C. C. Horvitz. 1991. "Adaptations for a two-phase seed dispersal system involving vertebrates and ants in a hemiepiphytic fig (*Ficus microcarpa:* Moraceae)." *American Journal of Botany,* 78: 971–977.

Keiffer, S. W. 1981. "Mount St. Helens lateral blast: Explanation of acoustic observations." *Annual Meeting, Geological Society of America, Abstracts,* 13: 487.

Kimball, H. H. 1913. "The effect of the atmospheric turbidity of 1912 on solar radiation intensities and skylight polarization." *Bulletin of the Mount Weather Observatory,* 5 (5): 295–312.

King, B. F., E. C. Dickinson, and M. W. Woodcock. 1975. *A Field Guide to the Birds of South-East Asia.* London: Collins.

Kingsley, C. 1863. *The Water-babies.* 1885 ed. London: Macmillan.

Kjellberg, F., and S. Maurice. 1989. "Seasonality in the reproductive phenology of *Ficus:* Its evolution and consequences." *Experientia,* 45: 653–660.

La Marche, V. C., and K. K. Hirschboeck. 1984. "Frost rings in trees as records of major volcanic eruptions." *Nature* (London), 307: 121–126.

Lack, D. 1976. *Island Biology, Illustrated by the Land Birds of Jamaica.* Oxford: Blackwell Scentific Publications.

Lam, H. J. 1935. "Een Halve eeuw Krakatau ondezoek." *Vakblad voor Biologen,* 16: 161.

Lamb, H. H. 1983. "Update of the chronology of assessments of the volcanic Dust Veil Index." *Climate Monitor,* 12: 79–80.

Lambert, F. R. 1989a. "Fig-eating by birds in a Malaysian lowland rain forest." *Journal of Tropical Ecology,* 5: 401–412.

————1989b. "Pigeons as seed predators and dispersers of figs in a Malaysian lowland forest." *Ibis*, 131: 521–527.

————1992. "Fig dimorphism in bird-dispersed gynodioecious *Ficus*." *Biotropica*, 24: 214–216.

Lambert, F. R., and A. G. Marshall. 1992. "Keystone characteristics of bird-dispersed *Ficus* in a Malaysian lowland rain forest." *Journal of Ecology*, 79: 793–809.

Larsen, T. B., and D. E. Pedgley. 1985. "Indian migrant butterflies displaced to Arabia by monsoon storm 'Aurora' in August 1983." *Ecological Entomology*, 10: 235–238.

Latter, J. H. 1981. "Tsunamis of volcanic origin: Summary of causes, with particular reference to Krakatoa, 1883." *Bulletin of Volcanology*, 44: 467–490.

Lawton, J. H. 1987. "Are there assembly rules for successional communities?" In A. J. Gray, M. J. Crawley, and P. J. Edwards (eds.), *Colonization, Succession and Stability*, pp. 225–244. Oxford: Blackwell Scientific Publications.

Lee, M. A. B. 1985. "The dispersal of *Pandanus tectorius* by the land crab *Cardisoma carnifex*." *Oikos*, 45: 169–173.

Leps, J., and K. Spitzer. 1990. "Ecological determinants of butterfly communities (Papilionoidea, Lepidoptera) in the Tam Dao Mountains, Vietnam." *Acta Entomologica Bohemoslovaca*, 87: 182–194.

Lindroth, C. H., H. Andersson, H. Bödvarsson, and S. H. Richter. 1973. "Surtsey, Iceland: The development of a new fauna 1963–1970. Terrestrial invertebrates." *Entomologica Scandinavica, Supplementum*, 5: 1–280.

Lloyd Praeger, R. 1915. "Clare Island survey. Part X." *Proceedings of the Royal Irish Academy*, 31 (10): 92–94.

Lotsy, J. P. 1908. *Vorlesungen ueber Deszendenztheorien*, vol 2. Jena.

Louda, S. V., and P. H. Zedler. 1985. "Predation in insular plant dynamics: An experimental assessment of postdispersal fruit and seed survival, Eniwetok Atoll, Marshall Islands." *American Journal of Botany*, 72 (3): 438–445.

MacArthur, R. H. 1972. *Geographical Ecology*. New York: Harper and Row.

MacArthur, R. H., and E. O. Wilson. 1963. "An equilibrium theory of insular zoogeography." *Evolution*, 17: 373–387.

————1967. *The Theory of Island Biogeography*. London: John Wiley.

MacKinnon, J., and K. Phillipps. 1994. *A Field Guide to the Birds of Borneo, Sumatra, Java and Bali, the Greater Sunda Islands*. Oxford: Oxford University Press.

Maeto, K., and I. W. B. Thornton. 1993. "A preliminary appraisal of the braconid (Hymenoptera) fauna of the Krakatau Islands (Indonesia) in 1984–1986, with comments on the colonizing abilities of parasitoid modes." *Japanese Journal of Entomology*, 61 (4): 787–801.

Main, B. Y. 1976. *Spiders*. Sydney: Collins.

————1982. "Some zoogeographic considerations of families of spiders occurring in New Guinea." In J. L. Gressitt (ed.), *Biogeography and Ecology of New Guinea*, pp. 583–602. The Hague: Junk.

Marshall, A. G. 1985. "Old world phytophagous bats (Megachiroptera) and their food plants: A survey." *Zoological Journal of the Linnean Society of London*, 83: 351–369.

Martens, E. C. von. 1867. *Die Preussische Expedition nach Ost-Asien. Zoologisher Theil 2. Die Landschnecken*. Berlin: R. von Decker.

McClure, H. E. 1966. "Flowering, fruiting and animals in the canopy of a tropical rain forest." *Malayan Forester,* 29: 182–203.

M'Closkey, R. T. 1978. "Niche separation and assembly in four species of Sonoran Desert rodents." *American Naturalist,* 112: 683–694.

——1985. "Species pools and combinations of heteromyid rodents." *Journal of Mammalogy,* 66: 132–134.

Mees, G. F. 1986. "A list of the birds recorded from Bangka Island, Indonesia." *Zoologische Verhandelingen* (Leiden), 232: 83–84.

Meinertzhagen R. 1955. "The speed and altitude of bird flight (with notes on other animals)." *Ibis,* 97: 81–117.

Michaelsen, W. 1924. "Oligochäten von Niederländisch-Indien." *Treubia,* 5: 379–401.

Midya, S., and R. L. Brahmachary. 1991. "The effect of birds upon the germination of banyan *(Ficus bengalensis)* seeds." *Journal of Tropical Ecology,* 7: 537–538.

Morgan, G. J. 1988. "Notes on terrestrial crabs (Decapoda: Anomura, Brachyura) of the Krakatau Islands, Indonesia." *Indo-Malayan Zoology,* 5: 307–309.

Muir, F. 1930. "On a small collection of fulgorids from the islands of Krakatau, Verlaten and Sebesi." *Treubia,* 12: 29–35.

Muller, J. J. A. 1897. "De eilanden Krakatau en Langeiland." *Tidschrift Koninklijk Nederlandsch Aardrijkskundig Genootschap,* 14: 118–122.

Neill, W. T. 1958. "The occurrence of reptiles and amphibians in saltwater areas." *Bulletin of Marine Science of the Gulf and Caribbean,* 8 (1): 1–95.

Nelson, J. E. 1965. "Movements of Australian flying foxes (Pteropodidae: Megachiroptera)." *Australian Journal of Zoology,* 13: 53–73.

Nève, G. A. de. 1982. "Anak Krakatau (1930–1980): Fifty years of geomorphological development and growth with the petrographically derived consequences." In *Proceedings PIT X Ikatan Ahli Geologi Indonesia, Bandung, 8–10 Desember 1981,* pp. 7–40. Bandung: Indonesian Association of Geologists.

——1983. "Krakatau's earliest known activity: Was it prehistoric?" *Berita Geologi* (Bandung), 15: 39–44.

——1985a. "Geovolcanology of the Krakatau group in the Sunda Strait region: Review of a hundred years development (1883–1983)." In Sastrapradja et al. (1985), pp. 20–34.

——1985b. "Earler eruptive activities of Krakatau in historic time and during the Quaternary." In Sastrapradja et al. (1985), pp. 35–46.

New, T. R., M. B. Bush, I. W. B. Thornton, and H. K. Sudarman. 1988. "The butterfly fauna of the Krakatau Islands after a century of recolonization." *Philosophical Transactions of the Royal Society of London* B, 322: 445–457.

New, T. R., and I. W. B. Thornton. 1988. "A pre-vegetation population of crickets subsisting on allochthonous aeolian debris on Anak Krakatau." *Philosophical Transactions of the Royal Society of London* B, 322: 481–485.

——1992a. "Colonization of the Krakatau Islands by invertebrates." In Thornton (1992), pp. 219–224.

——1992b. "The butterflies of Anak Krakatau, Indonesia: Faunal development in early succession." *Journal of the Lepidopterists' Society,* 46 (2): 83–96.

Newsome, D., K. Richards, and J. R. Flenley. 1982. "A carbonised wood sample from Rakata, and its radiocarbon assay. I. Stratigraphic observations." In Flenley and Richards (1982), pp. 174–177.

Ng, R., and S. S. Lee. 1982. "The vertical distribution of insects in a tropical primary lowland dipterocarp forest in Malaysia." *Malaysian Journal of Science*, 7: 37–52.

Ninkovich, D. 1976. "Late Cenozoic clockwise rotation of Sumatra." *Earth and Planetary Science Letters*, 29: 269–275.

———1979. "Distribution, age and chemical composition of tephra layers in deep-sea sediments off Western Indonesia." *Journal of Volcanology and Geothermal Research*, 5: 67–86.

Nishimura, S., H. Harjono, and S. Suparka. 1992. "The Krakatau Islands: The geotectonic setting." In Thornton (1992), pp. 87–98.

Noort, S. van, A. B. Ware, and S. G. Compton. 1989. "Release of pollinator-specific volatile attractants from figs of *Ficus burtt-davyi*." *South African Journal of Science*, 85: 323–324.

Oba, N., K. Tomita, and M. Yamamoto. 1992. "An interpretation of the 1883 cataclysmic eruption of Krakatau from geochemical studies on the partial melting of granite." In Thornton (1992), pp. 99–108.

O'Dowd, D. J., and P. S. Lake. 1989. "Red crabs in rain forest, Christmas Island: Removal and relocation of leaf-fall." *Journal of Tropical Ecology*, 5: 337–348.

———1991. "Red crabs in rain forest, Christmas Island: Removal and fate of fruits and seeds." *Journal of Tropical Ecology*, 7: 113–122.

Pacific Science Congress, Fourth. 1929. "The case of Krakatau." *Proceedings of the Fourth Pan-Pacific Congress, Java, 1929*, 1: 214–219.

Paine, R. T. 1966. "Food web complexity and species diversity." *American Naturalist*, 100: 65–75.

Partomihardjo, T., E. Mirmanto, S. Riswan, and E. Suzuki. 1993. "Drift fruit and seeds on Anak Krakatau beaches, Indonesia." *Tropics*, 2 (3): 143–156.

Partomihardjo, T., E. Mirmanto, and R. J. Whittaker. 1992. "Anak Krakatau's vegetation and flora circa 1991, with observations on a decade of development and change." In Thornton (1992), pp. 233–248.

Penzig, O. 1902. "Die fortschritte der flora der Krakatau." *Annales du Jardin Botanique de Buitenzorg*, 18: 92–113.

Pijl, L. van der. 1936. "Dubbele zaadverspreiding (diplochorie) bij *Mimosa, Terminalia, Calophyllum* en *Hernandia*." *De Tropische Natuur*, 25: 97–100.

———1941. "Flagelliflory and cauliflory as adaptations to bats in *Mucuna* and other plants." *Annals of the Botanical Gardens, Buitenzorg*, 51: 83–93.

———1949. "Botanische notities over een bezoek aan Anak Krakatau in Augustus 1949." *Chronica Naturae*, 105 (11): 283–285.

———1957. "The dispersal of plants by bats (chiropterochory)." *Acta Botanica Nederlandica*, 6: 291–315.

———1982. *Principles of Dispersal in Higher Plants*. Berlin: Springer-Verlag.

Proctor, V. W. 1968. "Long-distance dispersal of seeds by retention in digestive tract of birds." *Science*, 160: 321–322.

Ramirez, W. 1969. "Host specificity of fig wasps (Agaonidae)." *Evolution,* 24: 680–691.

Rampino, M. R., and S. Self. 1982. "Historic eruptions of Tambora (1815), Krakatau (1883) and Agung (1963), their stratospheric aerosols, and climatic impact." *Quaternary Research,* 18: 127–143.

———1984. "Sulphur-rich volcanic eruptions and stratospheric aerosols." *Nature* (London), 310: 677–679.

Rampino, M. R., S. Self, and R. W. Fairbridge. 1979. "Can rapid climatic change cause volcanic eruptions?" *Science,* 206: 826–829.

Raven, H. C. 1946. "Predators eating turtle eggs in the East Indies." *Copeia,* 1: 48.

Rawlinson, P. A., A. H. T. Widjoya, M. N. Hutchinson, and G. W. Brown. 1990. "The terrestrial vertebrate fauna of the Krakatau Islands, Sunda Strait, 1883–1986." *Philosophical Transactions of the Royal Society of London* B, 328: 3–28.

Rawlinson, P. A., R. A. Zann, S. A. van Balen, and I. W. B. Thornton. 1992. "Colonization of the Krakatau Islands by vertebrates." In Thornton (1992), pp. 225–231.

Recher, H. F. 1969. "Bird species diversity and habitat diversity in Australia and North America." *American Naturalist,* 103: 75–80.

Reck, H. 1936. *Santorin—Der Werdegang eines Inselvulcans und sein Ausbruch 1925–1928.* 3 vols. Berlin: Verlag D. Reimer.

Richards, K., and R. J. Whittaker. 1990. "A revised vegetation map of Krakatau." In Whittaker et al. (1990), pp. 16–20.

Richards, P. W. 1952. *The Tropical Rain Forest.* London: Cambridge University Press.

Rickart E. A., L. R. Heaney, P. D. Heideman, and R. C. B. Utzurrum. 1993. "The distribution and ecology of mammals on Leyte, Biliran, and Maripipi islands, Philippines." *Fieldiana: Zoology,* 72: 1–62.

Ridley, H. N. 1930. *The Dispersal of Plants throughout the World.* Ashford, U.K.: L. Reeve.

Robinson, J. V., and J. E. Dickerson Jr. 1987. "Does invasion sequence affect community structure?" *Ecology,* 68 (3): 587–589.

Robinson, J. V., and M. A. Edgemon. 1988. "An experimental evaluation of the effect of invasion history on community structure." *Ecology,* 69: 1410–1417.

Robinson, M. H. 1982. "The ecology and biogeography of spiders in Papua New Guinea." In J. L. Gressitt (ed.), *Biogeography and Ecology of New Guinea,* pp. 557–581. The Hague: Junk.

Rosengren, N. J. 1985. "The changing outlines of Sertung, Krakatau Islands, Indonesia." *Zeitschrift für Geomorphologie,* new ser., *Suppl.,* 57: 105–119.

Rosengren, N. J., and A. Suwardi. 1985. "Coastal geomorphology of Anak Krakatau and Sertung." In Sastrapradja et al. (1985), pp. 1–17.

Roughgarden, J. 1989. "The structure and assembly of communities." In J. Roughgarden, R. M. May, and S. A. Levin (eds.), *Perspectives in Ecological Theory,* pp. 203–226. Princeton: Princeton University Press.

Russell, F. A. R., and E. D. Archibald. 1888. "On the unusal optical phenomena of the atmosphere, 1883–1886, including twilight effects, coronal appearances, sky haze, coloured suns, moons." In Symons (1888), pp. 151–463.

Sastrapradja, D., S. Soemodihardjo, A. S. Soemartadipura, A. Soegiarto, S. Adisoe-marto, K. Kusumadinata, S. Birowo, S. Riswan, and A. B. Lapian, eds. 1985. *Symposium on 100 Years Development of Krakatau and Its Surroundings.* Jakarta: LIPI.

Scharff, R. F. 1925. "Sur le problem de l'ile de Krakatau." *Comptes-Rendus du Congrès de l'Association Française pour l'Avancement des Sciences (AFAS),* 49: 746–750.

Schedvin, N., S. Cook, and I. W. B. Thornton. 1995. "The diversity of bats on the Krakatau Islands, Indonesia." *Biodiversity Letters,* 2: 87–92.

Schoener, A., and T. W. Schoener. 1984. "Experiments on dispersal: Short-term floatation of insular anoles, with a review of similar abilities in other terrestrial animals." *Oecologia* (Berlin), 63: 289–294.

Schmidt, E. R., I. W. B. Thornton, and K. Hancock. 1994. "Tropical fruitflies (Diptera: Tephritidae) of the Krakatau Archipelago in 1990 and comments on faunistic changes since 1982." *Ecological Research,* 9: 1–8.

Selenka, E., and L. Selenka. 1905. *Sonnige Welten.* Rev. 2d ed. Wiesbaden: C. W. Kneidels Verlag.

Self, S. 1992 "Krakatau revisited: The course of events and interpretation of the 1883 eruption." In Thornton (1992), pp. 109–121.

Self, S., and M. R. Rampino. 1981. "The 1883 eruption of Krakatau." *Nature* (London), 294: 699–704.

Shinagawa, A., N. Miyauchi, and T. Higashi. 1992. "Cumulic soils on Rakata, Sertung and Panjang (Krakatau Is.) and properties of each solum." In Thornton (1992), pp. 139–151.

Sigurdsson, H., S. Carey, and C. Mandeville. 1991. "Krakatau: Submarine pyroclastic flows of the 1883 eruption of Krakatau volcano." *National Geographic Research and Exploration,* 7 (3): 310–327.

Sigurdsson, H., S. Carey, C. Mandeville, and S. Bronto. 1991. "Pyroclastic flows of the 1883 Krakatau eruption." *EOS Transactions, American Geophysical Union,* 72 (377): 380–381.

Simberloff, D. S. 1974. "Equilibrium theory of island biogeography and ecology." *Annual Review of Ecology and Systematics,* 5: 161–182.

————1978a. "Using island biogeographic distributions to determine if colonization is stochastic." *American Naturalist,* 112 (986): 713–726.

————1978b. "Colonization of islands by insects: Immigration, extinction, and diversity." In L. A. Mound and N. Waloff (eds.), *Diversity of Insect Faunas,* pp. 139–153. Oxford: Blackwell.

Simberloff, D., and E. O. Wilson. 1969. "Experimental zoogeography of islands: The colonization of empty islands." *Ecology,* 50: 278–296.

————1970. "Experimental zoogeography of islands: A two-year record of colonization." *Ecology,* 51: 934–937.

Simkin, T., and R. S. Fiske. 1983a. *Krakatau 1883: The Volcanic Eruption and Its Effects.* Washington, D.C.: Smithsonian Institution.

————1983b. "Krakatau 1883: A classic geophysical event." *EOS Transactions, American Geophysical Union,* 64 (34): 513–514.

Siswowidjoyo, S. 1983. "The renewed activity of Krakatau volcano after its catastrophic eruption in 1883." In Sastrapradja et al. (1983), pp. 192–198.

Slater, J. A. 1972. "Lygaeid bugs (Hemiptera: Lygaeidae) as seed predators of figs." *Biotropica,* 4: 145–151.

Smith, B. J., and M. Djajasasmita. 1988. "The land molluscs of the Krakatau Islands, Indonesia." *Philosophical Transactions of the Royal Society of London* B, 322: 379–400.

Smith, M. A. 1932. "Some notes on the monitors." *Journal of the Bombay Natural History Society,* 35 (3): 615–619.

Start, A. N. 1974. "The feeding biology in relation to food sources of nectarivorous bats (Chiroptera: Macroglossinae) in Malaysia." Thesis, University of Aberdeen.

Start, A. N., and A. G. Marshall. 1976. "Nectarivorous bats as pollinators of trees in West Malaysia." In J. Burley and B. T. Styles (eds.), *Variation, Breeding and Conservation of Tropical Forest Trees,* pp. 141–150. London: Academic Press.

Steadman, D. 1989. "Fossil birds and biogeography in Polynesia." *Acta XIX Congressus Internationalis Ornithologici,* II: 1526–1534.

———1993. "Biogeography of Tongan birds before and after human impact." *Proceedings of the National Academy of Sciences, USA,* 90: 818–822.

Steenis, C. G. G. J. van. 1931. "De niewe Flora van Krakatau na 1883." *Vakblad voor Biologen,* 13: 64–67.

———1938. "The botanical side of the Krakatoa problem." *Annales du Jardin Botanique de Buitenzorg,* 45: 94–96.

———1951. "Notes. Revegetation of Krakatau." *Flora Malesiana Bulletin,* 2: 292.

Stehn, C. E. 1929. "The geology and volcanism of the Krakatau-Group." *Proceedings of the Fourth Pan-Pacific Science Congress, Batavia,* 3–55.

Stoiber, R. E., and S. N. Williams. 1985. "Volcanic gas from the Krakatau eruptions of 1883." In Sastrapradja et al. (1985), pp. 107–110.

Strachey, R. 1884. "The Krakatau air-wave." *Nature* (London), 29 (738): 181–183.

———1888. "On the air waves and sounds caused by the eruption of Krakatoa in August, 1883." In Symons (1888), pp. 57–88.

Sudrajat (Sumartadipura), A. 1982. "The morphological development of Anak Krakatau Volcano, Sunda Strait." *Geologi Indonesia,* 9: 1–11.

Sudrajat, A., and S. Siswowidjoyo. 1987. "Merapi." *Bulletin of Volcanic Eruptions,* 24: 23.

Sukardjono, Lubis, and S. H. Sukorahardjono. 1985. "Shallow seismic reflection investigation in the northern part of the Sunda Strait." In Sastrapradja et al. (1985), pp. 123–131.

Sussman, R. W., and P. H. Raven. 1978. "Pollination by lemurs and marsupials: An archaic coevolutionary system." *Science,* 200 (4343): 731–736.

Sutawidjaja, I. S. 1993. "Erupsi G. Anak Krakatau 1992–1993." *Pertambangan dan Energi,* 4 (18): 44–50.

Switsur, V. R. 1982. "A carbonised wood sample from Rakata, and its radiocarbon assay. II. Radiocarbon assay." In Flenley and Richards (1982), pp. 177–180.

Symons, G. J., ed. 1888. *The Eruption of Krakatoa and Subsequent Phenomena. Report of the Krakatoa Committee of the Royal Society.* London: Trubner & Co.

Tagawa, H. 1992. "Primary succession and the effect of first arrivals on the subsequent development of forest types." In Thornton (1992), pp. 175–183.

Tagawa, H., E. Suzuki, T. Partomihardjo, and A. Suriadarma. 1985. "Vegetation and succession on the Krakatau Islands, Indonesia." *Vegetatio*, 60: 131–145.

Tan, E. M., M. E. Hanson, and C. P. Richter. 1954. "Swimming time of rats with relation to water temperature." *Federation Proceedings of the American Physiological Society*, 13 (498): 150–151.

Tedman, R. H., and L. S. Hall. 1985. "The morphology of the gastrointestinal tract and food transit time in the fruit bats *Pteropus alecto* and *P. poliocephalus* (Megachiroptera)." *Australian Journal of Zoology*, 33: 625–640.

Teijsmann, J. E. 1857. "Dagverhaal eener botanische reis over de westkust van Sumatra." *Natuurkundig Tijdschrift voor Nederlandsh-Indie*, 14: 249–376.

Thakur, M. L., and R. K. Thakur. 1992. "Termite fauna of Krakatau and associated islands, Sunda Straits, Indonesia." *Treubia* 30 (3): 213–317.

Thornton, I. W. B. 1984. "Krakatau—the development and repair of a tropical ecosystem." *Ambio*, 13: 217–225.

———1991. "Krakatau—studies on the origin and development of a fauna." In E. C. Dudley, ed., *The Unity of Evolutionary Biology* (Proceedings of the Fourth International Congress of Systematic and Evolutionary Biology), pp. 396–408. Portland: Dioscorides Press.

———ed. 1992a. *Krakatau—A Century of Change. GeoJournal* 28 (2): 81–302.

———1992b. "K. W. Dammerman—Fore-runner of island equilibrium theory?" *Global Ecology and Biogeography Letters*, 2:145–148.

———1994. "Figs, frugivores and falcons: An aspect of the assembly of mixed tropical forest on the emergent volcanic island, Anak Krakatau." *South Australian Geographical Journal*, 93: 3–21.

Thornton, I. W. B., S. G. Compton, and C. N. Wilson. 1996. "The role of animals in the colonization of the Krakatau Islands by fig trees (*Ficus* species)." *Journal of Biogeography*, in press.

Thornton, I. W. B., and T. R. New. 1988a. "Freshwater communities on the Krakatau Islands." *Philosophical Transactions of the Royal Society of London* B, 322: 487–492.

———1988b. "Krakatau invertebrates: The 1980s fauna in the context of a century of colonization." *Philosophical Transactions of the Royal Society of London* B, 322: 493–522.

Thornton, I. W. B., T. R. New, D. A. McLaren, H. K. Sudarman, and P. J. Vaughan. 1988. "Air-borne arthropod fall-out on Anak Krakatau and a possible pre-vegetation pioneer community." *Philosophical Transactions of the Royal Society of London* B, 322: 471–479.

Thornton, I. W. B., T. R. New, and P. J. Vaughan. 1988. "Colonization of the Krakatau Islands by Psocoptera (Insecta)." *Philosophical Transactions of the Royal Society of London* B, 322: 427–443.

Thornton, I. W. B., T. R. New, R. A. Zann, and P. A. Rawlinson. 1990. "Colonization of the Krakatau Islands by animals: A perspective from the 1980s." *Philosophical Transactions of the Royal Society of London* B, 328: 132–165.

Thornton, I. W. B., T. Partomihardjo, and J. Yukawa. 1994. "Observations on the effects, up to July 1993, of the current eruptive episode of Anak Krakatau." *Global Ecology and Biogeography Letters,* 4: 88–94.

Thornton, I. W. B., and N. J. Rosengren. 1988. "Zoological expeditions to the Krakatau Islands, 1984 and 1985: General Introduction." *Philosophical Transactions of the Royal Society of London,* ser. B, 322: 273–316.

Thornton, I. W. B., and D. Walsh. 1992. "Photographic evidence of rate of development of plant cover on the emergent island Anak Krakatau from 1971 to 1991 and implications for the effect of volcanism." In Thornton (1992), pp. 249–259.

Thornton, I. W. B., S. A. Ward, R. A. Zann, and T. R. New. 1992. "Anak Krakatau—a colonization model within a colonization model?" In Thornton (1992), pp. 271–286.

Thornton, I. W. B., R. A. Zann, and S. van Balen. 1993. "Colonization of Rakata (Krakatau Is.) by non-migrant land birds from 1883 to 1992 and implications for the value of island equilibrium theory." *Journal of Biogeography,* 20: 441–452.

Thornton, I. W. B., R. A. Zann, P. A. Rawlinson, C. R. Tidemann, A. S. Adhikerana, and A. H. J. Widjoya. 1989. "Colonization of the Krakatau Islands by vertebrates: Equilibrium, succession, and possible delayed extinction." *Proceedings of the National Academy of Sciences, USA,* 85: 515–518.

Thornton, I. W. B., R. A. Zann, and D. G. Stephenson. 1990. "Colonization of the Krakatau Islands by land birds, and the approach to an equilibrium number of species." *Philosophical Transactions of the Royal Society of London* B, 328: 55–93.

Tidemann, C. R., D. J. Kitchener, R. A. Zann, and I. W. B. Thornton. 1990. "Recolonization of the Krakatau Islands and adjacent areas of West Java, Indonesia, by bats (Chiroptera) 1883–1886." *Philosophical Transactions of the Royal Society of London* B, 328: 123–130.

Tol, J. van. 1990. "Zoological expeditions to the Krakataus, 1984 and 1985. Odonata." *Tijdschrift voor Entomologie,* 133: 273–279.

Toxopeus, L. J. 1950. "Over de pioneer-fauna van Anak Krakatau, met enige beschouwingen over het onstaat van de Krakatau-fauna." *Chronica Naturae,* 106: 27–34.

Treub, M. 1888. "Notice sur la nouvelle flore de Krakatau." *Annales du Jardin Botanique de Buitenzorg,* 7: 213–223.

Turner, B. 1992. "The colonization of Anak Krakatau: Interactions between wild sugar cane, *Saccharum spontaneum,* and the ant lion, *Myrmeleon frontalis." Journal of Tropical Ecology,* 8: 435–449.

Turrill, W. B. 1935. "Krakatau and its problems." *New Phytologist,* 34: 442–443.

Underwood, A. J., E. J. Denley, and M. J. Moran. 1983. "Experimental analyses of the structure and dynamics of mid-shore rocky intertidal communities in New South Wales." *Oecologia,* 56: 202–219.

Utzurrum, R. C. B., and P. D. Heideman. 1991. "Differential ingestion of viable vs nonviable *Ficus* seeds by fruit bats." *Biotropica,* 23 (3): 311–312.

Verbeek, R. D. M. 1881. "Topographischeen geologische beschrijving van Zuid-Sumatra." *Jahrbuch van het Mijnwezen in N.O.I.,* 1881 (1): 154–156, 179–181, 214–215.

———1884. "The Krakatau eruption." *Nature* (London), 30: 10–15.

———1885. *Krakatau.* Batavia: Landsdrukkerij.

Vincent, P. M., and G. Camus. 1983. "The 1883 Krakatau eruption initiated by a Mount St. Helens type event?" *EOS Transactions, American Geophysical Union,* 64: 872.

———1986. "The origin of the 1883 Krakatau tsunamis, by P. W. Francis: Discussion." *Journal of Volcanology and Geothermal Research,* 30: 169–177.

Vincent, P. M., G. Camus, and B. M. Larue. 1984. "Origine du grand tsunami du Krakatoa (27 Aout 1883) par immersion d'une coulée de débris." *Bulletin PIRPSEV,* 89: 1–19.

Vogel, J. W. 1690. *Journal einer Reise aus Holland nach Ost-Indien.* Frankfurt and Leipzig.

Wallace, A. R. 1880. *Island Life.* London: Richard Clay & Sons.

Ware, A. B., P. T. Kaye, S. G. Compton, and S. van Noort. 1993. "Fig volatiles: Their role in attracting pollinators and maintaining pollinator specificity." *Plant Systematics and Evolution,* 186: 147–156.

Wells, D. R. 1988. "Birds." In Cranbrook (1988), pp. 167–195.

Wexler, H. H. 1951. "On the effects of volcanic dust on insolation and weather (I)." *Bulletin of the American Meteorological Society,* 32: 10–15.

Wharton, W. J. L., and F. J. Evans. 1888. "On the seismic sea waves caused by the eruption of Krakatau, August 26 and 27, 1883." In Symons (1888), pp. 89–150.

Whitehead, D. R., and C. E. Jones. 1969. "Small islands and the equilibrium theory of insular biogeography." *Evolution,* 23: 171–179.

Whittaker, R. H., and S. A. Levin. 1977. "The role of mosaic phenomena in natural communities." *Theoretical Population Biology,* 12: 117–139.

Whittaker, R. J. 1982. "Aspects of the vegetation of Rakata." In Flenley and Richards (1982), pp. 54–82.

Whittaker, R. J., N. M. Asquith, M. B. Bush, and T. Partomihardjo, eds. 1990. *Krakatau Research Project 1989 Expedition Report.* Oxford: School of Geography, University of Oxford.

Whittaker, R. J., M. B. Bush, T. Partomihardjo, N. M. Asquith, and K. Richards. 1992. "Ecological aspects of plant colonization of the Krakatau Islands." In Thornton (1992), pp. 201–211.

Whittaker, R. J., M. B. Bush, and K. Richards. 1989. "Plant recolonization and vegetation succession on the Krakatau Islands, Indonesia." *Ecological Monographs,* 59: 59–123.

Whittaker, R. J., and S. H. Jones. 1994. "The role of frugivorous bats and birds in the rebuilding of a tropical forest ecosystem, Krakatau, Indonesia." *Journal of Biogeography,* 21: 246–258.

Whittaker, R. J., T. Partomihardjo, and S. Riswan. 1995. "Surface and buried seed banks from Krakatau, Indonesia: Implications for the sterilization hypothesis." *Biotropica,* 27 (3): 345–354.

Whittaker, R. J., and B. D. Turner. 1994. "Dispersal, fruit utilization and seed predation of *Dysoxylum gaudichaudianum* in early successional rain forest, Krakatau, Indonesia." *Journal of Tropical Ecology,* 10: 167–181.

Whittaker, R. J., J. Walden, and J. Hill. 1992. "Post-1883 ash fall on Panjang and Sertung and its ecological impact." In Thornton (1992), pp. 153–171.

Wiebes, J. T. 1986. "The association of figs and fig-insects." *Revue de Zoologie Afrique,* 100: 63–71.

Williams, H. 1941. "Calderas and their origins." *Bulletin of the Department of Geological Science, University of California,* 25–26: 238–346.

Williamson, M. 1982. *Island Populations.* Oxford: Oxford University Press.

Wilson, E. O. 1992. *The Diversity of Life.* Cambridge, MA: Harvard University Press

Wing Easton, N. 1929. "Review of C. A. Backer, The Problem of Krakatao." *Tidschrift Koninklijk Nederlandsch Aardrijkskundig Genootschap,* ser. II, 46: 871–872.

Winoto Suatmadji, R., A. Coomans, F. Rashid, E. Gevaert, and D. A. McLaren. 1988. "Nematodes of the Krakatau archipelago, Indonesia: A preliminary overview." *Philosophical Transactions of the Royal Society of London* B, 322: 369–378.

Wissel, C., and B. Maier. 1992. "A stochastic model for the species-area relationship." *Journal of Biogeography,* 19: 335–362.

Wodzicki, K., and H. Felten. 1975. "The Peka, or fruit bat *(Pteropus tonganus tonganus)* (Mammalia, Chiroptera), of Niue Island, South Pacific." *Pacific Science,* 29 (2): 131–138.

Womersley, H. 1932. "Collembola from Krakatau." *Entomologist's Monthly Magazine,* 68: 88.

Wood-Jones, F. 1909. "The fauna of Cocos-Keeling Atoll, collected by F. Wood-Jones." *Proceedings of the Zoological Society of London,* 1909: 132–160.

Wyrtki, K. 1961. *Physical Oceanography of the Southeast Asian Waters.* NAGA Report, vol. 2. San Diego: Scripps Institute of Oceanography, University of California.

Yamane, Sk. 1988. "The aculeate fauna of the Krakatau Islands (Insecta, Hymenoptera)." *Reports of the Faculty of Science, Kagoshima University (Earth Science and Biology),* 16: 75–107.

Yamane, Sk., T. Abe, and J. Yukawa. 1992. "Recolonization of the Krakataus by Hymenoptera and Isoptera (Insecta)." In Thornton (1992), pp. 213–218.

Yokoyama, I. 1981. "A geophysical interpretation of the 1883 Krakatau eruption." *Journal of Volcanology and Geothermal Research,* 9: 359–378.

———1982. "Author's reply to the comments by S. Self and M. R. Rampino." *Journal of Volcanology and Geothermal Research,* 13: 384–386.

———1987. "A scenario of the 1883 Krakatau tsunami." *Journal of Volcanology and Geothermal Research,* 34: 123.

Yokoyama, I., and O. Hadikusumo. 1969. "Volcanological survey of Indonesian volcanoes. Part 3. A gravity survey on the Krakatau Island, Indonesia." *Bulletin of the Earthquake Research Institute,* 47: 991–1001.

Yokoyama, T., A. Dharma, K. Hirooka, and S. Nishimura. 1985. "Paleomagnetic measurement and radiometric dating of the volcanic material at Krakatau volcano and neighbouring areas." In Sastrapradja et al. (1985), pp. 204–209.

Yukawa, J. 1984a. "An outbreak of *Cryptocerya jacobsoni* (Green) (Homoptera: Margarodidae) on Rakata Besar of the Krakatau Islands in Indonesia." *Applied Entomology and Zoology,* 19: 179–180.

———1984b. "Fruit flies of the genus *Dacus* (Diptera: Tephritidae) on the Krakatau Islands in Indonesia, with special reference to an outbreak of *Dacus albistrigatus* De Meijere." *Japanese Journal of Ecology,* 34: 281–288.

———1984c. "Geographical ecology of the butterfly fauna of the Krakatau Islands, Indonesia." *Tyô to Ga,* 35: 47–74.

Yukawa, J., T. Abe, T. Iwamoto, and Y. Seiki. 1984. "The fauna of the Krakatau, Peucang and Panaitan Islands." In H. Tagawa (ed.), *Researches on the Ecological Succession and the Formation Process of Volcanic Ash Soils on the Krakatau Islands,* pp. 91–114. Kagoshima University.

Yukawa, J., and S. Yamane. 1985. "Odonata and Hemiptera collected from the Krakataus and the surrounding islands, Indonesia." *Kontiû,* 53: 690–698.

Zann, R. A., and Darjono. 1992. "The birds of Anak Krakatau: The assembly of an avian community." In Thornton (1992), pp. 261–270.

Zann, R. A., E. B. Male, and Darjono. 1990. "Bird colonization of Anak Krakatau, an emergent volcanic island." *Philosophical Transactions of the Royal Society of London* B, 328: 95–121.

Zann, R. A., M. V. Walker, A. S. Adhikerana, G. W. Davison, E. B. Male, and Darjono. 1990. "The birds of the Krakatau Islands (Indonesia) 1984–86." *Philosophical Transactions of the Royal Society of London* B, 328: 29–54.

Zen, M. T. 1985. "Krakatau and the tectonic importance of the Sunda Strait." In Sastrapradja et al. (1985), pp. 100–106.

Zen, M. T., and A. Sudradjat. 1983. "History of the Krakatau volcanic complex in Strait Sunda and the mitigation of its future hazards." *Bulletin Jurusan Geologi,* 10: 1–28.

Zwaluwenburg, R. H. van. 1942. "Notes on the temporary establishment of insect and plant species on Canton Island." *Hawaiian Planters' Record,* 46: 49–52.

CREDITS

Figure 1 Modified from M. B. Bush, R. J. Whittaker, and T. Partomihardjo, "Forest development on Rakata, Panjang and Sertung: Contemporary dynamics (1979–1989)," pp. 185–199 in *Krakatau—A Century of Change,* ed. I. W. B. Thornton, *GeoJournal,* vol. 28, no. 2 (1992).

Figures 2 and 3 Modified from R. Strachey, "On the air waves and sounds caused by the eruption of Krakatoa in August, 1883," pp. 57–88 in *The Eruption of Krakatoa and Subsequent Phenomena: Report of the Krakatoa Committee of the Royal Society,* ed. G. J. Symons (London: Trubner & Co., 1888).

Figures 4 and 7 Modified from S. Self, "Krakatau revisited: The course of events and interpretation of the 1883 eruption," pp. 109–121 in *Krakatau—A Century of Change,* ed. I. W. B. Thornton, *GeoJournal,* vol. 28, no. 2 (1992).

Figures 5, 6, 11, 14, 16, 22, 23, 36, 38 Photograph by D. Walsh; copyright La Trobe University.

Figure 8 Modified from G. Camus and P. M. Vincent, "Un siècle pour comprendre l'éruption du Krakatoa," *La Recherche,* 14 (149): 1452–1457 (1983).

Figure 9 Modified from D. Ninkovich, "Late Cenozoic clockwise rotation of Sumatra," *Earth and Planetary Science Letters,* 29: 269–275 (1976).

Figure 10 Modified from P. Francis and S. Self, "The eruption of Krakatau," *Scientific American,* 249: 172–198 (1983). Copyright © 1983 by Scientific American, Inc. All rights reserved.

Figure 12 From I. W. B. Thornton, S. A. Ward, R. A. Zann, and T. R. New, "Anak Krakatau—A colonization model within a colonization model?" pp. 271–286 in *Krakatau—A Century of Change,* ed. I. W. B. Thornton, *GeoJournal,* vol. 28, no. 2 (1992).

Figure 13 Modified from N. J. Rosengren, "The changing outlines of Sertung, Krakatau Islands, Indonesia," *Zeitschrift für Geomorphologie,* new ser., *Suppl.,* 57: 105–119 (1983).

Figure 15 Photograph from The British Library, London (Add MS 15514 R51); by permission of The British Library.

Figure 17 Photograph by D. McG. Ewart.

Figure 18 Photograph by M. N. Hutchinson.

Figures 19–21 Photograph by the author.

Figure 24 Dating and outlines of flows prior to 1990 follow I. S. Sutawidjaja, "Erupsi G. Anak Krakatau 1992–1993," *Pertambangan dan Energi,* 4 (18): 44–50 (1993); island outline in 1986 and 1991 from N. Hadjar, S. H. Purwadi, and Dijardjana, "The remotely sensed data acquisition above Krakatau by using 12S camera and thermal infra-red scanner," pp. 316–322 in *Proceedings of the Symposium on 100 Years Development of Krakatau and Its Surroundings,* ed. D. Sastrapradja et al. (Jakarta: L.I.P.I., 1985).

Figure 25 Photograph by Igan S. Sutawidjaja.

Figures 27 and 39 Modified from I. W. B. Thornton, S. A. Ward, R. A. Zann, and T. R. New, "Anak Krakatau—A colonization model within a colonization model?" pp. 271–286 in *Krakatau—A Century of Change,* ed. I. W. B. Thornton, *GeoJournal,* vol. 28, no. 2 (1992).

Figure 28 Surtsey photograph by Sturla Fridriksson.

Figure 29 Modified from I. W. B. Thornton, "Krakatau—Studies on the origin and development of a fauna," pp. 396–408; reprinted with permission from *The Unity of Evolutionary Biology—Proceedings of the Fourth International Congress of Systematic and Evolutionary Biology,* vol. 1, ed. Elizabeth C. Dudley. Copyright 1991 by Dioscorides Press, an imprint of Timber Press, Inc. All rights reserved.

Figure 30 Modified from T. Partomihardjo, E. Mirmanto, and R. J. Whittaker, "Anak Krakatau's vegetation and flora circa 1991, with observations on a decade of development and change," pp. 233–248 in *Krakatau—A Century of Change,* ed. I. W. B. Thornton, *GeoJournal,* vol. 28, no. 2 (1992).

Figure 31 Modified from I. W. B. Thornton, "Figs, frugivores and falcons: An aspect of the assembly of mixed tropical forest on the emergent volcanic island, Anak Krakatau," *South Australian Geographical Journal,* 93: 3–21 (1994).

Figures 32 and 34 Photograph courtesy of Healesville Sanctuary, Victoria, Australia.

Figure 33 Photograph by P. A. Rawlinson, courtesy of Mrs. M. Rawlinson.

Figure 35 Modified from I. W. B. Thornton, R. A. Zann, and S. van Balen, "Colonization of Rakata (Krakatau Is.) by non-migrant land birds from 1883 to 1992 and implications for the value of island equilibrium theory," *Journal of Biogeography,* 20: 441–452 (1993).

Figure 37 Photograph by N. K. Schedvin.

Figure 40 Photograph by N. J. Rosengren.

Figure 41 Photograph by N. M. Hutchinson.

Photographs in biographical notes Courtesy of Mr. K. Lut, Rijksherbarium, Leiden, Netherlands.

AUTHOR INDEX

GENERAL INDEX